Optimal Design of Experiments

A Case Study Approach

Optimal Design of Experiments
A Case Study Approach

Peter Goos

University of Antwerp and Erasmus University Rotterdam

Bradley Jones

JMP Division of SAS

A John Wiley & Sons, Ltd., Publication

This edition first published 2011
© 2011 John Wiley & Sons, Ltd

Registered office
John Wiley & Sons Ltd, The Atrium, Southern Gate, Chichester, West Sussex, PO19 8SQ, United
Kingdom

For details of our global editorial offices, for customer services and for information about how to apply
for permission to reuse the copyright material in this book please see our website at www.wiley.com.

Library of Congress Cataloging-in-Publication Data

Goos, Peter.
   Optimal design of experiments: a case study approach / Peter Goos and Bradley Jones.
      p. cm.
   Includes bibliographical references and index.
   ISBN 978-0-470-74461-1 (hardback)
1. Industrial engineering–Experiments–Computer-aided design.   2. Experimental design–Data
processing.   3. Industrial engineering–Case studies.   I. Jones, Bradley.   II. Title.
   T57.5.G66 2011
   670.285–dc22

                                                                    2011008381

A catalogue record for this book is available from the British Library.

Print ISBN: 978-0-470-74461-1
ePDF ISBN: 978-1-119-97400-0
oBook ISBN: 978-1-119-97401-7
ePub ISBN: 978-1-119-97616-5
Mobi ISBN: 978-1-119-97617-2

Set in 10/12pt Times by Aptara Inc., New Delhi, India.

To Marijke, Bas, and Loes

To Roselinde

Contents

Preface **xiii**

Acknowledgments **xv**

1   A simple comparative experiment **1**
  1.1   Key concepts   1
  1.2   The setup of a comparative experiment   2
  1.3   Summary   8

2   An optimal screening experiment **9**
  2.1   Key concepts   9
  2.2   Case: an extraction experiment   10
      2.2.1   Problem and design   10
      2.2.2   Data analysis   14
  2.3   Peek into the black box   21
      2.3.1   Main-effects models   21
      2.3.2   Models with two-factor interaction effects   22
      2.3.3   Factor scaling   24
      2.3.4   Ordinary least squares estimation   24
      2.3.5   Significance tests and statistical power calculations   27
      2.3.6   Variance inflation   28
      2.3.7   Aliasing   29
      2.3.8   Optimal design   33
      2.3.9   Generating optimal experimental designs   35
      2.3.10 The extraction experiment revisited   40
      2.3.11 Principles of successful screening: sparsity, hierarchy,
           and heredity   41
  2.4   Background reading   44
      2.4.1   Screening   44
      2.4.2   Algorithms for finding optimal designs   44
  2.5   Summary   45

3   Adding runs to a screening experiment   47
   3.1   Key concepts   47
   3.2   Case: an augmented extraction experiment   48
      3.2.1   Problem and design   48
      3.2.2   Data analysis   55
   3.3   Peek into the black box   59
      3.3.1   Optimal selection of a follow-up design   60
      3.3.2   Design construction algorithm   65
      3.3.3   Foldover designs   66
   3.4   Background reading   67
   3.5   Summary   67

4   A response surface design with a categorical factor   69
   4.1   Key concepts   69
   4.2   Case: a robust and optimal process experiment   70
      4.2.1   Problem and design   70
      4.2.2   Data analysis   79
   4.3   Peek into the black box   82
      4.3.1   Quadratic effects   82
      4.3.2   Dummy variables for multilevel categorical factors   83
      4.3.3   Computing D-efficiencies   86
      4.3.4   Constructing Fraction of Design Space plots   87
      4.3.5   Calculating the average relative variance of prediction   88
      4.3.6   Computing I-efficiencies   90
      4.3.7   Ensuring the validity of inference based on ordinary least squares   90
      4.3.8   Design regions   91
   4.4   Background reading   92
   4.5   Summary   93

5   A response surface design in an irregularly shaped design region   95
   5.1   Key concepts   95
   5.2   Case: the yield maximization experiment   95
      5.2.1   Problem and design   95
      5.2.2   Data analysis   103
   5.3   Peek into the black box   108
      5.3.1   Cubic factor effects   108
      5.3.2   Lack-of-fit test   109
      5.3.3   Incorporating factor constraints in the design construction algorithm   111
   5.4   Background reading   112
   5.5   Summary   112

6 A "mixture" experiment with process variables   **113**
  6.1 Key concepts   113
  6.2 Case: the rolling mill experiment   114
      6.2.1 Problem and design   114
      6.2.2 Data analysis   121
  6.3 Peek into the black box   123
      6.3.1 The mixture constraint 123
      6.3.2 The effect of the mixture constraint on the model 123
      6.3.3 Commonly used models for data from mixture
             experiments   125
      6.3.4 Optimal designs for mixture experiments 127
      6.3.5 Design construction algorithms for mixture experiments 130
  6.4 Background reading   132
  6.5 Summary   133

7 A response surface design in blocks   **135**
  7.1 Key concepts   135
  7.2 Case: the pastry dough experiment   136
      7.2.1 Problem and design   136
      7.2.2 Data analysis   144
  7.3 Peek into the black box   151
      7.3.1 Model   151
      7.3.2 Generalized least squares estimation 153
      7.3.3 Estimation of variance components 156
      7.3.4 Significance tests   157
      7.3.5 Optimal design of blocked experiments 157
      7.3.6 Orthogonal blocking   158
      7.3.7 Optimal versus orthogonal blocking 160
  7.4 Background reading   160
  7.5 Summary   161

8 A screening experiment in blocks   **163**
  8.1 Key concepts   163
  8.2 Case: the stability improvement experiment   164
      8.2.1 Problem and design   164
      8.2.2 Afterthoughts about the design problem 169
      8.2.3 Data analysis   175
  8.3 Peek into the black box   179
      8.3.1 Models involving block effects 179
      8.3.2 Fixed block effects   182
  8.4 Background reading   184
  8.5 Summary   185

9 Experimental design in the presence of covariates   **187**
9.1   Key concepts   187
9.2   Case: the polypropylene experiment   188
      9.2.1   Problem and design   188
      9.2.2   Data analysis   197
9.3   Peek into the black box   206
      9.3.1   Covariates or concomitant variables   206
      9.3.2   Models and design criteria in the presence of covariates   206
      9.3.3   Designs robust to time trends   211
      9.3.4   Design construction algorithms   215
      9.3.5   To randomize or not to randomize   215
      9.3.6   Final thoughts   216
9.4   Background reading   216
9.5   Summary   217

10 A split-plot design   **219**
10.1   Key concepts   219
10.2   Case: the wind tunnel experiment   220
       10.2.1   Problem and design   220
       10.2.2   Data analysis   232
10.3   Peek into the black box   240
       10.3.1   Split-plot terminology   240
       10.3.2   Model   242
       10.3.3   Inference from a split-plot design   244
       10.3.4   Disguises of a split-plot design   247
       10.3.5   Required number of whole plots and runs   249
       10.3.6   Optimal design of split-plot experiments   250
       10.3.7   A design construction algorithm for optimal
                split-plot designs   251
       10.3.8   Difficulties when analyzing data from
                split-plot experiments   253
10.4   Background reading   253
10.5   Summary   254

11 A two-way split-plot design   **255**
11.1   Key concepts   255
11.2   Case: the battery cell experiment   255
       11.2.1   Problem and design   255
       11.2.2   Data analysis   263
11.3   Peek into the black box   267
       11.3.1   The two-way split-plot model   269
       11.3.2   Generalized least squares estimation   270
       11.3.3   Optimal design of two-way split-plot experiments   273

11.3.4   A design construction algorithm for D-optimal two-way
         split-plot designs   273
11.3.5   Extensions and related designs   274
11.4   Background reading   275
11.5   Summary   276

Bibliography   **277**

Index   **283**

Preface

Design of experiments is a powerful tool for understanding systems and processes. In practice, this understanding often leads immediately to improvements. We present optimal design of experiments as a general and flexible method for applying design of experiments. Our view is that optimal design of experiments is an appropriate tool in virtually any situation that suggests the possible use of design of experiments.

Books on application areas in statistics or applied mathematics, such as design of experiments, can present daunting obstacles to the nonexpert. We wanted to write a book on the practical application of design of experiments that would appeal to new practitioners and experts alike. This is clearly an ambitious goal and we have addressed it by writing a different kind of book.

Each chapter of the book contains a case study. The presentation of the case study is in the form of a play where two consultants, Brad and Peter, of the (fictitious) Intrepid Stats consulting firm, help clients in various industries solve practical problems. We chose this style to make the presentation of the core concepts of each chapter both informal and accessible.

This style is by no means unique. The use of dialogs dates all the way back to the Greek philosopher Plato. More recently, Galileo made use of this style to introduce scientific ideas. His three characters were: the teacher, the experienced student, and the novice.

Though our case studies involve scripted consulting sessions, we advise readers not to copy our consulting style when collaborating on their own design problems. In the interest of a compact exposition of the key points of each case, we skip much of the necessary information gathering involved in competent statistical consulting and problem solving.

We chose our case studies to show just how general and flexible the optimal design of experiments approach is. We start off by a chapter dealing with a simple comparative experiment. The next two chapters deal with a screening experiment and a follow-up experiment in a biotechnology firm. In Chapter 4, we show how a designed response surface experiment contributes to the development of a robust production process in food packaging. In Chapter 5, we set up a response surface experiment to maximize the yield of a chemical extraction process. Chapter 6 deals with an experiment, similar in structure to mixture experiments in the chemical and pharmaceutical industries, aimed at improving the finishing of aluminum sheets. In Chapters 7 and 8, we apply the optimal design of experiments approach to a vitamin stability experiment and a pastry dough experiment run over different days, and we demonstrate that this offers protection against day-to-day variation in the outcomes. In Chapter 9, we show how to take into account a priori information about the experimental units and how to deal with a time trend in the experimental results. In Chapter 10, we set up a wind tunnel experiment that involves factors whose levels are hard to change. Finally, in Chapter 11, we discuss the design of a battery cell experiment spanning two production steps.

Because our presentation of the case studies is often light on mathematical and statistical detail, each chapter also has a section that we call a "Peek into the black box." In these sections, we provide a more rigorous underpinning for the various techniques we employ in our case studies. The reader may find that there is not as much material in these sections on data analysis as might be expected. Many books on design of experiments are mostly about data analysis rather than design generation, evaluation, and comparison. We focus much of our attention in these peeks into the black box on explaining what the reader can anticipate from the analysis, before actually acquiring the response data. In nearly every chapter, we have also included separate frames, which we call "Attachments," to discuss topics that deserve special attention.

We hope that our book will appeal to the new practitioner as well as providing some utility to the expert. Our fondest wish is to empower more experimentation by more people. In the words of Cole Porter, "Experiment and you'll see!"

Acknowledgments

We would like to express our gratitude to numerous people who helped us in the process of writing this book.

First, we would like to thank Chris Nachtsheim for allowing us to use the scenario for the "mixture" experiment in Chapter 6, and Steven Gilmour for providing us with details about the pastry dough experiment in Chapter 7. We are also grateful to Ives Bourgonjon, Ludwig Claeys, Pascal Dekoninck, Tim De Rydt, Karen Dewilde, Heidi Dufait, Toine Machiels, Carlo Mol, and Marc Pauwels whose polypropylene project in Belgium, sponsored by Flanders' Drive, provided inspiration for the case study in Chapter 9.

A screening experiment described in Bie et al. (2005) provided inspiration for the case study in Chapters 2 and 3, while the work of Brenneman and Myers (2003) stimulated us to work out the response surface study involving a categorical factor in Chapter 4. We adapted the case study involving a constrained experimental region in Chapter 5 from an example in Box and Draper (1987). The vitamin stability experiment in Loukas (1997) formed the basis of the blocked screening experiment in Chapter 8. We turned the wind tunnel experiment described in Simpson et al. (2004) and the battery cell experiment studied in Vivacqua and Bisgaard (2004) into the case studies in Chapters 10 and 11.

Finally, we would like to thank Marjolein Crabbe, Marie Gaudard, Steven Gilmour, J. Stuart Hunter, Roselinde Kessels, Kalliopi Mylona, John Sall, Eric Schoen, Martina Vandebroek, and Arie Weeren for proofreading substantial portions of this book. Of course, all remaining errors are our own responsibility.

Heverlee,
Peter Goos

Cary,
Bradley Jones

January 2011

1

A simple comparative experiment

1.1 Key concepts

1. Good experimental designs allow for precise estimation of one or more unknown quantities of interest. An example of such a quantity, or parameter, is the difference in the means of two treatments. One parameter estimate is more precise than another if it has a smaller variance.

2. Balanced designs are sometimes optimal, but this is not always the case.

3. If two design problems have different characteristics, they generally require the use of different designs.

4. The best way to allocate a new experimental test is at the treatment combination with the highest prediction variance. This may seem counterintuitive but it is an important principle.

5. The best allocation of experimental resources can depend on the relative cost of runs at one treatment combination versus the cost of runs at a different combination.

Is A different from B? Is A better than B? This chapter shows that doing the same number of tests on A and on B in a simple comparative experiment, while seemingly sensible, is not always the best thing to do. This chapter also defines what we mean by the best or optimal test plan.

Optimal Design of Experiments: A Case Study Approach, First Edition. Peter Goos and Bradley Jones.
© 2011 John Wiley & Sons, Ltd. Published 2011 by John Wiley & Sons, Ltd.

1.2   The setup of a comparative experiment

Peter and Brad are drinking Belgian beer in the business lounge of Brussels Airport. They have plenty of time as their flight to the United States is severely delayed due to sudden heavy snowfall. Brad has just launched the idea of writing a textbook on tailor-made design of experiments.

[Brad] I have been playing with the idea for quite a while. My feeling is that design of experiments courses and textbooks overemphasize standard experimental plans such as full factorial designs, regular fractional factorial designs, other orthogonal designs, and central composite designs. More often than not, these designs are not feasible due to all kinds of practical considerations. Also, there are many situations where the standard designs are not the best choice.

[Peter] You don't need to convince me. What would you do instead of the classical approach?

[Brad] I would like to use a case-study approach. Every chapter could be built around one realistic experimental design problem. A key feature of most of the cases would be that none of the textbook designs yields satisfactory answers and that a flexible approach to design the experiment is required. I would then show that modern, computer-based experimental design techniques can handle real-world problems better than standard designs.

[Peter] So, you would attempt to promote optimal experimental design as a flexible approach that can solve any design of experiments problem.

[Brad] More or less.

[Peter] Do you think there is a market for that?

[Brad] I am convinced there is. It seems strange to me that, even in 2011, there aren't any books that show how to use optimal or computer-based experimental design to solve realistic problems without too much mathematics. I'd try to focus on how easy it is to generate those designs and on why they are often a better choice than standard designs.

[Peter] Do you have case studies in mind already?

[Brad] The robustness experiment done at Lone Star Snack Foods would be a good candidate. In that experiment, we had three quantitative experimental variables and one categorical. That is a typical example where the textbooks do not give very satisfying answers.

[Peter] Yes, that is an interesting case. Perhaps the pastry dough experiment is a good candidate as well. That was a case where a response surface design was run in blocks, and where it was not obvious how to use a central composite design.

[Brad] Right. I am sure we can find several other interesting case studies when we scan our list of recent consulting jobs.

[Peter] Certainly.

[Brad] Yesterday evening, I tried to come up with a good example for the introductory chapter of the book I have in mind.

[Peter] Did you find something interesting?

[Brad] I think so. My idea is to start with a simple example. An experiment to compare two population means. For example, to compare the average thickness of cables produced on two different machines.

[Peter] So, you'd go back to the simplest possible comparative experiment?

[Brad] Yep. I'd do so because it is a case where virtually everybody has a clear idea of what to do.

[Peter] Sure. The number of observations from the two machines should be equal.

[Brad] Right. But only if you assume that the variance of the thicknesses produced by the two machines is the same. If the variances of the two machines are different, then a 50–50 split of the total number of observations is no longer the best choice.

[Peter] That could do the job. Can you go into more detail about how you would work that example?

[Brad] Sure.

Brad grabs a pen and starts scribbling key words and formulas on his napkin while he lays out his intended approach.

[Brad] Here we go. We want to compare two means, say μ_1 and μ_2, and we have an experimental budget that allows for, say, $n = 12$ observations, n_1 observations from machine 1 and $n - n_1$ or n_2 observations from machine 2. The sample of n_1 observations from the first machine allows us to calculate a sample mean $\overline{X}_1$ for the first machine, with variance σ^2/n_1. In a similar fashion, we can calculate a sample mean $\overline{X}_2$ from the n_2 observations from the second machine. That second sample mean has variance σ^2/n_2.

[Peter] You're assuming that the variance in thickness is σ^2 for both machines, and that all the observations are statistically independent.

[Brad] Right. We are interested in comparing the two means, and we do so by calculating the difference between the two sample means, $\overline{X}_1 - \overline{X}_2$. Obviously, we want this estimate of the difference in means to be precise. So, we want its variance

$$\text{var}(\overline{X}_1 - \overline{X}_2) = \frac{\sigma^2}{n_1} + \frac{\sigma^2}{n_2} = \sigma^2\left(\frac{1}{n_1} + \frac{1}{n_2}\right)$$

or its standard deviation

$$\sigma_{\overline{X}_1 - \overline{X}_2} = \sqrt{\frac{\sigma^2}{n_1} + \frac{\sigma^2}{n_2}} = \sigma\sqrt{\frac{1}{n_1} + \frac{1}{n_2}}$$

to be small.

[Peter] Didn't you say you would avoid mathematics as much as possible?

[Brad] Yes, I did. But we will have to show a formula here and there anyway. We can talk about this later. Stay with me for the time being.

Brad empties his Leffe, draws the waiter's attention to order another, and grabs his laptop.

[Brad] Now, we can enumerate all possible experiments and compute the variance and standard deviation of $\overline{X}_1 - \overline{X}_2$ for each of them.

Table 1.1   Variance of sample mean difference for different
sample sizes n_1 and n_2 for $\sigma^2 = 1$.

n_1	n_2	$\text{var}(\overline{X}_1 - \overline{X}_2)$	$\sigma_{\overline{X}_1 - \overline{X}_2}$	Efficiency (%)
1	11	1.091	1.044	30.6
2	10	0.600	0.775	55.6
3	9	0.444	0.667	75.0
4	8	0.375	0.612	88.9
5	7	0.343	0.586	97.2
6	6	0.333	0.577	100.0
7	5	0.343	0.586	97.2
8	4	0.375	0.612	88.9
9	3	0.444	0.667	75.0
10	2	0.600	0.775	55.6
11	1	1.091	1.044	30.6

Before the waiter replaces Brad's empty glass with a full one, Brad has produced Table 1.1. The table shows the 11 possible ways in which the $n = 12$ observations can be divided over the two machines, and the resulting variances and standard deviations.

[Brad] Here we go. Note that I used a σ^2 value of one in my calculations. This exercise shows that taking n_1 and n_2 equal to six is the best choice, because it results in the smallest variance.

[Peter] That confirms traditional wisdom. It would be useful to point out that the σ^2 value you use does not change the choice of the design or the relative performance of the different design options.

[Brad] Right. If we change the value of σ^2, then the 11 variances will all be multiplied by the value of σ^2 and, so, their relative magnitudes will not be affected. Note that you don't lose much if you use a slightly unbalanced design. If one sample size is 5 and the other is 7, then the variance of our sample mean difference, $\overline{X}_1 - \overline{X}_2$, is only a little bit larger than for the balanced design. In the last column of the table, I computed the efficiency for the 11 designs. The design with sample sizes 5 and 7 has an efficiency of $0.333/0.343 = 97.2\%$. So, to calculate that efficiency, I divided the variance for the optimal design by the variance of the alternative.

[Peter] OK. I guess the next step is to convince the reader that the balanced design is not always the best choice.

Brad takes a swig of his new Leffe, and starts scribbling on his napkin again.

[Brad] Indeed. What I would do is drop the assumption that both machines have the same variance. If we denote the variances of machines 1 and 2 by σ_1^2 and σ_2^2, respectively, then the variances of $\overline{X}_1$ and $\overline{X}_2$ become σ_1^2/n_1 and σ_2^2/n_2. The variance of our sample mean difference $\overline{X}_1 - \overline{X}_2$ then is

$$\text{var}(\overline{X}_1 - \overline{X}_2) = \frac{\sigma_1^2}{n_1} + \frac{\sigma_2^2}{n_2},$$

Table 1.2   Variance of sample mean difference for different
sample sizes n_1 and n_2 for $\sigma_1^2 = 1$ and $\sigma_2^2 = 9$.

n_1	n_2	$\text{var}(\overline{X}_1 - \overline{X}_2)$	$\sigma_{\overline{X}_1 - \overline{X}_2}$	Efficiency (%)
1	11	1.818	1.348	73.3
2	10	1.400	1.183	95.2
3	9	1.333	1.155	100.0
4	8	1.375	1.173	97.0
5	7	1.486	1.219	89.7
6	6	1.667	1.291	80.0
7	5	1.943	1.394	68.6
8	4	2.375	1.541	56.1
9	3	3.111	1.764	42.9
10	2	4.600	2.145	29.0
11	1	9.091	3.015	14.7

so that its standard deviation is

$$\sigma_{\overline{X}_1 - \overline{X}_2} = \sqrt{\frac{\sigma_1^2}{n_1} + \frac{\sigma_2^2}{n_2}}.$$

[Peter] And now you will again enumerate the 11 design options?

[Brad] Yes, but first I need an a priori guess for the values of σ_1^2 and σ_2^2. Let's see what happens if σ_2^2 is nine times σ_1^2.

[Peter] Hm. A variance ratio of nine seems quite large.

[Brad] I know. I know. I just want to make sure that there is a noticeable effect on the design.

Brad pulls his laptop a bit closer and modifies his original table so that the thickness variances are $\sigma_1^2 = 1$ and $\sigma_2^2 = 9$. Soon, he produces Table 1.2.

[Brad] Here we are. This time, a design that requires three observations from machine 1 and nine observations from machine 2 is the optimal choice. The balanced design results in a variance of 1.667, which is 25% higher than the variance of 1.333 produced by the optimal design. The balanced design now is only $1.333/1.667 = 80\%$ efficient.

[Peter] That would be perfect if the variance ratio was really as large as nine. What happens if you choose a less extreme value for σ_2^2? Can you set σ_2^2 to 2?

[Brad] Sure.

A few seconds later, Brad has produced Table 1.3.

[Peter] This is much less spectacular, but it is still true that the optimal design is unbalanced. Note that the optimal design requires more observations from the machine with the higher variance than from the machine with the lower variance.

[Brad] Right. The larger value for n_2 compensates the large variance for machine 2 and ensures that the variance of $\overline{X}_2$ is not excessively large.

Table 1.3   Variance of sample mean difference for different sample sizes n_1 and n_2 for $\sigma_1^2 = 1$ and $\sigma_2^2 = 2$.

n_1	n_2	var($\overline{X}_1 - \overline{X}_2$)	$\sigma_{\overline{X}_1 - \overline{X}_2}$	Efficiency (%)
1	11	1.182	1.087	41.1
2	10	0.700	0.837	69.4
3	9	0.556	0.745	87.4
4	8	0.500	0.707	97.1
5	7	0.486	0.697	100.0
6	6	0.500	0.707	97.1
7	5	0.543	0.737	89.5
8	4	0.625	0.791	77.7
9	3	0.778	0.882	62.4
10	2	1.100	1.049	44.2
11	1	2.091	1.446	23.2

[Peter, pointing to Table 1.3] Well, I agree that this is a nice illustration in that it shows that balanced designs are not always optimal, but the balanced design is more than 97% efficient in this case. So, you don't lose much by using the balanced design when the variance ratio is closer to 1.

Brad looks a bit crestfallen and takes a gulp of his beer while he thinks of a comeback line.

[Peter] It would be great to have an example where the balanced design didn't do so well. Have you considered different costs for observations from the two populations? In the case of thickness measurements, this makes no sense. But imagine that the two means you are comparing correspond to two medical treatments. Or treatments with two kinds of fertilizers. Suppose that an observation using the first treatment is more expensive than an observation with the second treatment.

[Brad] Yes. That reminds me of Eric Schoen's coffee cream experiment. He was able to do twice as many runs per week with one setup than with another. And he only had a fixed number of weeks to run his study. So, in terms of time, one run was twice as expensive as another.

[Peter, pulling Brad's laptop toward him] I remember that one. Let us see what happens. Suppose that an observation from population 1, or an observation with treatment 1, costs twice as much as an observation from population 2. To keep things simple, let the costs be 2 and 1, and let the total budget be 24. Then, we have 11 ways to spend the experimental budget I think. One extreme option takes one observation for treatment 1 and 22 observations for treatment 2. The other extreme is to take 11 observations for treatment 1 and 2 observations for treatment 2. Each of these extreme options uses up the entire budget of 24. And, obviously, there are a lot of intermediate design options.

Peter starts modifying Brad's table on the laptop, and a little while later, he produces Table 1.4.

Table 1.4   Variance of sample mean difference for different designs when treatment 1 is twice as expensive as treatment 2 and the total cost is fixed.

n_1	n_2	$\mathrm{var}(\overline{X}_1 - \overline{X}_2)$	$\sigma_{\overline{X}_1 - \overline{X}_2}$	Efficiency (%)
1	22	1.045	1.022	23.2
2	20	0.550	0.742	44.2
3	18	0.389	0.624	62.4
4	16	0.313	0.559	77.7
5	14	0.271	0.521	89.5
6	12	0.250	0.500	97.1
7	10	0.243	0.493	100.0
8	8	0.250	0.500	97.1
9	6	0.278	0.527	87.4
10	4	0.350	0.592	69.4
11	2	0.591	0.769	41.1

[Peter] Take a look at this.

[Brad] Interesting. Again, the optimal design is not balanced. Its total number of observations is not even an even number.

[Peter, nodding] These results are not quite as dramatic as I would like. The balanced design with eight observations for each treatment is still highly efficient. Yet, this is another example where the balanced design is not the best choice.

[Brad] The question now is whether these examples would be a good start for the book.

[Peter] The good thing about the examples is that they show two key issues. First, the standard design is optimal for at least one scenario, namely, in the scenario where the number of observations one can afford is even, the variances in the two populations are identical and the cost of an observation is the same for both populations. Second, the standard design is often no longer optimal as soon as one of the usual assumptions is no longer valid.

[Brad] Surely, our readers will realize that it is unrealistic to assume that the variances in two different populations are exactly the same.

[Peter] Most likely. But finding the optimal design when the variances are different requires knowledge concerning the magnitude of σ_1^2 and σ_2^2. I don't see where that knowledge might come from. It is clear that choosing the balanced design is a reasonable choice in the absence of prior knowledge about σ_1^2 and σ_2^2, as that balanced design was at least 80% efficient in all of the cases we looked at.

[Brad] I can think of a case where you might reasonably expect different variances. Suppose your study used two machines, and one was old and one was new. There, you would certainly hope the new machine would produce less variable output. Still, an experimenter usually knows more about the cost of every observation than about its variance. Therefore, the example with the different costs for the two populations is possibly more convincing. If it is clear that observations for treatment 1 are twice as expensive as observations for treatment 2, you have just shown that the experimenter should drop the standard design, and use the unbalanced one instead. So, that sounds like a good example for the opening chapter of our book.

[Peter, laughing] I see you have already lured me into this project.

[Brad] Here is a toast to our new project!

They clink their glasses, and turn their attention toward the menu.

1.3 Summary

Balanced designs for one experimental factor at two levels are optimal if all the runs have the same cost, the observations are independent and the error variance is constant. If the error variances are different for the two treatments, then the balanced design is no longer best. If the two treatments have different costs, then, again, the balanced design is no longer best.

A general principle is that the experimenter should allocate more runs to the treatment combinations where the uncertainty is larger.

2

An optimal screening experiment

2.1 Key concepts

1. Orthogonal designs for two-level factors are also optimal designs. As a result, a computerized-search algorithm for generating optimal designs can generate standard orthogonal designs.

2. When a given factor's effect on a response changes depending on the level of a second factor, we say that there is a two-factor interaction effect. Thus, a two-factor interaction is a combined effect on the response that is different from the sum of the individual effects.

3. Active two-factor interactions that are not included in the model can bias the estimates of the main effects.

4. The alias matrix is a quantitative measure of the bias referred to in the third key concept.

5. Adding any term to a model that was previously estimated without that term removes any bias in the estimates of the factor effects due to that term.

6. The trade-off in adding two-factor interactions to a main-effects model after using an orthogonal main-effect design is that you may introduce correlation in the estimates of the coefficients. This correlation results in an increase in the variances of the effect estimates.

Screening designs are among the most commonly used in industry. The idea of screening is to explore the effects of many experimental factors in one relatively

Optimal Design of Experiments: A Case Study Approach, First Edition. Peter Goos and Bradley Jones.
© 2011 John Wiley & Sons, Ltd. Published 2011 by John Wiley & Sons, Ltd.

small study to find the few factors that most affect the response of interest. This methodology is based on the Pareto or sparsity-of-effects principle that states that most real processes are driven by a few important factors.

In this chapter, we generate an optimal design for a screening experiment and analyze the resulting data. As in many screening experiments, we are left with some ambiguity about what model best describes the underlying behavior of the system. This ambiguity will be resolved in Chapter 3. As it also often happens, even though there is some ambiguity about what the best model is, we identify new settings for the process that substantially improve its performance.

2.2   Case: an extraction experiment

2.2.1   Problem and design

Peter and Brad are taking the train to Rixensart, southeast of Brussels, to visit GeneBe, a Belgian biotech firm.

[Brad] What is the purpose of our journey?

[Peter] Our contact, Dr. Zheng, said GeneBe is just beginning to think about using designed experiments as part of their tool set.

[Brad] So, we should probably keep things as standard as possible.

[Peter] I guess you have a point. We need to stay well within their comfort zone. At least for one experiment.

[Brad] Do you have any idea what they plan to study?

[Peter] Dr. Zheng told me that they are trying to optimize the extraction of an antimicrobial substance from some proprietary cultures they have developed in house. He sketched the extraction process on the phone, but reproducing what he told me would be a bit much to ask. Microbiology is not my cup of tea.

[Brad] Likewise. I am sure Dr. Zheng will supply all the details we need during our meeting.

They arrive at GeneBe and Dr. Zheng meets them in the reception area.

[Dr. Zheng] Peter, it is good to see you again. And this must be. . . .

[Peter] Brad Jones, he is a colleague of mine from the States. He is the other principal partner in our firm, Intrepid Stats.

[Dr. Zheng] Brad, welcome to GeneBe. Let's go to a conference room and I will tell you about the study we have in mind.

In the conference room, Brad fires up his laptop, while Dr. Zheng gets coffee for everyone. After a little bit of small talk, the group settles in to discuss the problem at hand.

[Dr. Zheng] Some of our major customers are food producers. They are interested in inhibiting the growth of various microbes that are common in most processed foods. You know, *Escherichia coli*, *Salmonella typhimurium*, etc. In the past they have used chemical additives in food to do this, but there is some concern about the long-term effects of this practice. We have found a strong microbial inhibitor, a certain lipopeptide, in strains of *Bacillus subtilis*. If we can improve the yield of extraction of this inhibitor from our cultures, we may have a safer alternative than the current chemical agents. The main goal of the experiment we want to perform is to increase the yield of the extraction process.

[Brad] Right.

[Dr. Zheng] The problem is that we know quite a lot already about the lipopeptide, but not yet what affects the extraction of that substance.

[Brad] Can you tell us a bit about the whole process for producing the antimicrobial substance?

[Dr. Zheng] Sure. I will keep it simple though, as it is not difficult to make it sound very complicated. Roughly speaking, we start with a strain of *B. subtilis*, put it in a flask along with some medium, and cultivate it at 37°C for 24 hours while shaking the flask the whole time. The next step is to put the resulting culture in another flask, with some other very specific medium, and cultivate it at a temperature between 30°C and 33°C for some time. The culture that results from these operations is then centrifuged to remove bacterial cells, and then it is ready for the actual extraction.

[Peter] How does that work?

[Dr. Zheng] We start using 100 ml of our culture and add various solvents to it. In the extraction process, we can adjust the time in solution and the pH of the culture.

[Peter] Do you have an idea about the factors you would like to study? The time in solution and the pH seem ideal candidates.

[Dr. Zheng] Yes, we did our homework. We identified six factors that we would like to investigate. We want to look at the presence or absence of four solvents: methanol, ethanol, propanol, and butanol. The two other factors we want to investigate are indeed pH and the time in solution.

[Peter, nodding] Obviously, the response you want to study is the yield. How do you measure it?

[Dr. Zheng] The yield is expressed in milligrams per 100 ml. We determine the yield of a run of the extraction process by means of high-performance liquid chromatography or HPLC.

[Peter] That does not sound very simple either. What is the yield of your current extraction process?

[Dr. Zheng] We have been using methanol at neutral pH for 2 hours and getting about a 25 mg yield per batch. We need something higher than 45 mg to get management interested in taking the next step.

[Brad] That sounds like quite a challenge.

[Peter] How many processing runs can you afford for this study?

[Dr. Zheng] Design of experiments is not an accepted strategy here. This study is just a proof of concept. I doubt that I can persuade management to permit more than 15 runs. Given the time required to prepare the cultures, however, fewer than 15 trials would be better.

[Peter] Twelve is an ideal number for a screening experiment. Using large enough multiples of four allows you to estimate the main effects of your factors independently. You can then save three runs for doing a confirmatory experiment later on. If you think that is a good idea, then I think Brad will have a design for you in less than a minute.

Table 2.1 Brad's design for the extraction experiment at GeneBe.

Run	x_1	x_2	x_3	x_4	x_5	x_6
1	-1	-1	-1	$+1$	-1	-1
2	-1	$+1$	-1	-1	$+1$	-1
3	-1	$+1$	-1	$+1$	$+1$	$+1$
4	$+1$	$+1$	$+1$	-1	-1	-1
5	-1	-1	$+1$	-1	-1	$+1$
6	-1	$+1$	$+1$	$+1$	-1	-1
7	$+1$	$+1$	-1	-1	-1	$+1$
8	$+1$	-1	-1	-1	$+1$	-1
9	$+1$	-1	$+1$	$+1$	$+1$	-1
10	-1	-1	$+1$	-1	$+1$	$+1$
11	$+1$	$+1$	$+1$	$+1$	$+1$	$+1$
12	$+1$	-1	-1	$+1$	-1	$+1$

In a few seconds, Brad turns his laptop so that Dr. Zheng and Peter can see the screen.

[Brad] In generating this 12-run design, I used the generic names, x_1–x_6, for your six experimental factors. I also coded the absence and presence of a solvent using a -1 and a $+1$, respectively. For the factors pH and time, I used a -1 for their low levels and a $+1$ for their high levels.

He shows Dr. Zheng and Peter the design in Table 2.1.

[Dr. Zheng] That was fast! How did you create this table? Did you just pick it from a catalog of designs?

[Brad] In this case, I could have done just that, but I didn't. I created this design ex nihilo by using a computer algorithm that generates custom-built optimal designs.

[Dr. Zheng] That sounds fancy. I was hoping that, for our first project, we could just do something uncontroversial and use a design from a book or a catalog.

[Peter] I have been looking at this design while you two were talking and I would say that, for a two-level 12-run design, this is about as uncontroversial as you can get.

[Dr. Zheng] How so?

[Peter] This design has perfect structure in one sense. Notice that each column only has two values, -1 and $+1$. If we sum each column, we get zero, which means that each column has the same number of -1s and $+1$s. There is even more balance than that. Each pair of columns has four possible pairs of values: $++, +-, -+, --$. Each of these four possibilities appears three times.

[Brad] In the technical jargon, a design that has all these properties is an orthogonal design. In fact, I said earlier that I could have taken this design from a catalog of designs. That is because the optimal design of experiments algorithm in my software generated a design that can be derived from a Plackett–Burman design.

[Peter] A key property of orthogonal designs is that they allow independent estimation of the main effects.

[Dr. Zheng] I have heard of orthogonal designs as well as Plackett–Burman designs before. It seems they are very popular in food science research. This should make it easy to sell this design to management.

[Brad] That's good news.

[Dr. Zheng] If you could have selected this design from a catalog of designs, why did you build it from scratch?

[Brad] It's a matter of principle. We create each design according to the dictates of the problem we are trying to solve.

[Peter] As opposed to choosing a design from a catalog and then force-fitting it to the problem. Of course, in this case, force-fitting was not an issue. The design from the catalog turns out to be the same as the design that we create from scratch.

[Brad] I knew that it would turn out this way, but I always like to show that custom-built optimal designs are not necessarily fancy or exotic or complicated.

[Peter] True. We feel that it is appropriate to recommend custom-built optimal designs whether the problem is routine or extraordinary from a design point of view.

[Dr. Zheng] I think we can run your design this week. It would be helpful if you could replace the coded factor levels in your table with the actual factor levels we intend to use.

[Brad] Sure. What are they?

[Dr. Zheng] For each of the solvents, we will use 10 ml whenever it is used in the extraction process. So, the first four factors should range from 0 to 10 ml. For the pH, we will most likely use a value of 6 as the low level and a value of 9 as the high level. Finally, we were thinking of a range from 1 to 2 hours for the time in solution.

Soon, Brad has produced Table 2.2.

[Brad] Here you go.

Table 2.2 Design for the extraction experiment at GeneBe, using factor levels expressed in engineering units.

Run	Methanol (ml) x_1	Ethanol (ml) x_2	Propanol (ml) x_3	Butanol (ml) x_4	pH x_5	Time (h) x_6
1	0	0	0	10	6	1
2	0	10	0	0	9	1
3	0	10	0	10	9	2
4	10	10	10	0	6	1
5	0	0	10	0	6	2
6	0	10	10	10	6	1
7	10	10	0	0	6	2
8	10	0	0	0	9	1
9	10	0	10	10	9	1
10	0	0	10	0	9	2
11	10	10	10	10	9	2
12	10	0	0	10	6	2

[Dr. Zheng, handing Brad his memory stick] Can you copy this table to my stick? I hope to send you an e-mail with the data in the course of next week.

Later, on the train back, Peter and Brad are discussing the meeting they had with Dr. Zheng. One of the issues they discuss is related to the design in Tables 2.1 and 2.2.

[Peter] Did you realize how lucky you were when constructing the design for Dr. Zheng?

[Brad] How so?

[Peter] Well, your design could have had a row with the first four factors at their low level. That would have meant doing a run without any of the solvents.

[Brad] That would have been awkward. Replacing all the $+1$s with -1s and all -1s with $+1$s would have solved the problem. That operation gives a design that is equally good, in statistical terms, as the one I generated.

[Peter] But much better suited for Dr. Zheng's problem. It is always good to know that, very often, there is more than one design that is optimal. We can then pick the design that best fits our problem.

2.2.2 Data analysis

A week later Peter gets the following e-mail from Dr. Zheng with Table 2.3 enclosed.

Peter:

It looks like we have a big success already. Notice that run number 7 in the enclosed table has a yield of more than 45 mg. I could not resist taking the liberty of analyzing the data myself. It appears that the effects of methanol, ethanol, pH, and time are all statistically significant.

Table 2.3   Design and response data for the extraction experiment at GeneBe.

Run	Methanol x_1	Ethanol x_2	Propanol x_3	Butanol x_4	pH x_5	Time x_6	Yield (mg)
1	0	0	0	10	6	1	10.94
2	0	10	0	0	9	1	15.79
3	0	10	0	10	9	2	25.96
4	10	10	10	0	6	1	35.92
5	0	0	10	0	6	2	22.92
6	0	10	10	10	6	1	23.54
7	10	10	0	0	6	2	47.44
8	10	0	0	0	9	1	19.80
9	10	0	10	10	9	1	29.48
10	0	0	10	0	9	2	17.13
11	10	10	10	10	9	2	43.75
12	10	0	0	10	6	2	40.86

If I construct a prediction equation using the -1 to $+1$ coding that Brad used when he created the design, I get this:

$$\text{Yield} = 27.79 + 8.41x_1 + 4.27x_2 - 2.48x_5 + 5.21x_6, \qquad (2.1)$$

where

$$x_1 = (\text{Methanol} - 5)/5,$$
$$x_2 = (\text{Ethanol} - 5)/5,$$
$$x_5 = (\text{pH} - 7.5)/1.5,$$

and

$$x_6 = (\text{time} - 1.5)/0.5.$$

My approach was to fit a model with all the main effects first, and then remove all the nonsignificant ones from the model. I used a 5% significance level.

Can you and Brad take a look at this and come by tomorrow at 10am to present your results and recommendations?

Thanks,
-Zheng

The next morning Peter and Brad present their results to Dr. Zheng and his manager, Bas Ritter, in the same conference room as before.

[Peter] We verified the results you sent us in your e-mail, Dr. Zheng. Using your prediction equation, your best yield will average around 48 mg when you use methanol and ethanol as solvents at a pH of 6 for 2 hours.

[Bas] I was hoping you would recommend a setting with higher predicted yield. We would really prefer to have yields of 50 mg or more.

[Brad] We felt the same way, so we did a little extra digging. We used all subsets regression to fit all the models with up to seven effects. We were looking for possible large two-factor interaction effects.

[Bas] What is a two-factor interaction effect?

[Peter] Sometimes the effect on the response of changes in one factor depends on the value of another factor. When that happens, we say that the two factors interact in their effect on the response.

[Brad] In this case, along with a model containing all the main effects, we may have found a two-factor interaction involving ethanol and propanol. It appears that, when ethanol is not in the solvent system, adding propanol to the system raises the yield. On the other hand, when ethanol is in the solvent system, then adding propanol too decreases the yield. Here are two plots that show what I mean.

Brad turns his laptop to show Bas the graphs in Figures 2.1 and 2.2.

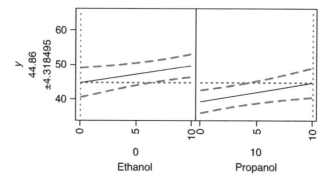

Figure 2.1   Prediction for a solvent system without ethanol but with propanol.

[Bas, scratching the back of his head] How does this show what you mean? How do I interpret these figures?

[Brad] Let me explain all the features of this plot. The left panel shows the effect on the response, Y, of the factor ethanol, while the right panel shows the effect of the factor propanol. In each of the two panels, the solid line represents the factor effects, as estimated from the data, for a given setting of all other factors. The two dashed lines in each panel form a 95% confidence band. The vertical dotted lines indicate the selected settings for the factors, and the horizontal dotted lines tell us what the predicted response is.

[Peter, pointing at Figure 2.1] In this figure's left panel, you can see that ethanol is at its low level, 0. In the right panel of the figure, we can see that, for that setting of the factor ethanol, increasing the level of the propanol factor has a positive impact on the response. So, if we fix ethanol at 0, the best thing we can do is set the factor propanol to its high level, 10. This gives a predicted yield of 44.86.

[Brad] Plus or minus 4.32.

[Dr. Zheng] I think I have it. Let me try to interpret the other figure. In the left panel, we can see that ethanol is now fixed at its high level. In the right panel, we see that propanol has a slightly negative impact on the yield.

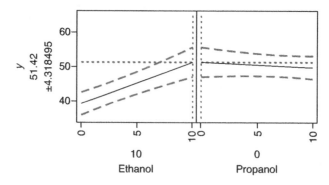

Figure 2.2   Prediction for a solvent system with ethanol but without propanol.

[Bas] So, if we fix ethanol at 10, we should use no propanol, in which case the predicted yield is 51.42. Again, $+4.32$ or -4.32.

[Brad] Perfect!

[Bas, pensively] So, in one case, adding propanol improves yield. In the other, it makes the yield worse. That seems strange.

[Peter] Two-factor interactions often seem strange, but they are quite common in our experience. There is some potential good news here. With all the main effects and the two-factor interaction in the model, we get a maximum predicted yield of 51.4 mg.

[Bas] What do we have to do to get that yield?

[Peter] Use a solvent system with methanol, ethanol, and butanol at a pH of 6 for 2 hours. But, we want to caution you that the two-factor interaction we found is only barely statistically significant. We also found many alternative models that are competitive with our preferred model in terms of their predictive capability. We think it would be wise to augment your original experiment with several extra runs to distinguish between the various competitive models and to nail down the two-factor interaction involving ethanol and propanol, assuming our preferred model is correct.

[Bas] I'll consider that. In the meantime, Dr. Zheng, let's do a few extractions at the settings Peter just suggested and see what kind of yield we get.

[Brad] That was going to be our other recommendation, but you beat us to it.

[Dr. Zheng] Dr. Ritter, I would like to ask Peter and Brad a few technical questions.

[Bas] I don't think I need to be here for this. Peter and Brad, thanks for your help. I will have Dr. Zheng get in touch with you if we decide to go with your follow-up design suggestion.

Bas leaves.

[Dr. Zheng] Can you show me the prediction equation for the model with the two-factor interaction?

[Peter] Sure. Here is the prediction equation. In other words, our estimated model. Note that I use coded units.

Peter writes the following equation on the white board:

$$\text{Yield} = 27.79 + 9.00x_1 + 4.27x_2 + 1.00x_3 + 1.88x_4 \\ - 3.07x_5 + 4.63x_6 - 1.77x_2x_3. \tag{2.2}$$

[Dr. Zheng] Your custom-built optimal design was really derived from a Plackett–Burman design, right?

[Peter] Yes.

[Dr. Zheng] Then, I am a little confused. The literature I have read says that these designs assume that two-factor interactions are negligible. I thought that you were only supposed to fit main-effects models with these designs.

[Brad] There is a reason for the usual assumption that the two-factor interactions are negligible. If there is a large two-factor interaction and you use a Plackett–Burman type of design, it will bias the estimates of the main effects of factors not involved in the interaction.

Dr. Zheng frowns.

[Brad, displaying Equation (2.1) on the screen of his laptop] Look at this. Here, I have the model that you estimated. That is a model involving only main effects. That model estimates the main effect for methanol to be 8.41. The estimate of the main effect of methanol in our prediction equation is 9.00.

[Dr. Zheng] That's a difference of 0.59, which isn't small. Why are the estimates different?

[Peter] Because the two-factor interaction involving ethanol and propanol is biasing your estimate of the main effect of methanol. Assuming this interaction is the only real two-factor interaction, then the expected value of the main-effect estimate for methanol, which I call $\hat{\beta}_1$, is the sum of the true (but unknown) effect of methanol, β_1, and one-third of the true (and also unknown) effect of the two-factor interaction involving ethanol and propanol, which I call β_{23}.

Peter writes the following equation on the white board:

$$E(\hat{\beta}_1) = \beta_1 + \frac{1}{3}\beta_{23}.$$

[Dr. Zheng] Why is the bias equal to one-third of β_{23}? Where does the one-third come from?

[Peter, who has taken over Brad's laptop and uses his favorite spreadsheet to demonstrate to Dr. Zheng where the one-third comes from] Normally, computing the bias requires inverting a matrix, but, because our design is orthogonal, the calculation is simpler. You start by creating a column for the interaction of the factors x_2 and x_3. Now, take sum of the product of x_1 with this column and divide the result by 12, which is the number of experimental runs. The result is one-third.

The screen of Brad's laptop now shows Table 2.4. When he sees Dr. Zheng is nodding, Peter continues.

[Peter] Now, when we include the two-factor interaction effect β_{23} involving ethanol and propanol in the model, its estimate is −1.77. The bias is $\beta_{23}/3 = -1.77/3 = -0.59$. So, the main-effect estimate for methanol in your main-effects model is 0.59 smaller than it should be. Once we find a real two-factor interaction and include it in the model, the estimates of the main effects are no longer biased by that interaction.

[Brad] That is why, when you run a Plackett–Burman type of design, it is a good idea to look for big two-factor interactions. If you find them, it can have a dramatic effect on your conclusions.

[Dr. Zheng] Hold on, the design is orthogonal for the main effects, right? But aren't your factor-effect estimates correlated once you add two-factor interactions to your model?

[Peter] Yes. And that complicates the model selection process a bit. But we think that the potential benefit of discovering a big two-factor interaction is worth the extra effort.

[Brad] I fit both the main-effects model and the model with the two-factor interaction. Here are the two sets of parameter estimates.

Brad turns his laptop so Dr. Zheng can see Tables 2.5 and 2.6.

Table 2.4   Calculation of the extent to which the two-factor interaction involving ethanol (x_2) and propanol (x_3) causes bias in the main-effect estimate of methanol (x_1).

Run	x_1	x_2	x_3	x_2x_3	$x_1x_2x_3$
1	−1	−1	−1	+1	−1
2	−1	+1	−1	−1	+1
3	−1	+1	−1	−1	+1
4	+1	+1	+1	+1	+1
5	−1	−1	+1	−1	+1
6	−1	+1	+1	+1	−1
7	+1	+1	−1	−1	−1
8	+1	−1	−1	+1	+1
9	+1	−1	+1	−1	−1
10	−1	−1	+1	−1	+1
11	+1	+1	+1	+1	+1
12	+1	−1	−1	+1	+1
				Average:	+1/3

[Brad, pointing at Table 2.5] Notice that all the estimates in the main-effects model have the same standard error. That is a consequence of the −1 to +1 coding scheme we use and also the fact that the design is orthogonal.

[Peter, drawing the attention to Table 2.6] Ah! But when we add the two-factor interaction, we get three different standard errors. The estimates of the intercept and the main effects of ethanol and propanol are uncorrelated with the estimate of the ethanol-by-propanol interaction effect. Therefore, these estimates have the smallest standard errors. The estimates of the other main effects are correlated with the estimate of the two-factor interaction effect by the same amount, so their standard errors are all the same but larger. Finally, the estimate for the two-factor interaction effect has

Table 2.5   Model with main effects only. Asterisks indicate effects that are significantly different from zero at a 5% level of significance.

Effect	Estimate	Standard error	t Ratio	p Value
Intercept	27.79	0.71	39.0	<.0001*
Methanol	8.41	0.71	11.8	<.0001*
Ethanol	4.27	0.71	6.0	0.0019*
Propanol	1.00	0.71	1.4	0.2214
Butanol	1.29	0.71	1.8	0.1292
pH	−2.48	0.71	−4.29	0.0178*
Time	5.22	0.71	8.37	0.0007*

Table 2.6   Model with main effects and ethanol-by-propanol interaction effect. Asterisks indicate effects that are significantly different from zero at a 5% level of significance.

Effect	Estimate	Standard error	t Ratio	p Value
Intercept	27.79	0.45	61.9	<.0001*
Methanol	9.00	0.49	18.3	<.0001*
Ethanol	4.27	0.45	9.5	0.0007*
Propanol	1.00	0.45	2.2	0.0908
Butanol	1.88	0.49	3.8	0.0186*
pH	−3.07	0.49	−6.2	0.0034*
Time	4.63	0.49	9.4	0.0007*
Ethanol × Propanol	−1.77	0.60	−2.9	0.0426*

the largest standard error of all due to its correlation with four other effects in the model.

[Dr. Zheng] I notice that in the model with the interaction term included, the main effect of butanol is highly significant. In the main-effects model, it is not significant at the 0.05 level. Is that also the result of the addition of the two-factor interaction term?

[Brad] It is. Adding the ethanol-by-propanol interaction effect to the model makes two things happen. First, the estimate of the main effect of butanol increased by 0.59 after adding the interaction to the model.

[Dr. Zheng] 0.59 again!

[Brad] Yes, this is because the aliasing of the butanol main effect with the ethanol-by-propanol interaction effect is one-third too.

[Dr. Zheng, nodding] Right. What is the second thing that happened by adding the interaction to the model?

[Brad] The root mean squared error of the model went down. This resulted in a decrease of the standard error of the factor-effect estimates. The two things that happened both made the t ratio of the significance test for the butanol main effect increase. The larger estimate caused the numerator of the t ratio to go up, and the smaller standard error (0.49 instead of 0.71) caused the denominator to go down.

[Dr. Zheng] To make sure I completely understand, can you explain why the root mean squared error decreased?

[Peter] The sum of squared errors measures the variation in the responses that is not explained by the model. By including the interaction effect of ethanol and propanol in the model, we have added an important extra explanatory variable to the model. The result is that there is less unexplained variation, so the sum of squared errors is smaller. If the decrease is large enough, then the mean squared error and the root mean squared error also go down.

[Dr. Zheng] And then the standard errors of the estimates go down too.

[Peter] Which, in the case of the main effect of butanol, leads to a significance probability, or p value, of less than 5%.

[Dr. Zheng] Thank you. I can definitely see the benefit of adding the interaction to the model. I also see why you are recommending a follow-up study to determine whether the interaction effect is real or not. I will do what I can to make that study happen.

[Peter] Great! We will be hoping to hear from you.

2.3   Peek into the black box

Usually, screening experiments involve two-level designs for fitting models with main effects or main effects and two-factor interaction effects. The benefit of restricting each factor to two levels and using simple models is that an experimental design with relatively few runs can accommodate many factors.

2.3.1   Main-effects models

Main-effects models contain only first-order terms in each experimental factor. We denote the k factors involved in an experiment $x_1, x_2, \ldots, x_k$. The model equation for the ith response is

$$Y_i = \beta_0 + \beta_1 x_{1i} + \cdots + \beta_k x_{ki} + \varepsilon_i, \tag{2.3}$$

where $\beta_0, \beta_1, \ldots, \beta_k$ are the unknown parameters of the model and ε_i is the deviation between the response, Y_i, and the unknown mean value of the response at the ith setting of the factors, $x_{1i}, x_{2i}, \ldots, x_{ki}$. The statistical interpretation of ε_i is that it is a random error. The standard assumption is that ε_i is normally distributed with mean zero and variance σ_ε^2 and that ε_i is independent of ε_j for all i and $j \neq i$. The variance σ_ε^2 is the error variance.

The parameter β_0 is the intercept or the constant. The parameters $\beta_1, \beta_2, \ldots, \beta_k$ represent the linear effects or the main effects of the factors $x_1, x_2, \ldots, x_k$ on the response Y.

The matrix form of the model, which we call a regression model, is more compact:

$$Y = X\beta + \varepsilon. \tag{2.4}$$

Here, Y and ε are both vectors with as many elements as there are runs in the experiment. We denote the number of runs, also named the sample size, by n. The vector Y contains the n responses, whereas the vector ε contains the n random errors. The matrix X is the model matrix. If there are k factors, then this matrix has $k + 1$ columns and n rows. The elements of the first column are all ones. The second column contains the n experimental values of the factor x_1. The third column contains the n experimental values of x_2, etc. The final column contains the n experimental values of the kth factor, x_k. The vector β has $k + 1$ elements corresponding to the intercept β_0 and the k main effects. For clarity, we show the building blocks of the model in detail in Attachment 2.1.

Attachment 2.1   Building blocks of the regression model in Equation (2.4)

In the regression model.

$$Y = X\beta + \varepsilon,$$

there are two vectors involving n elements:

$$\begin{bmatrix} Y_1 \\ Y_2 \\ \vdots \\ Y_n \end{bmatrix} \text{ and } \begin{bmatrix} \varepsilon_1 \\ \varepsilon_2 \\ \vdots \\ \varepsilon_n \end{bmatrix}.$$

For a main-effects model, the model matrix X has n rows and $k + 1$ columns:

$$X = \begin{bmatrix} 1 & x_{11} & x_{21} & \cdots & x_{k1} \\ 1 & x_{12} & x_{22} & \cdots & x_{k2} \\ \vdots & \vdots & \vdots & \ddots & \vdots \\ 1 & x_{1n} & x_{2n} & \cdots & x_{kn} \end{bmatrix}.$$

Finally, the unknown model parameters in the main-effects model (the intercept and the main effects) are contained within a vector β that has $k + 1$ elements:

$$\begin{bmatrix} \beta_0 \\ \beta_1 \\ \beta_2 \\ \vdots \\ \beta_k \end{bmatrix}.$$

For factor screening, many investigators use the main-effects model as the a priori model. In doing so, they assume that higher-order effects, such as two-factor interaction effects, can be ignored.

2.3.2   Models with two-factor interaction effects

A two-factor interaction effect is one kind of second-order effect. Models with two-factor interaction effects involve terms of the form $\beta_{ij}x_ix_j$. A main-effects-plus-two-factor-interaction model for an experiment with k factors is given by

$$\begin{aligned} Y_i = {} & \beta_0 + \beta_1 x_{1i} + \cdots + \beta_k x_{ki} \\ & + \beta_{12} x_{1i} x_{2i} + \beta_{13} x_{1i} x_{3i} + \cdots + \beta_{1k} x_{1i} x_{ki} \\ & + \beta_{23} x_{2i} x_{3i} + \cdots + \beta_{k-1,k} x_{k-1,i} x_{ki} + \varepsilon_i, \end{aligned} \tag{2.5}$$

where β_{ij} is the interaction effect of the factors x_i and x_j. Two-factor interaction effects are sometimes included in the a priori model of a screening experiment. Because there are $k(k-1)/2$ two-factor interaction effects, including all two-factor interactions in the a priori model requires at least $k(k-1)/2$ extra experimental runs.

When interaction effects are present, the impact of a factor on the response depends on the level of one or more other factors. Figures 2.1 and 2.2 show that, when a system has a two-factor interaction, the slope of either factor's relationship with the response depends on the setting of the other factor. In extreme cases, the slope can change from negative to positive, or vice versa. Therefore, in the presence of interaction effects, it is no longer meaningful to interpret the main effects $\beta_1, \beta_2, \ldots, \beta_k$ separately. Generally, designed experiments are a powerful tool for detecting interaction effects.

The matrix form of the regression model in Equation (2.4) remains valid in the presence of two-factor interaction effects. However, the contents of the model matrix $\mathbf{X}$ and of the vector $\boldsymbol{\beta}$ change when compared to the main-effects model. In Attachment 2.2, we show the structure of the model matrix and the corresponding vector $\boldsymbol{\beta}$ for the model with two-factor interactions.

Attachment 2.2   Building blocks of the regression model in Equation (2.4) when the model contains main effects and two-factor interactions.

For a model containing all main effects and all two-factor interactions, the model matrix $\mathbf{X}$ has n rows and $k+1+k(k-1)/2$ columns:

$$\mathbf{X} = \begin{bmatrix} 1 & x_{11} & x_{21} & \cdots & x_{k1} & x_{11}x_{21} & \cdots & x_{11}x_{k1} & x_{21}x_{31} & \cdots & x_{k-1,1}x_{k1} \\ 1 & x_{12} & x_{22} & \cdots & x_{k2} & x_{12}x_{22} & \cdots & x_{12}x_{k2} & x_{22}x_{32} & \cdots & x_{k-1,2}x_{k2} \\ \vdots & \vdots & \vdots & \ddots & \vdots & \vdots & \ddots & \vdots & \vdots & \ddots & \vdots \\ 1 & x_{1n} & x_{2n} & \cdots & x_{kn} & x_{1n}x_{2n} & \cdots & x_{1n}x_{kn} & x_{2n}x_{3n} & \cdots & x_{k-1,n}x_{kn} \end{bmatrix}.$$

Finally, the unknown model parameters in the main-effects-plus-two-factor-interaction model (representing the intercept, the main effects and the interaction effects) are contained within a vector $\boldsymbol{\beta}$ that has $k+1+k(k-1)/2$ elements:

$$\begin{bmatrix} \beta_0 \\ \beta_1 \\ \beta_2 \\ \vdots \\ \beta_k \\ \beta_{12} \\ \vdots \\ \beta_{1k} \\ \beta_{23} \\ \vdots \\ \beta_{k-1,k} \end{bmatrix}.$$

2.3.3 Factor scaling

For a continuous experimental factor, it is possible to smoothly change the level over an interval, from a lower end point, L, to an upper end point, U. It is common to scale the levels of such a factor so that they lie on the interval $[-1, +1]$. This can be done as follows. The midpoint, M, of the unscaled interval $[L, U]$ is

$$M = \frac{L + U}{2},$$

and the half of the range of that interval, Δ, is

$$\Delta = \frac{U - L}{2}.$$

The scaled level x_k of a factor with unscaled level l_k then is

$$x_k = \frac{l_k - M}{\Delta}. \tag{2.6}$$

The scaled value x_k is -1 when the factor is at its low level, L, and $+1$ when the factor is at its high level, U.

This scaling convention has a couple of benefits. First, it allows for the direct comparison of the sizes of the factor effects. If, for instance, an effect β_i is twice as large in magnitude as another effect β_j, then the ith factor's effect on the response is twice as big as the jth factor's effect on the response. Second, models containing both main effects and two-factor interactions will always have the main effects correlated with two-factor interactions if the factors are not scaled. This makes model selection more difficult.

If factors are categorical and have only two levels, it is also standard to encode one of the levels as -1 and the other as $+1$. This way of coding, named effects-type coding, creates a correspondence between the effects of categorical factors and those of continuous factors.

2.3.4 Ordinary least squares estimation

For fitting the regression model in Equation (2.4) to the data, we generally use ordinary least squares estimation. The ordinary least squares estimator of the vector of unknown model coefficients, $\boldsymbol{\beta}$, is

$$\hat{\boldsymbol{\beta}} = (\mathbf{X}'\mathbf{X})^{-1}\mathbf{X}'\mathbf{Y}. \tag{2.7}$$

The variance–covariance matrix of this estimator is

$$\mathrm{var}(\hat{\boldsymbol{\beta}}) = \sigma_\varepsilon^2 (\mathbf{X}'\mathbf{X})^{-1}. \tag{2.8}$$

The diagonal elements of this matrix are the variances of the estimates $\hat{\beta}_0, \hat{\beta}_1, \ldots, \hat{\beta}_k$ of the individual model coefficients, the intercept β_0 and the main effects $\beta_1, \ldots, \beta_k$

of the k experimental factors. The off-diagonal elements are the covariances between pairs of estimates. In general, for the main-effects model in Equation (2.3), the variance–covariance matrix is a $(k + 1) \times (k + 1)$ matrix with the following structure:

$$
\text{var}(\hat{\boldsymbol{\beta}}) = \begin{bmatrix}
\text{var}(\hat{\beta}_0) & \text{cov}(\hat{\beta}_0, \hat{\beta}_1) & \text{cov}(\hat{\beta}_0, \hat{\beta}_2) & \cdots & \text{cov}(\hat{\beta}_0, \hat{\beta}_k) \\
\text{cov}(\hat{\beta}_0, \hat{\beta}_1) & \text{var}(\hat{\beta}_1) & \text{cov}(\hat{\beta}_1, \hat{\beta}_2) & \cdots & \text{cov}(\hat{\beta}_1, \hat{\beta}_k) \\
\text{cov}(\hat{\beta}_0, \hat{\beta}_2) & \text{cov}(\hat{\beta}_1, \hat{\beta}_2) & \text{var}(\hat{\beta}_2) & \cdots & \text{cov}(\hat{\beta}_2, \hat{\beta}_k) \\
\vdots & \vdots & \vdots & \ddots & \vdots \\
\text{cov}(\hat{\beta}_0, \hat{\beta}_k) & \text{cov}(\hat{\beta}_1, \hat{\beta}_k) & \text{cov}(\hat{\beta}_2, \hat{\beta}_k) & \cdots & \text{var}(\hat{\beta}_k)
\end{bmatrix}. \quad (2.9)
$$

The square roots of the diagonal elements of this matrix are the standard errors of the estimates of the intercept and the factor effects. It is desirable that the variances on the diagonal of this matrix be as small as possible, because this allows for powerful significance tests and narrow confidence intervals about the unknown factor effects.

Note that the variance–covariance matrix of the estimator is directly proportional to the error variance, σ_ε^2, which is unknown. We can estimate the error variance using the mean squared error:

$$
\hat{\sigma}_\varepsilon^2 = \frac{1}{n - p}(\mathbf{Y} - \mathbf{X}\hat{\boldsymbol{\beta}})'(\mathbf{Y} - \mathbf{X}\hat{\boldsymbol{\beta}}), \quad (2.10)
$$

where p denotes the number of parameters in $\boldsymbol{\beta}$. For the main-effects model in Equation (2.3), p equals $k + 1$. For the model with two-factor interaction effects in Equation (2.5), p is $k + 1 + k(k - 1)/2$. The root mean squared error is just the square root of $\hat{\sigma}_\varepsilon^2$.

When planning an experiment, we set σ_ε^2 to one, and consider only the elements of $(\mathbf{X}'\mathbf{X})^{-1}$. We call the diagonal elements of that matrix the relative variances of the estimates of the model's parameters, because they tell us how large the variances are compared to the error variance. We denote the relative variance of the estimate $\hat{\beta}_i$ of β_i by v_i:

$$
v_i = \frac{1}{\sigma_\varepsilon^2}\text{var}(\hat{\beta}_i).
$$

Using the $[-1, +1]$ factor-scaling convention described earlier, and assuming that all the factors are two-level factors, then the minimum possible value for the relative variance of an estimate is $1/n$, where n is the number of experimental runs. All of the estimates attain this minimum value when the sum of the element-wise products of all the pairs of columns in the model matrix, $\mathbf{X}$, is zero. For a main-effects model, this requires that

$$
\sum_{i=1}^{n} x_{ki} = 0,
$$

for every experimental factor x_k, and that

$$\sum_{i=1}^{n} x_{ki} x_{li} = 0,$$

for every pair of factors x_k and x_l. For a model including all the two-factor interactions too, it is also required that

$$\sum_{i=1}^{n} x_{ki} x_{li} x_{mi} = 0,$$

for every set of three factors x_k, x_l, and x_m, and that

$$\sum_{i=1}^{n} x_{ki} x_{li} x_{mi} x_{qi} = 0,$$

for every set of four factors x_k, x_l, x_m, and x_q.

When these relationships hold, the matrix $\mathbf{X'X}$ is diagonal and all of its diagonal elements are n. Then, $(\mathbf{X'X})^{-1}$ is also diagonal and its diagonal elements are $1/n$. When an experimental design has the properties that the matrix $\mathbf{X'X}$ is diagonal and all the diagonal elements are n, we say that the design is orthogonal for its associated model.

When a design is orthogonal for a given model, all the parameters in the model can be estimated independently. The covariance between each pair of estimates $\hat{\beta}_i$ and $\hat{\beta}_j$, $\text{cov}(\hat{\beta}_i, \hat{\beta}_j)$, is zero in that case. We say that the estimates are independent or uncorrelated. Dropping one or more terms from the model then has no impact on the estimates of the remaining parameters. If a design is not orthogonal, then re-fitting the model after dropping a term leads to changes in some of the estimates.

The inverse of the variance–covariance matrix in Equation (2.8),

$$\mathbf{M} = \frac{1}{\sigma_\varepsilon^2} \mathbf{X'X}, \tag{2.11}$$

is called the information matrix for the model's parameter vector $\boldsymbol{\beta}$. This matrix summarizes the available information on the model parameters. It is diagonal whenever the variance–covariance matrix in Equation (2.8) is diagonal, and vice versa. So, the information matrix is diagonal for designs that are orthogonal for a given model.

The key goal of screening experiments is to detect factors or interactions that have large effects on the response. This is done via significance tests. We say that factors or interactions corresponding to significant effects are *active*.

2.3.5   Significance tests and statistical power calculations

2.3.5.1   Significance tests

A statistical test of significance is a simple procedure. In the screening situation of this chapter, we estimated the factor effects in the model using ordinary least squares. In the model given by Equation (2.3), each observed response is subject to error. The errors in the responses are transmitted to the factor-effect estimates, so that they also have variability. The diagonal elements of the matrices in Equations (2.8) and (2.9) give the variances of the factor-effect estimates. Now, if we divide the factor-effect estimate by the square root of its variance, we obtain the signal-to-noise ratio

$$t = \frac{\hat{\beta}_i}{\sqrt{\text{var}(\hat{\beta}_i)}}. \tag{2.12}$$

The signal is the factor-effect estimate (here the estimate of a main effect). The noise is the square root of the estimate's variance.

We call the signal-to-noise ratio the t statistic because, under the assumptions of the model in Equation (2.3), the ratio is a t-distributed random variable *if the true value of β_i is actually* 0. It is unlikely that a t-distributed random variable, which has a zero mean, takes very negative or very positive values. Therefore, if the t ratio is very negative or very positive, we regard it as an indication that the true value of β_i is different from 0.

To quantify how unlikely a given value for the t ratio is given a zero value for the factor effect β_i, we use the p value. Roughly speaking, the p value is the probability that, by chance alone, we obtain a t ratio as big as the observed one or bigger. We regard a small p value as a strong indication that the true value of β_i is different from zero. We declare factor effects with small p values significantly different from zero, or, more informally, we say these factors effects are significant or active.

How small does the p value have to be before we decide that a factor-effect estimate is significantly different from zero? If we require the p value to be really small, like 0.0001, then we will not make many false positive errors —roughly one in every 10,000 significance tests we do. However, at the same time we will fail to detect many small but real signals. So, the choice of a cutoff value for the p value is a trade-off between having false positive or false negative results. Fisher (1926), one of the founding fathers of statistics, chose a cutoff value of 5% or 0.05, and decided to "ignore entirely all results which fail to reach this level." The value 0.05 has since become standard, but there is nothing absolute about this number that merits such favored treatment. In any case, we call whatever cutoff value we choose the significance level α of the test.

A key feature of a significance test is that we depart from the hypothesis that the factor effect β_i is zero. Under this assumption, we check whether the factor-effect estimate we obtained is likely. If the p value is small, this indicates our factor-effect estimate is unlikely. We then conclude that the hypothesis we started from is wrong and that β_i must be different from zero.

2.3.5.2   Power calculations

The statistical power of a significance test is the probability of detecting a signal of a certain size. The starting point for calculating the power of a test is a hypothesis that β_i has a certain nonzero value, the signal. To calculate the power we must know three things:

1. the cutoff value, α, for the significance test we will perform;

2. the error variance, σ_ε^2, of the a priori model; and

3. the size of the factor effect, β_i, we wish to detect.

Generally, the error variance, σ_ε^2, is unknown before the experiment. This immediately makes prospective power calculations somewhat dubious. We can get around this problem by expressing the size of the effect β_i we wish to detect in terms of its ratio with σ_ε. We can say, for instance, that we want to detect any signal β_i that is as big as the noise level σ_ε. Now, we do not have to know either the size of the effect that is of practical importance or the noise level. We only have to specify their ratio, $\beta_i/\sigma_\varepsilon$.

Given the ratio $\beta_i/\sigma_\varepsilon$, statistical software can calculate the probability that a significance test for the effect β_i will lead to the conclusion that it is significantly different from zero. We call this probability the *power* of the test. Obviously, we want this probability to be close to one, because this indicates that our analysis will identify effects that are as large as the noise level, by declaring them significant, almost with certainty. If this is the case, it means that our experiment allows for powerful significance tests. In general, the power of the significance tests is larger for larger numbers of runs and for larger factor effects β_i.

We recommend doing power calculations before running an experiment. If the power of the tests turns out unsatisfactory, this can be an argument to convince management to increase the experimental budget and perform more runs. A possible alternative is to identify one or more sources of variability, so that you can block your experiment and hereby reduce the noise level σ_ε. The use of blocking is discussed in Chapters 7 and 8. If none of these solutions is possible, it might still be worth conducting the experiment, because factor effects are sometimes larger than expected and the variance σ_ε^2 is sometimes smaller than expected, and because designed experiments are the only way to establish causal relationships.

2.3.6   Variance inflation

Ideally, an experimenter will choose a design that is orthogonal for the model under investigation. However, in practice, budget constraints on the number of runs or constraints on the factor-level combinations often make it impossible to run an orthogonal design. In the case where the design is not orthogonal, the relative variance of at least one of the estimates of the model parameters will be bigger than $1/n$. We then say that the variance of that estimate is inflated, and we call the factor by which the variance is inflated the variance inflation factor (VIF). For two-level designs, the

VIF is

$$VIF = nv_i. \tag{2.13}$$

This is equivalent to the standard definition of VIFs in textbooks on regression models given the factor scaling we advocate.

The minimum value of a VIF is one. For orthogonal designs, the VIF of every factor-effect estimate is 1. When a design is not orthogonal, at least one VIF is greater than one. If the VIF for an individual factor effect is four, then the variance of that factor effect's estimate is four times as large as it would be if it were possible to find an orthogonal alternative for the given design. The length of a confidence interval for a factor effect is proportional to the square root of its variance. Therefore, a VIF of four means that the factor effect's confidence interval is twice as long as it would be for an orthogonal design.

The VIF is a commonly used multicollinearity diagnostic in regression analysis. Many authors use a rule of thumb that a VIF of greater than five is indicative of possible collinearity problems.

2.3.7   Aliasing

When choosing an a priori model for an experiment, you implicitly assume that this model is adequate to describe the system you are studying. However, it is impossible to know for sure that this is true. For example, in many screening experiments investigators want to keep the number of runs near a bare minimum to fit a main-effects model. To do this, they assume that two-factor interactions are negligible. However, it may happen that one or more two-factor interactions are as large as the main effects. If this is the case, and the large two-factor interactions are not included in the model, then their existence in the physical system can bias the estimates of the main effects.

To formalize this mathematically, suppose, using the matrix representation introduced earlier, that the true model is

$$Y = \mathbf{X}_1 \boldsymbol{\beta}_1 + \mathbf{X}_2 \boldsymbol{\beta}_2 + \boldsymbol{\varepsilon}, \tag{2.14}$$

but that the investigator intends to fit the model

$$Y = \mathbf{X}_1 \boldsymbol{\beta}_1 + \boldsymbol{\varepsilon}. \tag{2.15}$$

The investigator thus assumes that the term $\mathbf{X}_2 \boldsymbol{\beta}_2$ is negligible. Then, the estimate of the vector $\boldsymbol{\beta}_1$, which contains the effects included in the estimated model, is biased by the values in the vector $\boldsymbol{\beta}_2$, i.e., the effects of the terms not included in the estimated model. In screening situations, the model initially fitted is often a main-effects model, and the fear is that active two-factor interactions cause bias in the main-effect estimates. In that case,

$$
\boldsymbol{\beta}_1 = \begin{bmatrix} \beta_0 \\ \beta_1 \\ \beta_2 \\ \vdots \\ \beta_k \end{bmatrix}, \quad
\boldsymbol{\beta}_2 = \begin{bmatrix} \beta_{12} \\ \vdots \\ \beta_{1k} \\ \beta_{23} \\ \vdots \\ \beta_{k-1,k} \end{bmatrix},
$$

$$
\mathbf{X}_1 = \begin{bmatrix}
1 & x_{11} & x_{21} & \cdots & x_{k1} \\
1 & x_{12} & x_{22} & \cdots & x_{k2} \\
\vdots & \vdots & \vdots & \ddots & \vdots \\
1 & x_{1n} & x_{2n} & \cdots & x_{kn}
\end{bmatrix}
$$

and

$$
\mathbf{X}_2 = \begin{bmatrix}
x_{11}x_{21} & \cdots & x_{11}x_{k1} & x_{21}x_{31} & \cdots & x_{k-1,1}x_{k1} \\
x_{12}x_{22} & \cdots & x_{12}x_{k2} & x_{22}x_{32} & \cdots & x_{k-1,2}x_{k2} \\
\vdots & \ddots & \vdots & \vdots & \ddots & \vdots \\
x_{1n}x_{2n} & \cdots & x_{1n}x_{kn} & x_{2n}x_{3n} & \cdots & x_{k-1,n}x_{kn}
\end{bmatrix}.
$$

The expected value of the least squares estimator of $\boldsymbol{\beta}_1$ is

$$
\mathrm{E}(\hat{\boldsymbol{\beta}}_1) = \boldsymbol{\beta}_1 + \mathbf{A}\boldsymbol{\beta}_2, \tag{2.16}
$$

where

$$
\mathbf{A} = (\mathbf{X}_1'\mathbf{X}_1)^{-1}\mathbf{X}_1'\mathbf{X}_2 \tag{2.17}
$$

is the alias matrix. To see this, we start from the ordinary least squares estimator of $\boldsymbol{\beta}_1$:

$$
\hat{\boldsymbol{\beta}}_1 = (\mathbf{X}_1'\mathbf{X}_1)^{-1}\mathbf{X}_1'\boldsymbol{Y}. \tag{2.18}
$$

The expected value of this estimator is

$$
\begin{aligned}
\mathrm{E}(\hat{\boldsymbol{\beta}}_1) &= \mathrm{E}[(\mathbf{X}_1'\mathbf{X}_1)^{-1}\mathbf{X}_1'\boldsymbol{Y}] \\
&= (\mathbf{X}_1'\mathbf{X}_1)^{-1}\mathbf{X}_1'\mathrm{E}[\boldsymbol{Y}] \\
&= (\mathbf{X}_1'\mathbf{X}_1)^{-1}\mathbf{X}_1'\mathrm{E}[\mathbf{X}_1\boldsymbol{\beta}_1 + \mathbf{X}_2\boldsymbol{\beta}_2 + \boldsymbol{\varepsilon}] \\
&= (\mathbf{X}_1'\mathbf{X}_1)^{-1}\mathbf{X}_1'(\mathbf{X}_1\boldsymbol{\beta}_1 + \mathbf{X}_2\boldsymbol{\beta}_2) \\
&= (\mathbf{X}_1'\mathbf{X}_1)^{-1}\mathbf{X}_1'\mathbf{X}_1\boldsymbol{\beta}_1 + (\mathbf{X}_1'\mathbf{X}_1)^{-1}\mathbf{X}_1'\mathbf{X}_2\boldsymbol{\beta}_2 \\
&= \boldsymbol{\beta}_1 + (\mathbf{X}_1'\mathbf{X}_1)^{-1}\mathbf{X}_1'\mathbf{X}_2\boldsymbol{\beta}_2, \\
&= \boldsymbol{\beta}_1 + \mathbf{A}\boldsymbol{\beta}_2.
\end{aligned}
$$

This derivation shows that, generally, the estimator $\hat{\boldsymbol{\beta}}_1$ is biased. In only two cases is the bias zero. First, if all the parameters in the vector $\boldsymbol{\beta}_2$ are zero, then $E(\hat{\boldsymbol{\beta}}_1) = \boldsymbol{\beta}_1$. This is the case where the factor effects represented by $\boldsymbol{\beta}_2$ are not active. Second, if $\mathbf{X}_1'\mathbf{X}_2$ is a zero matrix, then $\mathbf{A}\boldsymbol{\beta}_2$ is zero.

Each row of the alias matrix, $\mathbf{A}$, corresponds to an element of $\boldsymbol{\beta}_1$ in the estimated model. Nonzero elements in a row of $\mathbf{A}$ show the degree of biasing of the corresponding element of $\boldsymbol{\beta}_1$ due to two-factor-interaction effects not included in the estimated model but present in the true model in the matrix $\mathbf{X}_2$. A zero element in the ith row and jth column of the alias matrix indicates that the effect corresponding to the jth column does not bias the estimate of the parameter corresponding to the ith row. This is the ideal situation. Failing this, we would like the elements of the alias matrix, $\mathbf{A}$, to be as small as possible.

Table 2.7 shows the alias matrix for the design used in the extraction experiment, assuming that a main-effects model was fit and that all two-factor interaction effects are active. The table shows that the estimate of the main effect of methanol, β_1, is potentially biased by the interaction effects β_{23}, β_{24}, β_{25}, β_{26}, β_{34}, β_{35}, β_{36}, β_{45}, β_{46}, and β_{56}. More specifically, the expected value of the estimator of β_1 is

$$E(\hat{\beta}_1) = \beta_1 + \frac{1}{3}\beta_{23} - \frac{1}{3}\beta_{24} - \frac{1}{3}\beta_{25} + \frac{1}{3}\beta_{26} + \frac{1}{3}\beta_{34}$$
$$+ \frac{1}{3}\beta_{35} - \frac{1}{3}\beta_{36} + \frac{1}{3}\beta_{45} + \frac{1}{3}\beta_{46} - \frac{1}{3}\beta_{56}.$$

Assuming that the only important interaction effect is that involving the factors ethanol and propanol, this expression can be simplified to

$$E(\hat{\beta}_1) = \beta_1 + \frac{1}{3}\beta_{23}.$$

The case where elements of the alias matrix equal -1 or $+1$ is special. In this case, there are one or more columns in $\mathbf{X}_1$ that also appear in $\mathbf{X}_2$ (either as the identical columns or their negatives), and there is no data-driven way to distinguish between the corresponding effects or any linear combination of them. When this happens, then we say that the corresponding effects are *confounded* or *aliased*.

Consider, for example, the four-run two-level design in Table 2.8 for estimating a three-factor main-effects model. The alias matrix for this design is given in Table 2.9. It shows, for instance, that the expectation of the estimate of the main effect β_1 is

$$E(\hat{\beta}_1) = \beta_1 + \beta_{23}.$$

In this case, we say that β_1 and β_{23} are confounded or aliased. There is no way to distinguish, based on this design, between the effect of x_1 and the interaction effect involving x_2 and x_3.

Table 2.7 Alias matrix, **A**, for the extraction experiment, indicating the potential bias in the main-effect estimates due to two-factor interactions.

	β_{12}	β_{13}	β_{14}	β_{15}	β_{16}	β_{23}	β_{24}	β_{25}	β_{26}	β_{34}	β_{35}	β_{36}	β_{45}	β_{46}	β_{56}
β_0	0	0	0	0	0	0	0	0	0	0	0	0	0	0	0
β_1	0	0	0	0	0	$1/3$	$-1/3$	$-1/3$	$1/3$	$1/3$	$1/3$	$-1/3$	$1/3$	$1/3$	$-1/3$
β_2	0	$1/3$	$-1/3$	$-1/3$	$1/3$	0	0	0	0	$1/3$	$-1/3$	$-1/3$	$1/3$	$1/3$	$1/3$
β_3	$1/3$	0	$1/3$	$1/3$	$-1/3$	0	$1/3$	$-1/3$	$-1/3$	0	0	0	$1/3$	$-1/3$	$1/3$
β_4	$-1/3$	$1/3$	0	$1/3$	$1/3$	$1/3$	0	$1/3$	$1/3$	0	$1/3$	$-1/3$	0	0	$1/3$
β_5	$-1/3$	$1/3$	$1/3$	0	$-1/3$	$-1/3$	$1/3$	0	$1/3$	$1/3$	0	$1/3$	0	$1/3$	0
β_6	$1/3$	$-1/3$	$1/3$	$-1/3$	0	$-1/3$	$1/3$	$1/3$	0	$-1/3$	$1/3$	0	$1/3$	0	0

Table 2.8 Four-run design with three two-level factors for estimating a main-effects model.

Run	x_1	x_2	x_3
1	-1	-1	$+1$
2	$+1$	-1	-1
3	-1	$+1$	-1
4	$+1$	$+1$	$+1$

2.3.8 Optimal design

In Equations (2.8) and (2.9), we introduced the variance–covariance matrix of the ordinary least squares estimator. The diagonal elements of this matrix are the individual variances of the estimates of the model coefficients $\beta_0, \beta_1, \ldots, \beta_k$ for the main-effects model and of $\beta_0, \beta_1, \ldots, \beta_k, \beta_{12}, \beta_{13}, \ldots, \beta_{k-1,k}$ for the model including main effects and two-factor interaction effects. We would like these variances (individually and collectively) to be as small as possible. We know that, for two-level designs and using our factor-scaling convention, the minimum value the relative variance can take for any coefficient is $1/n$. This occurs when the design is orthogonal for the model, so that the matrix $(\mathbf{X}'\mathbf{X})^{-1}$ is diagonal. In this case, the determinant of $(\mathbf{X}'\mathbf{X})^{-1}$, $|(\mathbf{X}'\mathbf{X})^{-1}|$, is easy to calculate. It is the product of the diagonal elements, namely $(1/n)^p$, where p is the number of terms in the model. This is the smallest possible value for $|(\mathbf{X}'\mathbf{X})^{-1}|$. So, in a sense, the determinant can be used as an overall measure of the variance of the estimates in $\hat{\boldsymbol{\beta}}$.

Minimizing the determinant of $(\mathbf{X}'\mathbf{X})^{-1}$ is equivalent to maximizing the determinant of $\mathbf{X}'\mathbf{X}$, $|\mathbf{X}'\mathbf{X}|$. The matrix $\mathbf{X}'\mathbf{X}$ is proportional to the information matrix defined in Equation (2.11). Choosing the factor settings to maximize the determinant of $\mathbf{X}'\mathbf{X}$ is one way of maximizing the information in a design. We call a design that maximizes the determinant of the information matrix, a D-optimal design. The "D" in D-optimal stands for determinant. For two-level factors and main-effects models or models including main effects and two-factor interaction effects, orthogonal designs are D-optimal. However, if the number of experimental runs is not a multiple of 4, there are no orthogonal designs available for two-level factors.

Table 2.9 Alias matrix, $\mathbf{A}$, for the four-run design in Table 2.8, indicating the potential bias in the main-effect estimates due to two-factor interactions.

	β_{12}	β_{13}	β_{23}
β_0	0	0	0
β_1	0	0	1
β_2	0	1	0
β_3	1	0	0

The emphasis of classical textbooks on orthogonal designs has left many experimenters with the impression that using nonorthogonal designs makes no sense at all. At the same time, these experimenters face practical problems that simply do no allow the use of orthogonal designs. Common reasons for this are budget constraints that do not allow the number of runs to be a multiple of 4 and constraints on the combinations of factor levels. One of the things we want to demonstrate in this book is that experimental design remains a useful tool in these situations. It is just that the best possible designs for these situations are not orthogonal. With nonorthogonal designs, we can still estimate all the effects in the model of interest. As mentioned in Section 2.3.4, the estimates of the different factor effects will be correlated. Also, as we learned in Section 2.3.6, the variance of the estimates will generally be somewhat inflated. For any optimal design, however, variance inflation is usually not dramatic and the correlation between the factor effects' estimates is so small that it is no cause for concern.

In any case, regardless of the number of runs, we can find a design that maximizes the determinant of the information matrix for the model we want to use. This is a major strength of optimal experimental design: for any number of runs, you can find a design that maximizes the information on the model you want to estimate. In this respect, optimal experimental designs are completely different from full factorial or fractional factorial designs. When you want to use factorial or (regular) fractional factorial designs, you also start from an a priori model in which you include main effects or main effects and interaction effects. In the former case, you typically use a resolution III fractional factorial design. In the latter case, you use either a resolution IV design or even a resolution V design. In both situations, however, the use of a (regular) fractional factorial or a full factorial design requires that the number of runs be a power of 2. This offers very little flexibility to the researcher. In contrast, the optimal experimental design approach allows for any number of runs.

It turns out that the D-optimal design does not depend on the factor scaling, although the actual value of the determinant $|\mathbf{X}'\mathbf{X}|$ is different for differently scaled factors. That is, although the maximal value of the determinant is different if you scale the factors differently, a design that gives the maximal value for one factor scaling also gives the maximal value under another factor scaling. However, the D-optimal design may not be unique. For a specified number of runs and a priori model, there may be many designs having the maximal value for the determinant of the matrix $\mathbf{X}'\mathbf{X}$. It is also true that, for a specified number of runs and a priori model, there may be many different orthogonal designs.

We call the determinant of $\mathbf{X}'\mathbf{X}$ the *D-optimality criterion*. This optimality criterion is by far the most commonly used one in the design of experiments. For experiments in which the focus is on the estimation of factor effects and on significance testing, using the D-optimality criterion is certainly justified. It can be shown that a D-optimal design minimizes the volume of the joint confidence ellipsoid about the p parameters in $\boldsymbol{\beta}$. This confidence ellipsoid is a p-dimensional version of a confidence interval for one effect. So, roughly speaking, a D-optimal design guarantees the smallest possible confidence intervals for the model's parameters.

In Chapter 4, we introduce an alternative optimality criterion, the I-optimality criterion, that focuses on precise predictions.

2.3.9   Generating optimal experimental designs

If the information matrix, or $\mathbf{X}'\mathbf{X}$, is not diagonal, then computing the D-optimality criterion by hand is generally tedious and error-prone. In general, finding a design that maximizes the D-optimality criterion requires calculating this determinant (or, at least, multiplicative changes in the criterion) multiple times. Therefore, except in very special cases, generating optimal designs is an exercise for computers. Among the various optimality criteria for experimental designs, the D-optimality criterion is the fastest to calculate. This is another reason for the predominant use of the D-optimality criterion.

In this section, we describe one common method, the coordinate-exchange algorithm, for finding D-optimal designs using a computer. In our description, we assume that the interest is in a main-effects model or a model including main effects and two-factor interactions and that any categorical factors have two levels only. Before we get to the particulars of the coordinate-exchange algorithm, we need to introduce some terminology.

2.3.9.1   Preliminaries

We call the matrix of factor settings the design matrix $\mathbf{D}$. For an experiment with n runs and k factors,

$$\mathbf{D} = \begin{bmatrix} x_{11} & x_{21} & \cdots & x_{k1} \\ x_{12} & x_{22} & \cdots & x_{k2} \\ \vdots & \vdots & \ddots & \vdots \\ x_{1n} & x_{2n} & \cdots & x_{kn} \end{bmatrix}.$$

Whenever the model of interest contains the factors' main effects, the design matrix $\mathbf{D}$ is a submatrix of the model matrix $\mathbf{X}$. This will usually be the case. An exception is when a mixture-process variable experiment is performed. Such an experiment is discussed in Chapter 6. In any case, the number of rows in $\mathbf{D}$ and $\mathbf{X}$ is the same. The number of columns in $\mathbf{X}$, p, is usually larger than the number of columns of $\mathbf{D}$, k. Whenever a change is made to the factor settings in a given row of $\mathbf{D}$, this leads to one or more changes in the corresponding row of $\mathbf{X}$.

We call the rows of the design matrix $\mathbf{D}$ the design points. Every design point dictates at what factor settings the response has to be measured in the experiment.

It is possible for the D-optimality criterion to be zero. In that case, we say that the design is singular and the inverse of $\mathbf{X}'\mathbf{X}$ does not exist. As long as the number of runs, n, is larger than the number of parameters in the model, p, we can find a design with a positive D-optimality criterion value. We say that a design with a positive D-criterion value is nonsingular. A necessary condition for a design to be nonsingular is that the number of design points, i.e. the number of different factor-level combinations, is larger than or equal to the number of model parameters, p. Note that the number of runs may be larger than the number of design points or factor-level combinations because of the fact that certain combinations may be replicated in a design.

2.3.9.2   Outline of the coordinate-exchange algorithm

The coordinate-exchange algorithm proceeds element by element through the rows and columns of the design matrix, $\mathbf{D}$. In each iteration of the algorithm, we consider possible changes for every element in the design matrix. Because every element of the design matrix is essentially a coordinate of a point in the experimental space, this kind of algorithm is called a coordinate-exchange algorithm.

For all factors that are continuous, the algorithm starts by generating random values on the interval $[-1, +1]$, one for every cell of the design matrix $\mathbf{D}$ corresponding to a continuous experimental factor. For all factors that are categorical, the algorithm starts by generating random values from the set $\{-1, +1\}$ instead. For a design with n runs and k experimental factors, nk random values are needed in total. The resulting design is a random starting design. Almost always, a random starting design is nonsingular. If, exceptionally, the starting design is singular, then the algorithm starts all over again and generates a completely new random design.

The next step in the algorithm is to improve the starting design on an element-by-element basis. For the kinds of models discussed in this chapter, and using the factor scaling introduced in Section 2.3.3, we consider changing each element of the starting design, x_{ij}, to either -1 or $+1$ and evaluate the effect of the change on the D-optimality criterion. If the larger of these two effects increases the D-optimality criterion over its previous value, then we exchange the value of x_{ij} for -1 or $+1$, depending on which one had the larger effect.

After investigating possible changes for each element in the design, we repeat this process until no element changes within an entire iteration through the factor settings or until we have made a prespecified maximum number of iterations. The output of this procedure is a design that can no longer be improved by changing one coordinate at a time. It is the best design among the broad set of neighboring designs. There is, however, no guarantee that there exists no better design that is not a neighboring design. We, therefore, say that the exchange procedure leads to a locally optimal design.

In general, every random starting design may lead to a different locally optimal design. Some of these locally optimal designs will be better than others in terms of the D-optimality criterion. The best design among all locally optimal designs is the globally optimal design. To make the likelihood of finding the globally optimal design large, we repeat the exchange algorithm a large number of times. Except for trivial problems involving a few factors only and a small number of runs, we recommend using at least 1000 random starts of the algorithm. This ensures that either we find the globally optimal design or a design with a D-optimality criterion value that is very close to the global optimum.

The search for an optimal experimental design is similar to searching for the highest peak in a mountain range. Every mountain range has several peaks, each of which is a local maximum. Only one of the peaks is the highest, the global maximum. Algorithms for optimal experimental design are like a climber who is dropped at a random location in a mountain range and who starts climbing the nearest peak without seeing all of the potentially higher peaks in the range. To increase the chance of finding the highest peak, it is necessary to drop the climber at various random locations in the range and let him search for the highest peak in the immediate neighborhood of each drop-off point.

2.3.9.3   The coordinate-exchange algorithm in action

Suppose we have two continuous factors, x_1 and x_2, and we wish to fit a main-effects model using four experimental runs. The coordinate-exchange algorithm starts by creating a random starting design. Our starting design, which is graphically displayed in Figure 2.3(a), is

$$\mathbf{D} = \begin{bmatrix} x_{11} & x_{21} \\ x_{12} & x_{22} \\ x_{13} & x_{22} \\ x_{14} & x_{2n} \end{bmatrix} = \begin{bmatrix} 0.4 & -0.8 \\ 0.5 & 0.2 \\ -0.2 & -0.9 \\ 0.9 & 0.3 \end{bmatrix}.$$

The corresponding model matrix is

$$\mathbf{X} = \begin{bmatrix} 1 & x_{11} & x_{21} \\ 1 & x_{12} & x_{22} \\ 1 & x_{13} & x_{22} \\ 1 & x_{14} & x_{2n} \end{bmatrix} = \begin{bmatrix} 1 & 0.4 & -0.8 \\ 1 & 0.5 & 0.2 \\ 1 & -0.2 & -0.9 \\ 1 & 0.9 & 0.3 \end{bmatrix},$$

so that

$$\mathbf{X'X} = \begin{bmatrix} 4 & 1.60 & -1.20 \\ 1.60 & 1.26 & 0.23 \\ -1.20 & 0.23 & 1.58 \end{bmatrix}.$$

The determinant of this information matrix, $|\mathbf{X'X}|$, is 1.0092. This is the D-optimality criterion value that we will now improve by changing the design matrix $\mathbf{D}$ element by element. We start with the first element of the matrix, x_{11}, which equals 0.4.

Consider changing the value 0.4 to -1 and $+1$. The resulting design matrices are

$$\mathbf{D}_1 = \begin{bmatrix} -1 & -0.8 \\ 0.5 & 0.2 \\ -0.2 & -0.9 \\ 0.9 & 0.3 \end{bmatrix}$$

and

$$\mathbf{D}_2 = \begin{bmatrix} +1 & -0.8 \\ 0.5 & 0.2 \\ -0.2 & -0.9 \\ 0.9 & 0.3 \end{bmatrix}.$$

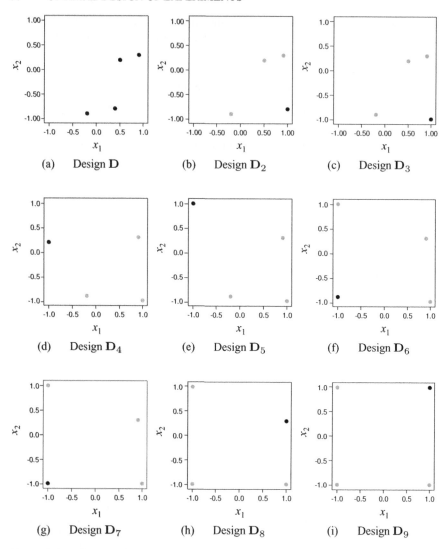

Figure 2.3 The coordinate-exchange algorithm: from random starting design to optimal design.

The determinants of the information matrices corresponding to $\mathbf{D}_1$ and $\mathbf{D}_2$ are 2.2468 and 3.6708, respectively. Both determinants are bigger than the initial determinant, but the one associated with a value of $+1$ for the first element is larger. Therefore, we replace 0.4 with $+1$, which is the value associated with the larger determinant. The resulting design is shown geometrically in Figure 2.3(b), where the point whose first coordinate has been modified is shown in black. The algorithm then shifts its attention to the second element in the first row of the design matrix $\mathbf{D}_2$, -0.8.

We consider changing this value to -1 and $+1$. The resulting determinants are 4.7524 and 0.6324, respectively. Here, only using -1 improves the determinant. Therefore, we replace -0.8 with -1 and obtain the design shown in Figure 2.3(c), with the design matrix

$$\mathbf{D}_3 = \begin{bmatrix} +1 & -1 \\ 0.5 & 0.2 \\ -0.2 & -0.9 \\ 0.9 & 0.3 \end{bmatrix}.$$

We then move to the first element of the next row, 0.5. We replace this element with -1 and $+1$, which results in the determinants 14.7994 and 4.5434, respectively. So, we replace 0.5 with -1 and obtain design matrix

$$\mathbf{D}_4 = \begin{bmatrix} +1 & -1 \\ -1 & 0.2 \\ -0.2 & -0.9 \\ 0.9 & 0.3 \end{bmatrix},$$

which is graphically shown in Figure 2.3(d).

We proceed to the element in the second column of the second row, 0.2. Replacing this value with $+1$ yields a larger determinant, 22.3050, than replacing it with -1. The resulting design matrix then is

$$\mathbf{D}_5 = \begin{bmatrix} +1 & -1 \\ -1 & +1 \\ -0.2 & -0.9 \\ 0.9 & 0.3 \end{bmatrix}.$$

Moving to the first column of the third row, we find that replacing -0.2 with -1 is better, resulting in a determinant of 39.9402. We, therefore, adjust the design matrix to

$$\mathbf{D}_6 = \begin{bmatrix} +1 & -1 \\ -1 & +1 \\ -1 & -0.9 \\ 0.9 & 0.3 \end{bmatrix}.$$

Moving to the second column of the third row, the best choice is to replace -0.9 with -1, yielding the determinant 42.96 and design matrix

$$\mathbf{D}_7 = \begin{bmatrix} +1 & -1 \\ -1 & +1 \\ -1 & -1 \\ 0.9 & 0.3 \end{bmatrix}.$$

Moving to the last row, we replace 0.9 with $+1$, yielding the determinant 45.52 and design matrix

$$\mathbf{D}_7 = \begin{bmatrix} +1 & -1 \\ -1 & +1 \\ -1 & -1 \\ +1 & 0.3 \end{bmatrix}.$$

Finally, considering the second column in the last row, we replace 0.3 with $+1$, yielding the design matrix

$$\mathbf{D}_8 = \begin{bmatrix} +1 & -1 \\ -1 & +1 \\ -1 & -1 \\ +1 & +1 \end{bmatrix},$$

model matrix

$$\mathbf{X} = \begin{bmatrix} 1 & +1 & -1 \\ 1 & -1 & +1 \\ 1 & -1 & -1 \\ 1 & +1 & +1 \end{bmatrix},$$

$$\mathbf{X'X} = \begin{bmatrix} 4 & 0 & 0 \\ 0 & 4 & 0 \\ 0 & 0 & 4 \end{bmatrix},$$

and

$$|\mathbf{X'X}| = 4^3 = 64.$$

We note that $\mathbf{X'X}$ is a diagonal matrix, so that this design is orthogonal and has the globally maximal information. Therefore, we can stop the coordinate-exchange algorithm without considering any further iterations and without any additional random starts. The designs corresponding to design matrices $\mathbf{D}_5$–$\mathbf{D}_9$ are shown in Figures 2.3(e)–(i).

For experiments with more factors and runs, and with more involved models, finding an optimal design is, of course, harder. Then, more than one iteration through the design matrix is generally required.

2.3.10   The extraction experiment revisited

The design for the extraction experiment in the case study is an orthogonal design for the model containing the intercept and all the main effects of the factors. Therefore, the VIFs for the main effects are all one. However, when we add the interaction involving ethanol and propanol to the model, the design is no longer orthogonal.

Table 2.10 Variance inflation factors (VIFs) for the design in Table 2.1 for a model with main effects and one two-factor interaction.

Effect	VIF
Intercept	1
Methanol	1.2
Ethanol	1
Propanol	1
Butanol	1.2
pH	1.2
Time	1.2
Ethanol × Propanol	1.8

Neither $\mathbf{X'X}$ nor $(\mathbf{X'X})^{-1}$ are then diagonal, causing some of the VIFs to be larger than 1. The VIFs for all the terms in the extended model are displayed in Table 2.10.

The variances of the main-effect estimates of all the factors except for ethanol and propanol increase by 20%. This implies roughly a 10% increase in the standard error of these estimates, and, equivalently, a 10% increase in the length of the corresponding confidence intervals compared to the model containing only the main effects. The two-factor interaction effect involving ethanol and propanol has a VIF of 1.8. Thus, the variance of this coefficient is 80% larger than it would be if there were a 12-run orthogonal design available for fitting this model. There is, however, no 12-run design with two-level factors and orthogonal columns for the six main effects and one interaction effect.

Suppose that, in advance of experimentation, the investigator suspected a two-factor interaction involving x_2 and x_3. Then, the investigator could add the two-factor interaction to the a priori model for the experiment. A 12-run D-optimal design for a model including six main effects and the interaction involving x_2 and x_3 appears in Table 2.11. As shown in Table 2.12, this design has VIFs of 1 for all of the main effects except those of x_2 and x_3. The intercept, the main effects of x_2 and x_3, and the interaction effect involving x_2 and x_3, all have a VIF of 1.25. The improvement of these VIFs compared to the values in Table 2.10 arises from the explicit inclusion of the interaction effect in the a priori model when generating the design.

2.3.11 Principles of successful screening: sparsity, hierarchy, and heredity

2.3.11.1 Sparsity

The Pareto principle states that most of the variability in any measured output of a system or process is due to variation in a small number of inputs. Screening experiments depend on this principle as axiomatic because they investigate the effects of many factors with a number of runs that may barely exceed the number of factors.

Table 2.11   D-optimal design for a model with main effects and the two-factor interaction involving x_2 and x_3.

Run	x_1	x_2	x_3	x_4	x_5	x_6
1	+1	+1	−1	+1	−1	+1
2	+1	−1	−1	+1	+1	−1
3	−1	+1	+1	+1	+1	−1
4	−1	−1	−1	+1	−1	+1
5	+1	+1	+1	−1	−1	+1
6	+1	−1	+1	+1	+1	+1
7	−1	−1	−1	−1	+1	+1
8	−1	+1	−1	−1	+1	+1
9	+1	+1	−1	−1	+1	−1
10	−1	+1	−1	+1	−1	−1
11	−1	−1	+1	−1	−1	−1
12	+1	−1	−1	−1	−1	−1

If most systems were driven by complicated functions of many factors, screening studies would have little hope of success.

In fact, screening studies are often extraordinarily successful. This fact is substantiated by thousands of published case studies, as a quick internet search will verify.

In screening experimentation, the Pareto principle is restated as the principle of factor (or effect) sparsity. In the same way that a screen lets small things pass through while capturing large things, a screening experiment signals the presence of large factor effects while ignoring the negligible ones.

Perhaps a better statement of the sparsity principle in screening experiments is captured by the 80-20 rule. That is, you expect to identify (at least) 80% of the variability in the response with 20% of the factors.

Table 2.12   Variance inflation factors (VIFs) for the design in Table 2.11 for a model with main effects and one two-factor interaction.

Effect	VIF
Intercept	1.25
Methanol	1
Ethanol	1.25
Propanol	1.25
Butanol	1
pH	1
Time	1
Ethanol × Propanol	1.25

The probability of correctly identifying all the large effects drops precipitously as the number of large effects passes half the number of runs.

2.3.11.2 Hierarchy

The principle of hierarchy in regression modeling states that first-order (linear or main) effects account for the largest source of variability in most systems or processes. Second-order effects composed of two-factor-interaction effects and quadratic effects of individual factors are the next largest source of variability (see Chapter 4 for a discussion of quadratic effects).

In model selection methods, the principle of model hierarchy suggests that the addition of main effects should precede the inclusion of second-order effects such as two-factor interaction effects or quadratic effects.

2.3.11.3 Heredity

Heredity is another screening principle that relates mainly to model selection. There are two types of model heredity, strong and weak. A model with strong heredity has the property that, if it includes a two-factor interaction involving the factors x_i and x_j, it also includes both the main effect of factor x_i and the main effect of factor x_j. Models with weak heredity only require one of the two main effects to be present in the model if their interaction is included.

A model based on the strong heredity principle has the attractive technical advantage that predictions made with the estimated model do not change when the model is fit with a different scaling of the factors.

2.3.11.4 Empirical evidence

Li et al. (2006) reanalyzed 113 data sets from published full factorial experiments with three to seven factors to validate the above principles. They report that the percentage of first-order effects is higher than the percentage of second-order effects, which is in turn higher than the percentage of third-order effects. This supports the idea of hierarchy. They also report that, for the active two-factor interactions, the conditional probability that both main effects are active is higher than the conditional probability that only one main effect is active, and much higher than the conditional probability that no main effects are active. This supports the principle of heredity.

However, Li et al. (2006) also point out that violations of the sparsity, hierarchy, and heredity principles are more common than the literature on screening experiments suggests. A remarkable finding is that 80% of the active two-factor interaction effects are synergistic. Therefore, when larger responses are desirable, exploiting positive main effects will lead to additional increases in response due to active two-factor interactions (even if those interactions have not been identified or estimated). By contrast, when smaller responses are desirable, procedures that exploit main effects to reduce the response are likely to be counteracted by active two-factor interactions.

2.4 Background reading

2.4.1 Screening

A comprehensive overview of the design and analysis of two-level screening experiments is given in Mee (2009). This work first discusses completely randomized full factorial designs and then continues with completely randomized fractional factorial designs. The initial focus is on regular fractional factorial designs; these are the most well-known fractional factorial designs and can be obtained using design generators (see also Box et al. (2005), Montgomery (2009), and Wu and Hamada (2009)). However, nonregular orthogonal two-level fractional factorial designs, including Plackett–Burman designs, Hadamard designs and other orthogonal arrays, also receive substantial attention. Nonregular orthogonal two-level fractional factorial designs exist whenever the number of available runs is a multiple of 4. Therefore, they offer more flexibility to experimenters than regular two-level fractional factorial designs, for which the sample size is necessarily a power of 2. It is important to realize that all these orthogonal two-level designs are optimal in many different ways, and that custom-built optimal experimental designs offer even more flexibility as regards the sample size of the experiment.

For screening purposes, besides nonregular two-level fractional factorial designs, two-level supersaturated designs have gained substantial popularity. Supersaturated designs have fewer experimental runs than experimental factors. Mee (2009) and Gilmour (2006) provide an overview of the recent developments in this area. Guidelines for analyzing data from two-level supersaturated designs, and for setting up these designs, are given in Marley and Woods (2010).

Jones and Nachtsheim (2011a) propose a new screening approach, involving three levels per factor. Their designs have one more run than twice the number of factors, and possess the attractive properties that the main-effect estimates are not biased by any two-factor interaction effect or quadratic effect, and that they allow an assessment of possible curvature in the factor-response relationship.

It is also possible to create designs for screening experiments that minimize aliasing. Jones and Nachtsheim (2011b) show how to create designs that minimize the squared elements of the alias matrix, while ensuring a certain minimum precision for the factor effects of primary interest.

For an overview of industrial screening experiments and screening experiments for medical applications, drug discovery and microarrays, including discussions of supersaturated designs, group screening, and screening in simulation studies and computer experiments, we refer to Dean and Lewis (2006).

2.4.2 Algorithms for finding optimal designs

The coordinate-exchange algorithm is due to Meyer and Nachtsheim (1995) and has been implemented in various software packages for design and analysis of experiments. The coordinate-exchange algorithm runs in polynomial time, which means that the time it needs to find an optimal design does not explode when the size of the design and the number of factors increases. This is different from point-exchange algorithms, which were introduced by Fedorov (1972) and modified by various researchers to speed them up (for instance, by Johnson and Nachtsheim (1983) and Atkinson and Donev (1989)). The main drawback of point-exchange algorithms is that they require a list of possible design points, called a candidate set, as an input. For design problems involving large numbers of experimental factors, this list becomes too large to handle.

In recent years, several researchers have investigated the use of genetic algorithms, simulating annealing algorithms and tabu search algorithms for finding optimal experimental designs. While these algorithms are reported to have led to better designs in some cases, they have never gained much popularity. The main reason for this is that they are much more complex than coordinate-exchange and point-exchange algorithms, and yet do not lead to designs that make a difference in practice.

2.5 Summary

Screening designs are usually designs that involve several factors, each set at two-levels. Then, whenever the number of test runs is a multiple of 4, it is always possible to generate an orthogonal design for a main-effects model. The most cost-efficient screening designs have only a few more test runs than factors.

Though screening designs are primarily meant for finding large main effects, it is important to keep in mind that two-factor interactions are fairly common. If a two-factor interaction is active, then its effect can bias the estimates of the main effects.

As in our example, it can be very useful to add one or more important two-factor interaction effects to the model. The resulting model matrix will generally not be orthogonal. However, the reduction in the root mean squared error due the inclusion of the interaction effect(s) in the model more than compensates for the variance inflation introduced by the nonorthogonality.

3

Adding runs to a screening experiment

3.1 Key concepts

1. Ideally, experimentation is a sequential process.

2. Screening experiments often leave some questions unanswered. Adding new runs to the original experiment can provide the necessary information to answer these questions.

3. An optimal choice of the runs to add allows the investigator to include extra terms in the fitted model and fit the new model without drastically increasing the total run size.

4. Including all the terms from a set of competing models in the updated a priori model when searching for the optimal runs to add can reveal the best of many potential models.

There are often many models that are equally (or nearly equally) good at explaining the results of a screening experiment. This ambiguity is a consequence of the extreme economy in the number of runs employed. When there are multiple feasible models, acquiring more information by adding runs to the screening design is a natural course of action.

In this chapter, we optimally augment the screening experiment from the previous chapter by adding eight new runs for the purpose of fitting a larger a priori model. The new model contains two-factor interaction terms that appeared in the best-fitting models for the original data. We analyze the combined data from the original screening experiment and the eight additional runs, and identify the best model from

Optimal Design of Experiments: A Case Study Approach, First Edition. Peter Goos and Bradley Jones.
© 2011 John Wiley & Sons, Ltd. Published 2011 by John Wiley & Sons, Ltd.

a list of competing models while dramatically reducing the variance of the model estimates.

3.2   Case: an augmented extraction experiment

3.2.1   Problem and design

Peter and Brad are boarding the train to Rixensart, to visit GeneBe, the Belgian biotech firm where they had run a screening experiment to study an extraction process. The goal of the experiment was to increase the yield of the process.

[Peter] I take it that GeneBe is interested in following up on that screening experiment they did earlier. Your text message was terse as usual. "Meet me at the train station at 8:30 AM."

[Brad] Sorry. I hate typing with my thumbs on my iPhone. Anyway, I got a call from Dr. Zheng last night. He has management approval to further study which of the various models we fit using all-subsets regression might best describe his extraction process. In particular, he is interested in whether ethanol and propanol actually do interact in their effect on yield.

[Peter] Are they running the process at the settings we recommended?

[Brad] I think so, and I imagine that they are getting the results we predicted or they would not have asked us to come back.

[Peter] Then, they are already realizing improved yields. Are they doing the extra experimentation just out of intellectual curiosity?

[Brad] I doubt it. I'm sure Dr. Zheng will tell us what their motivation is.

After a short walk from the Rixensart train station, they arrive at GeneBe and Dr. Zheng meets them in the reception area.

[Dr. Zheng] Hello guys. I am glad you were available today.

[Peter] We wouldn't pass up this chance to do sequential experimentation.

[Dr. Zheng] Excellent, let's go to the conference room and I will tell you what's going on.

In the conference room, Brad starts his laptop while Dr. Zheng orders coffee.

[Peter] We were curious about your motivation for this follow-up experiment. Have you been realizing the higher yield we predicted?

[Dr. Zheng] Yes. We have been using your recommendations and the yields have doubled. They are averaging more than 50 mg per 100 ml consistently, which was what management required for continuing to work on this project.

[Brad] That's great. I hope that management is ready to continue development of your antimicrobial product.

[Dr. Zheng] Maybe. When you were here, you told Bas Ritter and me that the interaction effect of ethanol and propanol was only barely significant, and that there were several alternative models that had similar predictive performance.

[Brad] That is true. But you have already demonstrated a dramatically improved yield. It seems as though that would be validation enough.

[Dr. Zheng] As I told you last time, design of experiments is not an accepted strategy here. The first study was a proof of concept. There is some political backlash saying that we have not proved anything yet. My supervisor—you remember Bas Ritter—and I think that, if we can present a sound model for the process, it would silence the backlash. You had mentioned that there were a number of competitive models. I have the go-ahead from Bas to do a follow-up experiment to sort these out and to arrive at a defensible model.

[Brad] Actually, this feels like an opportunity to influence some thinking at GeneBe.

[Dr. Zheng, sighing] I hope so. Do you still have the data from the screening experiment?

[Brad] Yes. I have a folder for GeneBe in the file system on my laptop. If you let me know how many new runs you can afford, I can augment the original runs with the new experimental runs in short order.

[Dr. Zheng] I am concerned about the time that has passed in between the last experiment and this one. How can we assume that everything has remained constant?

[Peter] You make a good point. It is often not wise to assume that everything has remained constant.

[Brad] As it turns out, there is a rather standard approach for dealing with this issue. You define a new factor that designates whether a given run was in the original experiment or in the new experiment. This factor, then, has two levels and is considered a blocking factor because it divides the full set of experimental runs into two groups, or blocks.

[Dr. Zheng] How does this procedure help in a practical sense?

[Peter] If the process yield has shifted upward or downward, the estimated effect of the blocking factor will capture the shift and keep it from contaminating the effects we want to investigate. In this case, the effects of interest are the union of the effects in all the competitive models.

[Dr. Zheng] Your preferred model had only one new effect of primary interest, namely, the two-factor interaction involving ethanol and propanol. How many effects are involved in the competing models? What is the minimum number of extra processing runs we need to cover all the competitive models?

Brad opens his briefcase and pulls out his GeneBe folder. After a short search, he finds a sheet of paper with Table 3.1 that displays the set of competitive models, along with several goodness-of-fit measures, and shows it to Dr. Zheng.

[Brad] We have nine models here. Ignoring the intercept, each has six or seven terms. All these models do a good job of fitting the data.

[Dr. Zheng] Wow! If I had known about all these different possibilities, I would have been more nervous about producing our last several batches. How did you come up with your recommendations given this much uncertainty about the model?

Brad pulls another sheet from his folder that displays Table 3.2.

[Peter] For each of the nine models we got, we determined the factor settings that would maximize the yield. It turns out that, even though the models do not have the same terms, they are in substantial agreement about which settings lead to the

Table 3.1   Nine competitive models from the all-subsets analysis of the data from the original extraction experiment, along with the number of terms in each of them and several goodness-of-fit measures: the coefficient of determination (R^2), the root mean squared error (RMSE) and the corrected Akaike information criterion (AICc).

Model	Number	R^2	RMSE	AICc
1 Methanol, Ethanol, Butanol, pH, Time, Ethanol × Propanol	6	0.986	2.08	105
2 Methanol, Ethanol, Propanol, Time, Methanol × Time, Propanol × Time	6	0.985	2.13	106
3 Methanol, Ethanol, Butanol, Time, Ethanol × Time, Butanol × Time	6	0.983	2.31	108
4 Methanol, Ethanol, Butanol, pH, Time, Butanol × Time	6	0.981	2.42	109
5 Methanol, Ethanol, Propanol, Butanol, pH, Time	6	0.980	2.47	109
6 Methanol, Ethanol, Propanol, pH, Time, Propanol × Time	6	0.979	2.53	110
7 Methanol, Ethanol, Butanol, pH, Time, Methanol × Butanol, Ethanol × Propanol	7	0.995	1.45	138
8 Methanol, Ethanol, Butanol, pH, Time, Ethanol × Propanol, Ethanol × Time	7	0.994	1.45	138
9 Methanol, Ethanol, Propanol, Butanol, pH, Time, Ethanol × Propanol	7	0.994	1.56	139

Table 3.2   Recommended factor settings obtained from the nine competing models in Table 3.1 to maximize yield, plus the "consensus" choice. NA means that the given factor does not appear in the model.

Model	Methanol	Ethanol	Propanol	Butanol	pH	Time
1	+	+	−	+	−	+
2	+	+	−	NA	NA	+
3	+	+	NA	−	NA	+
4	+	+	NA	+	−	+
5	+	+	+	+	−	+
6	+	+	−	NA	−	+
7	+	+	−	+	−	+
8	+	+	−	+	−	+
9	+	+	−	+	−	+
Consensus	+	+	−	+	−	+

maximum yield. After we produced this table, Brad and I felt reasonably confident about our recommendation to use 10 ml of methanol, ethanol, and butanol, with the low pH level and the long time in solution.

[Dr. Zheng, frowning] I suppose that makes sense. Still, you guys concealed a lot of complexity in your last presentation.

[Brad] We worried about that, but we knew that design of experiments was on trial at GeneBe. We wanted to make a strong recommendation about how to proceed without hedging. We did say that there were other competitive models. At the time, you were more interested in understanding the aliasing of main-effect terms with active two-factor interactions.

[Peter] Also, if you recall, we had two recommendations. One was to try a confirmatory run at what we thought would turn out to be the best factor-level combination. The original experiment did not have a run at that setting. So, no matter what, you would have learned something from that run. As it happens, that setting worked out rather well. Our other recommendation was to augment the original experiment to address the "multiple feasible models" issue.

[Dr. Zheng] What you are saying is true. Still, I wish you had told me more in that conversation we had after Bas Ritter left. I had the impression that there were just a few good competing models.

[Brad] I apologize for not laying it all out for you before. I will communicate with a goal of full disclosure in the future.

[Dr. Zheng, mollified] I accept your apology. Now, before going back to my question about how many more runs we need to do, can you tell me more about this table of models you showed me? I know that R^2 is the percentage of the total variation of the response explained by the model. I also know what the root mean squared error (RMSE) is. But what is AICc? Am I supposed to be looking for big values or small ones?

[Peter] That is the corrected Akaike information criterion. It is a way of comparing alternative models that penalizes models with more terms. For AICc, lower values are better. We included the R^2 value because it is the most popular goodness-of-fit criterion. But it always increases as you add more terms to the model, so we do not recommend it as a model selection criterion.

[Dr. Zheng] That is pretty well known. How about the RMSE?

[Peter] The RMSE, which, of course, we want to be small, is in the units of the response. That is nice. And, if you add a term of no explanatory value, the RMSE will increase. However, minimizing the RMSE as a model selection approach tends to result in models with too many terms. Minimizing the AICc results in more parsimonious models.

[Brad, pointing at Table 3.1] In our table, the first six models have low values of AICc, while the last three models have low RMSE values.

[Peter, interrupting] In the spirit of full disclosure, these are not the only models having six or seven terms with low RMSE and AICc. We found many six- or seven-term models with low RMSE or AICc. We did so by estimating all possible models with up to seven terms. We call this approach all-subsets regression.

[Dr. Zheng, pointing to Table 3.1] I assume, then, that you chose this set of models from the larger group based on some other principle.

[Peter] Yes, that is true. We tend to believe in models that exhibit a property called heredity.

[Dr. Zheng] And what makes a model exhibit heredity?

[Peter] A model exhibits strong heredity if, for every two-factor interaction in the model, the model also contains the main effects of both factors involved in that two-factor interaction.

[Brad] For example, if you have the interaction involving ethanol and propanol in the model, then, for the model to exhibit strong heredity, it must also include the main effect of ethanol and the main effect of propanol. Before you ask, weak heredity means that only one of the associated main effects needs to be present.

[Peter] We eliminated quite a few models having two-factor interactions with no corresponding main effects.

[Dr. Zheng] That seems like a common sense approach for reducing the number of competing models and simultaneously the complexity of the model selection effort. Let's hope it works . . . . Going back to my original question, how many new runs do we need to fit all these extra potential terms?

[Brad, referring to Table 3.1] There are six possible interactions to consider: ethanol × propanol, methanol × butanol, methanol × time, ethanol × time, propanol × time, and butanol × time. If you add the block effect that Peter mentioned earlier, you have seven effects that we did not include in the a priori model of our original screening experiment. I would recommend adding at least eight runs so that we can get estimates of all these effects. That should allow us to clear up our uncertainty about which of these models is actually driving the process.

[Dr. Zheng] I think we can get approval for eight new runs. I was concerned that you would want to "fold over" our original experiment. That would double our original budget and would be a harder sell.

[Brad] Wait a few minutes and I will create the design for optimally augmenting your 12 original runs with eight new ones.

[Peter] It is interesting that you mention the foldover idea. That is an augmentation approach that appears in many textbooks on design of experiments. If all your factors have only two levels you can fold over any design, and the resulting design comes with a guarantee that no active two-factor interactions will bias any main-effect estimate. Of course, you are right that this approach doubles the cost of the experiment. And in our case, this technique is not sufficient, because we want to estimate quite a few two-factor interaction effects, involving five of the six factors. Therefore, the foldover technique can't help us out.

[Dr. Zheng, interrupting] Peter, while Brad is working, can you tell me in what sense the new runs are an optimal augmentation of the original design?

[Peter] In the software application that Brad uses, he can specify an a priori model for use in the design optimality criterion. In our case, the a priori model is the main-effects model for the original screening experiment, plus all the two-factor interactions that Brad mentioned before, plus the block effect required to capture the possible shift in responses between the time of the original experiment and the follow-up experiment.

Dr. Zheng nods in agreement and Peter continues.

[Peter] The software makes sure that there are enough additional runs to estimate the new model. This can be useful if you are adding a lot of new effects to the original model. Of course, in our experiment, the number of added runs is governed by the interplay between economics and statistical power.

[Dr. Zheng] What do you mean?

[Peter] You can improve the power to detect an active effect—that is, the probability that you detect an effect that's really there—by adding runs to an experimental plan. So, the more runs the better. But economics dictates a practical limit on the number of runs.

[Dr. Zheng] That is certainly true, especially so in our case. But I still don't understand this concept of optimal augmentation.

[Peter] The original experimental runs are fixed because you have already done them, but they also provide a lot of information about the process. So, you want to use that information when choosing the factor settings for the follow-up experiment. In Brad's software, he adds the new runs to the old runs in such a way as to maximize the total information from the combined set of runs for fitting the new, updated a priori model you specify.

[Dr. Zheng] In what sense are you maximizing the information in the design?

[Peter] That is an interesting issue. When you estimate main effects and two-factor interaction effects, these estimates have variability because your responses have variability that gets transmitted to the estimates. You cannot change the fact that the responses are subject to random errors, but you can minimize the variance that is transmitted to the estimates of the effects. An omnibus measure of this transmitted variance is the determinant of the variance–covariance matrix of the effect estimates. Brad's software finds the design that minimizes this determinant. We call that design optimal.

[Brad] I have our new design.

Brad turns the screen of his laptop, so that the others can see the design shown in Table 3.3.

[Dr. Zheng] I notice that the eight new runs do not balance the number of high and low values of each factor. It seems a little disturbing that there are five runs with no methanol and only three runs with methanol. Doesn't that mean that we have introduced some correlation in the estimates? Can that be optimal?

[Brad] You are correct that our combined design is not orthogonal and, therefore, has some correlation among all the estimates of the effects. We should definitely look at some design diagnostics to see how much correlation there is, and how much the correlation is increasing the variance of the effect estimates from their theoretical minimum values.

[Dr. Zheng] So, what about these design diagnostics you mentioned. Are our eight runs going to be enough?

[Brad, displaying Table 3.4] Here is a table showing the relative variance for the estimates of the model parameters using the augmented design. It also has a column displaying the power to detect an effect that has the same size as the standard deviation of the random error.

Table 3.3 Original screening design with observed responses (upper panel) and additional runs for the follow-up experiment (lower panel).

Run	Methanol x_1	Ethanol x_2	Propanol x_3	Butanol x_4	pH x_5	Time x_6	Block x_7	Yield (mg)
1	0	0	0	10	6	1	1	10.94
2	0	10	0	0	9	1	1	15.79
3	0	10	0	10	9	2	1	25.96
4	10	10	10	0	6	1	1	35.92
5	0	0	10	0	6	2	1	22.92
6	0	10	10	10	6	1	1	23.54
7	10	10	0	0	6	2	1	47.44
8	10	0	0	0	9	1	1	19.80
9	10	0	10	10	9	1	1	29.48
10	0	0	10	0	9	2	1	17.13
11	10	10	10	10	9	2	1	43.75
12	10	0	0	10	6	2	1	40.86
13	10	0	10	0	9	2	2	.
14	0	0	10	0	6	1	2	.
15	10	10	0	10	6	1	2	.
16	0	10	10	0	6	2	2	.
17	0	0	0	0	9	2	2	.
18	0	10	10	10	9	1	2	.
19	0	0	10	10	6	2	2	.
20	10	0	0	0	6	1	2	.

[Peter] What we learn from this table is that, if the effect of any two-factor interaction is only the same size as the error's standard deviation, there is a 94% chance that a hypothesis test at a significance level of 0.05 will reject the null hypothesis that there is actually no interaction.

[Dr. Zheng] That is a mouthful. Can you tell me something that will make sense to management?

[Brad] Loosely speaking, we are quite confident that, if any two-factor interaction is really there, we will find it. Of course the probability that we will find it depends on its actual size. If it is only the same size as the standard deviation of the error, which I recall was estimated to be around 1.6 from our original analysis, then the probability of detection is 94%. If the real effect is twice as big, around 3.2, then we are virtually certain to detect it. The bottom line is that eight runs will do the trick.

[Dr. Zheng] Then, I will run these eight trials and e-mail you the table with the responses as soon as I have them.

[Peter] We will look forward to hearing from you.

Table 3.4   Relative variances of the factor-effect estimates obtained from the 12 original runs and the eight additional runs for the extraction experiment in Table 3.3, along with the power for detecting effects that have the same size as the error's standard deviation.

Effect	Variance	Power
Intercept	0.058	0.930
Methanol	0.056	0.938
Ethanol	0.056	0.938
Propanol	0.056	0.938
Butanol	0.056	0.938
pH	0.052	0.950
Time	0.055	0.939
Block	0.058	0.930
Ethanol × Propanol	0.055	0.939
Methanol × Time	0.056	0.938
Ethanol × Time	0.056	0.938
Propanol × Time	0.056	0.938
Butanol × Time	0.056	0.938
Methanol × Butanol	0.055	0.939

3.2.2   Data analysis

A week later Peter and Brad get the following e-mail from Dr. Zheng, with Table 3.5 enclosed.

Peter and Brad,

Management was not very happy about the fact that the yield for these eight runs was so low. I told them that the purpose of this experiment was not to maximize yield, but that it was to demonstrate the existence of any two-factor interaction effects and simultaneously improve our estimates of the main effects. I hope that was ok.

In any case, having you two come tomorrow morning to explain your analysis would help me out. Can you and Brad take a look at the new data table and come by tomorrow at 10 AM to present your results and recommendations? We have a meeting with the management to discuss the next steps in the development of our antimicrobial substance. Remember that the dress code here is casual, so please don't bother with a coat and tie.

Thanks,
-Zheng

Table 3.5   Design and response data for the original and follow-up extraction experiment at GeneBe.

	Methanol	Ethanol	Propanol	Butanol	pH	Time		Yield
Run	x_1	x_2	x_3	x_4	x_5	x_6	Block	(mg)
1	0	0	0	10	6	1	1	10.94
2	0	10	0	0	9	1	1	15.79
3	0	10	0	10	9	2	1	25.96
4	10	10	10	0	6	1	1	35.92
5	0	0	10	0	6	2	1	22.92
6	0	10	10	10	6	1	1	23.54
7	10	10	0	0	6	2	1	47.44
8	10	0	0	0	9	1	1	19.80
9	10	0	10	10	9	1	1	29.48
10	0	0	10	0	9	2	1	17.13
11	10	10	10	10	9	2	1	43.75
12	10	0	0	10	6	2	1	40.86
13	10	0	10	0	9	2	2	34.03
14	0	0	10	0	6	1	2	13.47
15	10	10	0	10	6	1	2	41.50
16	0	10	10	0	6	2	2	27.07
17	0	0	0	0	9	2	2	9.07
18	0	10	10	10	9	1	2	15.83
19	0	0	10	10	6	2	2	24.44
20	10	0	0	0	6	1	2	24.83

The next morning, Peter and Brad present their results to the management team at GeneBe. Dr. Zheng and his manager, Bas Ritter, are both there.

[Bas, introducing Peter as the presenter] Allow me to introduce Dr. Peter Goos. He and his partner at Intrepid Stats, Dr. Bradley Jones, are here to present the results of our augmentation experiment, which is a follow up to the successful screening experiment we ran a couple of months ago. Peter?

[Peter] Thanks for coming. We are excited to be here to present our new results. We understand there was some concern about the low yields in the most recent study. We plan to explain that, but, first, let me share some good news with you.

Peter shows Table 3.6 with the results of the previous study.

[Peter] This table shows the estimates of the factor effects on the yield of the extraction process of the microbial inhibitor from our analysis of the screening experiment performed here at GeneBe a couple of months ago. Notice that the last row of the table has a term "Ethanol × Propanol." This is the two-factor interaction effect involving ethanol and propanol in your antimicrobial product development effort. The last entry in that row is 0.0426. This is the significance probability for this effect. It is common practice to view any significance probability less than 0.05 as an indication that the observed effect is not due to chance. We thought that this effect

Table 3.6   Factor-effect estimates obtained from the original screening experiment at GeneBe, with marginally significant interaction effect involving ethanol and butanol. Asterisks indicate effects that are significantly different from zero at a 5% level of significance.

Effect	Estimate	Standard error	t Ratio	p Value
Intercept	27.79	0.45	61.9	<.0001*
Methanol	9.00	0.49	18.3	<.0001*
Ethanol	4.27	0.45	9.5	0.0007*
Propanol	1.00	0.45	2.2	0.0908
Butanol	1.88	0.49	3.8	0.0186*
pH	−3.07	0.49	−6.2	0.0034*
Time	4.63	0.49	9.4	0.0007*
Ethanol × Propanol	−1.77	0.60	−2.9	0.0426*

might indeed be real and recommended a follow-up study to confirm the presence of this effect.

Peter changes to a new slide, showing Table 3.7.

[Peter] Here are the results of a similar analysis we performed after receiving the data from the eight augmented design runs GeneBe recently completed. Note that the values of the estimates have hardly changed, but because of having the extra data, we have become much more confident about our results. Now, the significance probability for the two-factor interaction effect that we are interested in is less than 0.0001. This is so small that we regard it as strong evidence that the two-factor interaction involving ethanol and propanol is really there and that it is not due to random chance.

[Bas, raising his hand] I notice that you have an extra row in this table, labeled "Block." What does that mean?

[Peter] We added the block term to the model to investigate whether the process mean had shifted in the interim between the two experiments. The estimate of this

Table 3.7   Factor-effect estimates obtained from the reduced model for the combined screening and follow-up experiment.

Effect	Estimate	Standard error	t Ratio	p Value
Intercept	27.18	0.25	107.70	<.0001*
Methanol	9.08	0.25	36.71	<.0001*
Ethanol	4.41	0.25	17.83	<.0001*
Propanol	1.10	0.25	4.44	0.0010*
Butanol	1.85	0.25	7.47	<.0001*
pH	−3.08	0.24	−12.60	<.0001*
Time	4.44	0.25	17.94	<.0001*
Ethanol × Propanol	−1.87	0.25	−7.57	<.0001*
Block	0.61	0.25	2.42	0.0337*

term is positive indicating that the first block of runs had a higher estimated yield in the middle of each factor's range than the second block. This could mean that some other factors have changed in the meantime, leading to average yields that are a little over a milligram lower than they were a few weeks ago.

[Bas] A little over a milligram? You mean 0.61 mg?

[Peter] Because we coded the low and high levels of the experimental factors by −1 and +1, we need to double the estimates in our table to get the effect of changing a factor's level from low to high on the yield. Because we used a +1 to code runs from the original experiment and a −1 to code runs from the follow-up experiment, the difference in yield between the two blocks is two times 0.61 mg, or 1.22 mg.

[Bas] You promised to explain the low yields of the eight runs in the augmentation study. Can you say something about that?

Peter changes slides again, showing Table 3.8.

[Peter] Using our model estimates, we created a prediction formula that we used to predict the yield of the eight new responses. As you can see, the observed yields and their predicted values are quite close. This is especially true since your typical run-to-run variation, as indicated by the RMSE of the fitted model, is more than one and a half milligram. So, even though the observed yield is low for these runs, we now have established that the two-factor interaction involving ethanol and propanol exists and is reproducible. Our recommended process settings from the original experiment were based in part on the assumption that the two-factor interaction was real.

[Bas] How is that?

[Peter] We recommended that you leave propanol out of your solvent system. That is because our model indicated that, if ethanol was in the solvent system, then leaving propanol out of the system would result in higher yields. We understand that the current process does not include propanol and it is maintaining an average yield of 50 mg per 100 ml.

[Bas] That also seems like good evidence that your original recommendations were sound. Are there any other questions?

Table 3.8   Observed and predicted yields (in mg) for the eight factor-level combinations studied in the follow-up experiment.

Run	Observed yield	Predicted yield
13	34.03	33.72
14	13.47	12.85
15	41.50	41.32
16	27.07	26.79
17	9.07	9.62
18	15.83	15.46
19	24.44	25.41
20	24.83	25.07

Table 3.9 Factor-effect estimates for the model with six two-factor interactions obtained from the combined screening and follow-up experiment.

Effect	Estimate	Standard error	t Ratio	p Value
Intercept	27.22	0.20	134.30	<.0001
Methanol	9.14	0.20	45.98	<.0001
Ethanol	4.37	0.20	21.98	<.0001
Propanol	1.03	0.20	5.20	0.0020
Butanol	1.81	0.20	9.09	<.0001
pH	−3.11	0.19	−16.12	<.0001
Time	4.39	0.20	22.16	<.0001
Block	0.57	0.20	2.84	0.0297
Methanol × Butanol	0.44	0.20	2.21	0.0689
Methanol × Time	0.48	0.20	2.40	0.0533
Ethanol × Propanol	−1.83	0.20	−9.23	<.0001
Ethanol × Time	−0.22	0.20	−1.08	0.3210
Propanol × Time	0.03	0.20	0.16	0.8752
Butanol × Time	0.04	0.20	0.23	0.8288

[Dr. Zheng] I notice that you did not include the other terms from the a priori model of the augmentation experiment. Is that because the other potential two-factor interaction effects turned out not to be statistically significant?

Peter changes slides once more, showing Table 3.9.

[Peter] Dr. Zheng, that is exactly the case. Here are the estimates of all the terms in the full model for the augmentation experiment.

[Peter] As you can see, none of the other terms has an estimated effect larger than one half, and none of these effects is statistically significant. I removed these insignificant terms and refit the model to get a simpler prediction equation. That was what I showed you two slides ago.

[Dr. Zheng] Thank you. That makes sense.

[Bas] Are there any other questions?

No further questions emerge.

[Bas] In that case, Peter, thank you for your presentation. You have made some believers here in the utility of both designed experiments and the benefits of design augmentation studies. We will now discuss whether we will continue the development of this antimicrobial substance.

3.3   Peek into the black box

Design augmentation is the process of adding experimental runs to an existing experimental design. In other words, it is the process of creating a follow-up experiment. This is useful if the initial experiment gave insufficiently precise parameter estimates or, if some effect is large enough to be practically important but is still too small to be statistically significant. Design augmentation is also useful in screening situations where there is ambiguity regarding which effects are important due to complete or partial confounding or aliasing of effects. As a matter of fact, performing extra experimental runs is the only data-driven way to break confounding patterns and to disentangle confounded effects. The challenge is to find the best possible runs to add to the experiment.

3.3.1   Optimal selection of a follow-up design

We explain how to select the best possible runs for a follow-up experiment using the scenario for the extraction experiment. The original design for the experiment was given in Table 2.1. This design is optimal for the main-effects model

$$Y = \beta_0 + \beta_1 x_1 + \beta_2 x_2 + \beta_3 x_3 + \beta_4 x_4 + \beta_5 x_5 + \beta_6 x_6 + \varepsilon. \tag{3.1}$$

The model matrix corresponding to the design and the main-effects model is

$$\mathbf{X}^* = \begin{bmatrix} 1 & -1 & -1 & -1 & +1 & -1 & -1 \\ 1 & -1 & +1 & -1 & -1 & +1 & -1 \\ 1 & -1 & +1 & -1 & +1 & +1 & +1 \\ 1 & +1 & +1 & +1 & -1 & -1 & -1 \\ 1 & -1 & -1 & +1 & -1 & -1 & +1 \\ 1 & -1 & +1 & +1 & +1 & -1 & -1 \\ 1 & +1 & +1 & -1 & -1 & -1 & +1 \\ 1 & +1 & -1 & -1 & -1 & +1 & -1 \\ 1 & +1 & -1 & +1 & +1 & +1 & -1 \\ 1 & -1 & -1 & +1 & -1 & +1 & +1 \\ 1 & +1 & +1 & +1 & +1 & +1 & +1 \\ 1 & +1 & -1 & -1 & +1 & -1 & +1 \end{bmatrix}, \tag{3.2}$$

and the information contained within this matrix for model (3.1) is

$$\mathbf{X}^{*'}\mathbf{X}^* = \begin{bmatrix} 12 & 0 & 0 & 0 & 0 & 0 & 0 \\ 0 & 12 & 0 & 0 & 0 & 0 & 0 \\ 0 & 0 & 12 & 0 & 0 & 0 & 0 \\ 0 & 0 & 0 & 12 & 0 & 0 & 0 \\ 0 & 0 & 0 & 0 & 12 & 0 & 0 \\ 0 & 0 & 0 & 0 & 0 & 12 & 0 \\ 0 & 0 & 0 & 0 & 0 & 0 & 12 \end{bmatrix}. \tag{3.3}$$

This matrix is diagonal, which implies that all the parameters in the model, i.e., the intercept and the main effects, can be estimated independently. The determinant of this matrix, which we named the D-optimality criterion value in the previous chapter, equals 12^7. The fact that the determinant is positive signifies that the design contains sufficient information to estimate the main-effects model.

The purpose of the additional runs in the extraction experiment is to resolve the remaining ambiguity after analyzing the data from the original experiment. More specifically, the scenario requires adding runs so that a model involving six two-factor interaction terms,

$$Y = \beta_0 + \beta_1 x_1 + \beta_2 x_2 + \beta_3 x_3 + \beta_4 x_4 + \beta_5 x_5 + \beta_6 x_6 + \beta_{14} x_1 x_4$$
$$+ \beta_{16} x_1 x_6 + \beta_{23} x_2 x_3 + \beta_{26} x_2 x_6 + \beta_{36} x_3 x_6 + \beta_{46} x_4 x_6 + \varepsilon, \qquad (3.4)$$

can be estimated. This will allow us to test which of the potential interaction effects is active.

Often, quite a bit of time goes by between the original experiment and the follow-up experiment. It is, therefore, common to observe a shift in average response between the two experiments. We recommend adding a term, δx_7, to the model to capture this potential shift in the mean response:

$$Y = \beta_0 + \beta_1 x_1 + \beta_2 x_2 + \beta_3 x_3 + \beta_4 x_4 + \beta_5 x_5 + \beta_6 x_6 + \beta_{14} x_1 x_4$$
$$+ \beta_{16} x_1 x_6 + \beta_{23} x_2 x_3 + \beta_{26} x_2 x_6 + \beta_{36} x_3 x_6 + \beta_{46} x_4 x_6 + \delta x_7 + \varepsilon. \quad (3.5)$$

The parameter δ represents the shift in response, and the factor x_7 is a two-level factor that takes the value $+1$ for all the runs in the original experiment and -1 for all the runs in the follow-up experiment. In design of experiments, we name the parameter δ a block effect and the factor x_7 a blocking factor. Blocking is a fundamental principle in experimental design. In Chapters 7 and 8, we discuss blocking in more detail.

To construct the follow-up design, we first need to create the model matrix for the new model corresponding to the original experiment, $\mathbf{X}_1$. Like the original model matrix $\mathbf{X}^*$, the new model matrix has 12 rows. However, as the new model in Equation (3.5) involves 14 unknown parameters instead of 7, it will have 14 columns: one for the intercept, six for the main effects, six others for the two-factor interaction effects, and a final one for the block effect. For the original extraction experiment, the new model matrix is

$$\mathbf{X}_1 = \begin{bmatrix}
1 & -1 & -1 & -1 & +1 & -1 & -1 & -1 & +1 & +1 & +1 & +1 & -1 & +1 \\
1 & -1 & +1 & -1 & -1 & +1 & -1 & +1 & +1 & -1 & -1 & +1 & +1 & +1 \\
1 & -1 & +1 & -1 & +1 & +1 & +1 & -1 & -1 & -1 & +1 & -1 & +1 & +1 \\
1 & +1 & +1 & +1 & -1 & -1 & -1 & -1 & -1 & +1 & -1 & -1 & +1 & +1 \\
1 & -1 & -1 & +1 & -1 & -1 & +1 & +1 & -1 & -1 & -1 & +1 & -1 & +1 \\
1 & -1 & +1 & +1 & +1 & -1 & -1 & -1 & +1 & +1 & -1 & -1 & -1 & +1 \\
1 & +1 & +1 & -1 & -1 & -1 & +1 & -1 & +1 & -1 & +1 & -1 & -1 & +1 \\
1 & +1 & -1 & -1 & -1 & +1 & -1 & -1 & -1 & +1 & +1 & +1 & +1 & +1 \\
1 & +1 & -1 & +1 & +1 & +1 & -1 & +1 & -1 & -1 & +1 & -1 & -1 & +1 \\
1 & -1 & -1 & +1 & -1 & +1 & +1 & +1 & -1 & -1 & -1 & +1 & -1 & +1 \\
1 & +1 & +1 & +1 & +1 & +1 & +1 & +1 & +1 & +1 & +1 & +1 & +1 & +1 \\
1 & +1 & -1 & -1 & +1 & -1 & +1 & +1 & +1 & -1 & -1 & +1 & +1 \\
\end{bmatrix},$$

$$(3.6)$$

and the information contained within the original experiment on the parameters in the extended model (3.5) is given by

$$
\mathbf{X}_1'\mathbf{X}_1 =
\begin{bmatrix}
12 & 0 & 0 & 0 & 0 & 0 & 0 & 0 & 0 & 0 & 0 & 0 & 0 & +12 \\
0 & 12 & 0 & 0 & 0 & 0 & 0 & 0 & 0 & 4 & 4 & -4 & 4 & 0 \\
0 & 0 & 12 & 0 & 0 & 0 & 0 & -4 & 4 & 0 & 0 & -4 & 4 & 0 \\
0 & 0 & 0 & 12 & 0 & 0 & 0 & 4 & -4 & 0 & -4 & 0 & -4 & 0 \\
0 & 0 & 0 & 0 & 12 & 0 & 0 & 0 & 4 & 4 & 4 & -4 & 0 & 0 \\
0 & 0 & 0 & 0 & 0 & 12 & 0 & 4 & -4 & -4 & 4 & 4 & 4 & 0 \\
0 & 0 & 0 & 0 & 0 & 0 & 12 & 4 & 0 & -4 & 0 & 0 & 0 & 0 \\
0 & 0 & -4 & 4 & 0 & 4 & 4 & 12 & 0 & -4 & -4 & 4 & 0 & 0 \\
0 & 0 & 4 & -4 & 4 & -4 & 0 & 0 & 12 & 4 & 0 & 0 & 0 & 0 \\
0 & 4 & 0 & 0 & 4 & -4 & -4 & -4 & 4 & 12 & 0 & 0 & 4 & 0 \\
0 & 4 & 0 & -4 & 4 & 4 & 0 & -4 & 0 & 0 & 12 & 0 & 0 & 0 \\
0 & -4 & -4 & 0 & -4 & 4 & 0 & 4 & 0 & 0 & 0 & 12 & 0 & 0 \\
0 & 4 & 4 & -4 & 0 & 4 & 0 & 0 & 0 & 4 & 0 & 0 & 12 & 0 \\
+12 & 0 & 0 & 0 & 0 & 0 & 0 & 0 & 0 & 0 & 0 & 0 & 0 & 12
\end{bmatrix}.
\tag{3.7}
$$

The determinant of this matrix is zero, which indicates that the original design does not contain enough information to estimate the extended model involving the six additional two-factor interactions and the block effect. One reason for this is that the original design contains fewer experimental runs than there are parameters in the extended model. Another reason is that the last column of $\mathbf{X}_1$ is equal to the first column. The block effect is, therefore, confounded with the intercept. In mathematical jargon, the first and last columns of $\mathbf{X}_1$ are, therefore, linearly dependent. This can be seen from the information matrix $\mathbf{X}_1'\mathbf{X}_1$ too, by noting that the last element in the first row, $+12$, is as large as the first and the last diagonal elements of the matrix. If a design gives rise to an information matrix with zero determinant for a given model, we say that the design is singular for that model.

The next step in the design augmentation process is to add rows to the model matrix $\mathbf{X}_1$, one for each additional run. In the follow-up extraction experiment, there are eight additional runs. Therefore, if we denote the additional rows of the model matrix by $\mathbf{X}_2$ and the complete model matrix by $\mathbf{X}$, we have

$$
\mathbf{X} = \begin{bmatrix} \mathbf{X}_1 \\ \mathbf{X}_2 \end{bmatrix},
$$

$$
=
\left[
\begin{array}{cccccccccccccc}
1 & -1 & -1 & -1 & +1 & -1 & -1 & -1 & +1 & +1 & +1 & +1 & -1 & +1 \\
1 & -1 & +1 & -1 & -1 & +1 & -1 & +1 & +1 & -1 & +1 & +1 & +1 & +1 \\
1 & -1 & +1 & -1 & +1 & +1 & +1 & -1 & -1 & -1 & +1 & -1 & +1 & +1 \\
1 & +1 & +1 & +1 & -1 & -1 & -1 & -1 & -1 & +1 & -1 & -1 & +1 & +1 \\
1 & -1 & -1 & +1 & -1 & -1 & +1 & +1 & -1 & -1 & -1 & +1 & -1 & +1 \\
1 & -1 & +1 & +1 & +1 & -1 & -1 & -1 & +1 & +1 & -1 & -1 & -1 & +1 \\
1 & +1 & +1 & -1 & -1 & -1 & +1 & -1 & +1 & -1 & +1 & -1 & -1 & +1 \\
1 & +1 & -1 & -1 & -1 & +1 & -1 & -1 & -1 & +1 & +1 & +1 & +1 & +1 \\
1 & +1 & -1 & +1 & +1 & +1 & -1 & +1 & -1 & -1 & +1 & -1 & -1 & +1 \\
1 & -1 & -1 & +1 & -1 & +1 & +1 & +1 & -1 & -1 & -1 & +1 & -1 & +1 \\
1 & +1 & +1 & +1 & +1 & +1 & +1 & +1 & +1 & +1 & +1 & +1 & +1 & +1 \\
1 & +1 & -1 & -1 & +1 & -1 & +1 & +1 & +1 & +1 & -1 & -1 & +1 & +1 \\
\hline
1 & \cdot & \cdot & \cdot & \cdot & \cdot & \cdot & \cdot & \cdot & \cdot & \cdot & \cdot & \cdot & -1 \\
1 & \cdot & \cdot & \cdot & \cdot & \cdot & \cdot & \cdot & \cdot & \cdot & \cdot & \cdot & \cdot & -1 \\
1 & \cdot & \cdot & \cdot & \cdot & \cdot & \cdot & \cdot & \cdot & \cdot & \cdot & \cdot & \cdot & -1 \\
1 & \cdot & \cdot & \cdot & \cdot & \cdot & \cdot & \cdot & \cdot & \cdot & \cdot & \cdot & \cdot & -1 \\
1 & \cdot & \cdot & \cdot & \cdot & \cdot & \cdot & \cdot & \cdot & \cdot & \cdot & \cdot & \cdot & -1 \\
1 & \cdot & \cdot & \cdot & \cdot & \cdot & \cdot & \cdot & \cdot & \cdot & \cdot & \cdot & \cdot & -1 \\
1 & \cdot & \cdot & \cdot & \cdot & \cdot & \cdot & \cdot & \cdot & \cdot & \cdot & \cdot & \cdot & -1 \\
1 & \cdot & \cdot & \cdot & \cdot & \cdot & \cdot & \cdot & \cdot & \cdot & \cdot & \cdot & \cdot & -1
\end{array}
\right].
\tag{3.8}
$$

We now have to find values for the levels of the six experimental factors that maximize the determinant of the total information matrix

$$
\begin{aligned}
\mathbf{X'X} &= \begin{bmatrix} \mathbf{X}_1 \\ \mathbf{X}_2 \end{bmatrix}' \begin{bmatrix} \mathbf{X}_1 \\ \mathbf{X}_2 \end{bmatrix}, \\
&= \begin{bmatrix} \mathbf{X}'_1 & \mathbf{X}'_2 \end{bmatrix} \begin{bmatrix} \mathbf{X}_1 \\ \mathbf{X}_2 \end{bmatrix}, \\
&= \mathbf{X}'_1\mathbf{X}_1 + \mathbf{X}'_2\mathbf{X}_2.
\end{aligned}
\tag{3.9}
$$

This last expression shows that the new information matrix is the sum of the information matrix of the original experiment, $\mathbf{X}'_1\mathbf{X}_1$, and that of the follow-up experiment, $\mathbf{X}'_2\mathbf{X}_2$. A D-optimal follow-up design maximizes the determinant

$$
|\mathbf{X'X}| = |\mathbf{X}'_1\mathbf{X}_1 + \mathbf{X}'_2\mathbf{X}_2|
\tag{3.10}
$$

over all possible matrices $\mathbf{X}_2$.

It is common sense to take into account the design used in the original experiment when setting up the follow-up experiment. Not doing so, by just maximizing $|\mathbf{X}'_2\mathbf{X}_2|$, will lead to a follow-up experiment that pays equal attention to the main effects, about which we already have substantial information from the initial experiment, and the interaction effects, about which we have insufficient information.

In the extraction experiment, the design of the original study was singular for the extended model (3.5) involving the interaction effects. Therefore, its information matrix, $\mathbf{X}'_1\mathbf{X}_1$, cannot be inverted. However, in situations where $\mathbf{X}'_1\mathbf{X}_1$ is invertible, the determinant of the total experiment can be written as

$$
|\mathbf{X}'_1\mathbf{X}_1 + \mathbf{X}'_2\mathbf{X}_2| = |\mathbf{X}'_1\mathbf{X}_1| \times |\mathbf{I}_{n_2} + \mathbf{X}_2(\mathbf{X}'_1\mathbf{X}_1)^{-1}\mathbf{X}'_2|,
\tag{3.11}
$$

where n_2 is the number of runs in the follow-up experiment. As the model matrix $\mathbf{X}_1$ is given when designing the follow-up experiment, the determinant $|\mathbf{X}'_1\mathbf{X}_1|$ is a constant. Therefore, in a situation where $\mathbf{X}'_1\mathbf{X}_1$ is invertible, maximizing the total information that will be available after the follow-up experiment requires maximizing

$$
|\mathbf{I}_{n_2} + \mathbf{X}_2(\mathbf{X}'_1\mathbf{X}_1)^{-1}\mathbf{X}'_2|.
\tag{3.12}
$$

This expression shows explicitly that the optimal follow-up experiment is generally not found by maximizing $|\mathbf{X}'_2\mathbf{X}_2|$, but that $\mathbf{X}_1$ needs to be taken into account when designing the follow-up experiment with model matrix $\mathbf{X}_2$.

Optimizing the follow-up design for the extraction experiment, by maximizing the determinant in Equation (3.10), results in the following model matrix

$$
\mathbf{X} = \left[
\begin{array}{ccccccccccccccc}
1 & -1 & -1 & -1 & +1 & -1 & -1 & -1 & +1 & +1 & +1 & +1 & -1 & +1 \\
1 & -1 & +1 & -1 & -1 & +1 & -1 & +1 & +1 & -1 & -1 & +1 & +1 & +1 \\
1 & -1 & +1 & -1 & +1 & +1 & +1 & -1 & -1 & -1 & +1 & -1 & +1 & +1 \\
1 & +1 & +1 & +1 & -1 & -1 & -1 & -1 & -1 & +1 & -1 & -1 & +1 & +1 \\
1 & -1 & -1 & +1 & -1 & -1 & +1 & +1 & -1 & -1 & -1 & +1 & -1 & +1 \\
1 & -1 & +1 & +1 & +1 & -1 & -1 & -1 & +1 & +1 & -1 & -1 & -1 & +1 \\
1 & +1 & +1 & -1 & -1 & -1 & +1 & -1 & +1 & -1 & +1 & -1 & -1 & +1 \\
1 & +1 & -1 & -1 & -1 & +1 & -1 & -1 & -1 & +1 & +1 & +1 & +1 & +1 \\
1 & +1 & -1 & +1 & +1 & +1 & -1 & +1 & -1 & -1 & +1 & -1 & -1 & +1 \\
1 & -1 & -1 & +1 & -1 & +1 & +1 & +1 & -1 & -1 & -1 & +1 & -1 & +1 \\
1 & +1 & +1 & +1 & +1 & +1 & +1 & +1 & +1 & +1 & +1 & +1 & +1 & +1 \\
1 & +1 & -1 & -1 & +1 & -1 & +1 & +1 & +1 & +1 & -1 & -1 & +1 & +1 \\
\hline
1 & +1 & -1 & +1 & -1 & +1 & +1 & -1 & +1 & -1 & -1 & +1 & -1 & -1 \\
1 & -1 & -1 & +1 & -1 & -1 & -1 & +1 & +1 & -1 & +1 & -1 & +1 & -1 \\
1 & +1 & +1 & -1 & +1 & -1 & -1 & +1 & -1 & -1 & -1 & +1 & -1 & -1 \\
1 & -1 & +1 & +1 & -1 & -1 & +1 & +1 & -1 & +1 & +1 & +1 & -1 & -1 \\
1 & -1 & -1 & -1 & -1 & +1 & +1 & +1 & -1 & +1 & -1 & -1 & -1 & -1 \\
1 & -1 & +1 & +1 & +1 & +1 & -1 & -1 & +1 & +1 & -1 & -1 & -1 & -1 \\
1 & -1 & -1 & +1 & +1 & -1 & +1 & -1 & -1 & -1 & -1 & +1 & +1 & -1 \\
1 & +1 & -1 & -1 & -1 & -1 & -1 & -1 & -1 & +1 & +1 & +1 & +1 & -1 \\
\end{array}
\right], \qquad (3.13)
$$

with information matrix

$$
\mathbf{X} = \left[
\begin{array}{cccccccccccccc}
20 & -2 & -2 & 2 & -2 & -2 & 0 & 0 & -2 & 0 & -2 & 2 & -2 & 2 \\
-2 & 20 & 0 & -4 & 0 & 0 & -2 & -2 & 0 & 2 & 4 & 0 & 4 & 0 \\
-2 & 0 & 20 & 0 & 4 & 0 & -2 & -2 & 4 & 2 & 0 & -4 & 0 & 4 \\
2 & -4 & 0 & 20 & 0 & 0 & 2 & 2 & 0 & -2 & -4 & 0 & -4 & 0 \\
-2 & 0 & 4 & 0 & 20 & 0 & -2 & -2 & 4 & 2 & 0 & -4 & 0 & 0 \\
-2 & 0 & 0 & 0 & 0 & 20 & 2 & 2 & 0 & -2 & 0 & 0 & 0 & 4 \\
0 & -2 & -2 & 2 & -2 & 2 & 20 & 4 & -2 & -4 & -2 & 2 & -2 & 2 \\
0 & -2 & -2 & 2 & -2 & 2 & 4 & 20 & -2 & -4 & -2 & 2 & -2 & 2 \\
-2 & 0 & 4 & 0 & 4 & 0 & -2 & -2 & 20 & 2 & 0 & -4 & 0 & 0 \\
0 & 2 & 2 & -2 & 2 & -2 & -4 & -4 & 2 & 20 & 2 & -2 & 2 & 2 \\
-2 & 4 & 0 & -4 & 0 & 0 & -2 & -2 & 0 & 2 & 20 & 0 & 4 & 4 \\
2 & 0 & -4 & 0 & -4 & 0 & 2 & 2 & -4 & -2 & 0 & 20 & 0 & 0 \\
-2 & 4 & 0 & -4 & 0 & 0 & -2 & -2 & 0 & 2 & 4 & 0 & 20 & 0 \\
2 & 0 & 4 & 0 & 0 & 4 & 2 & 2 & 0 & 2 & 4 & 0 & 0 & 20 \\
\end{array}
\right]. \qquad (3.14)
$$

This information matrix is far from singular, as it has a determinant of 6.5541×10^{17}. So, the augmented design, i.e., the combination of the initial design and the follow-up design, contains sufficient information to estimate all the effects in the

Table 3.10   Relative variances of factor-effect estimates from the combined designs of the initial extraction experiment and the follow-up experiment.

Effect	Variance	VIF
Intercept	0.058	1.161
Methanol	0.056	1.117
Ethanol	0.056	1.117
Propanol	0.056	1.117
Butanol	0.056	1.117
pH	0.052	1.049
Time	0.055	1.110
Methanol × Butanol	0.055	1.110
Methanol × Time	0.056	1.117
Ethanol × Propanol	0.055	1.110
Ethanol × Time	0.056	1.117
Propanol × Time	0.056	1.117
Butanol × Time	0.056	1.117
Block	0.058	1.161

extended model. An interesting feature of the augmented design can be seen when inverting the information matrix and looking at the diagonal elements of the inverse, $(\mathbf{X}'\mathbf{X})^{-1}$. As explained in the previous chapter, these diagonal elements are the relative variances of the factor-effect estimates. They are shown in Table 3.10, along with the corresponding variance inflation factors.

The variances of all the effects in the extended model turn out to be roughly of the same size: they all lie between 0.052 and 0.058. This shows that maximizing the total information matrix can result in all effects in the extended model receiving roughly equal attention. None of the variances is as small as $1/20 = 0.05$, indicating that there is some variance inflation. This is due to the fact that the information matrix is not diagonal, which is in turn due to the columns of the model matrix $\mathbf{X}$ not being orthogonal. As we can see from the last column of Table 3.10, the maximum variance inflation factor for any estimated effect is 1.161. This is much smaller than the alarm level of five, which is used for variance inflation factors. So, there is no reason for concern about the correlation among the factor-effect estimates.

3.3.2   Design construction algorithm

For finding the design that maximizes the determinant of the total information matrix in Equation (3.10), or, equivalently, the determinant in Equation (3.12) when $\mathbf{X}_1'\mathbf{X}_1$ is invertible, we use a version of the coordinate-exchange algorithm, which was introduced in Section 2.3.9., tailored to design augmentation. It differs from the original coordinate-exchange algorithm in that it operates only on X_2 rather than on the complete model matrix X.

The algorithm starts by generating random values on the interval $[-1, +1]$, one for every cell of the model matrix X_2 corresponding to an experimental factor. If the follow-up design requires n_2 runs and k experimental factors are involved, then $n_2 k$ random values are needed.

The resulting starting design is then improved on a coordinate-by-coordinate basis. The algorithm goes through the $n_2 k$ coordinates one by one, and finds the value in the set $\{-1, +1\}$ that optimizes the D-optimality criterion value in Equation (3.10). If the D-optimality criterion value improves, the old value of the coordinate is replaced with the new value. This process is repeated until no more coordinate exchanges are made in an entire pass through the design matrix.

The process of generating a starting design and iteratively improving it coordinate by coordinate is repeated for a large number of starting designs to increase the likelihood of obtaining a design for which the determinant in Equation (3.10) is close to the global optimum.

3.3.3   Foldover designs

The best-known technique for creating designs for follow-up experiments is the foldover technique. This technique, in which the follow-up design is equal to the original with all the signs in one or more columns reversed, has the advantage of being so simple that one can obtain the design by hand.

The foldover technique has two main disadvantages. First, the foldover technique requires the follow-up experiment to have the same size as the initial experiment. This automatically doubles the total experimental cost, often unnecessarily. Second, the foldover technique is only effective for removing the ambiguity concerning specific sets of effects. For instance, folding over is effective for de-aliasing main effects from interaction effects and for de-aliasing the main effect of one factor and all two-factor interaction effects involving that factor. Often, however, the original experiment suggests quite different sets of effects to be of interest. For instance, after the initial extraction experiment, the interest was in all the main effects and six additional two-factor interactions that could not all be de-aliased by using a foldover design.

Therefore, a more flexible approach to designing follow-up screening experiments is required in most practical situations. Using optimal experimental design, as we did in the extraction experiment case study, is such an approach: it allows the researcher to create a follow-up experiment that will give the most precise estimates for any set of main effects, interaction effects, and/or polynomial effects (such as quadratic effects; see Chapter 4). The only requirement is that the number of experimental runs in the follow-up experiment is larger than or equal to the number of extra terms in the extended model. For instance, for the extraction experiment, a follow-up experiment with just seven runs could have been used.

3.4   Background reading

There is a substantial literature on the design of follow-up experiments. There can be many reasons for doing one or more follow-up experiments. Both Atkinsonet al. (2007) and Mee (2009) provide lists of reasons why performing follow-up experiments is useful.

Mee (2009) provides a thorough overview of approaches that can be taken to set up follow-up experiments. For instance, he shows that augmenting a regular fractional factorial resolution III design with its mirror image ensures that the main effects are no longer confounded with two-factor interaction effects. He also discusses the use of the foldover and semifoldover techniques after a resolution IV fractional factorial design. The semifoldover technique is similar to the foldover technique, but only half the runs of the foldover designs are actually performed. Finally, Mee (2009) shows the added value of the optimal design of follow-up experiments using an injection molding example. Another such example, using a leaf spring experiment, is given by Wu and Hamada (2009). Miller and Sitter (1997) demonstrate that the folded-over 12-run Plackett–Burman design is useful for considering up to 12 factors in 24 runs, even if one anticipates that some two-factor interactions may be significant.

Even though optimal design is a more flexible and more economical approach to the design of follow-up experiments, the literature on this subject is not nearly as extensive as that on foldover techniques. The first published example of a D-optimal follow-up experimental design appears in Mitchell (1974). A general account of the D-optimal design of follow-up experiments, along with several examples, is given in Atkinson et al. (2007).

A more advanced Bayesian approach for selecting designs for follow-up experiments is suggested by Meyer et al. (1996). In their discussion of this article, Jones and DuMouchel (1996) provide an alternative optimal design approach based on posterior probabilities of competing models being the true model. Ruggoo and Vandebroek (2004) propose a Bayesian strategy for optimally designing the sequence consisting of both the initial and follow-up experiments.

3.5   Summary

This chapter dealt with augmenting a screening design to resolve questions about which of many possible two-factor interaction effects were active. Algorithmic construction methods using an optimal design approach provide an extremely cost-effective way to discriminate among several highly likely models.

It often happens that an initial screening experiment points to a factor-level combination that solves the immediate engineering problem without requiring further experimentation other than verification runs. Of course, this is a highly desirable result but not one you can always expect.

We encourage practitioners to approach experimentation as an iterative activity, where each iteration builds on current knowledge. In our view, the optimal design approach simplifies this process.

4

A response surface design with a categorical factor

4.1 Key concepts

1. If there are only a few factors driving the performance of a product or process, you can experiment with the goal of finding optimal settings for these factors. For optimization, it is useful to model any curvature that exists in the relationships between the factors and the response.

2. The simplest way to model curvature in the relationship between a continuous factor and a response is to use quadratic effects. The inclusion of quadratic effects in the model requires that each continuous factor be set at a minimum of three levels. We call test plans for such experiments response surface designs.

3. While standard response surface designs apply only to continuous factors, there are many situations where categorical factors as well as continuous ones are of interest.

4. The I-optimal design criterion, with its focus on precise predictions, is appropriate for selecting response surface designs.

5. By contrast, the D-optimal design criterion, with its focus on precise parameter estimation, is more appropriate for screening experiments.

6. Modeling two-factor interactions involving categorical and continuous factors is necessary for discovering factor-level combinations that make the predicted response robust to changes in the levels of the categorical factors.

7. Fraction of Design Space (FDS) plots are a powerful diagnostic tool for comparing designs whose goal is precise prediction.

Optimal Design of Experiments: A Case Study Approach, First Edition. Peter Goos and Bradley Jones.
© 2011 John Wiley & Sons, Ltd. Published 2011 by John Wiley & Sons, Ltd.

The case study in this chapter deals with an experiment performed to make a process robust to raw material from multiple suppliers. We address the optimal design of experiments for second-order models that include categorical factors. We also show how to use FDS plots for diagnostic evaluation and comparison of alternative designs.

4.2   Case: a robust and optimal process experiment

4.2.1   Problem and design

Peter and Brad have just landed at the Dallas Fort Worth airport in Irving, Texas, on a consulting trip to the corporate headquarters of Lone Star Snack Foods. As they drive their rental car up I45, Brad briefs Peter on their upcoming meeting with a team of engineers and statisticians.

[Brad] Lone Star Snack Foods has gotten on the robust processing bandwagon recently. They have a new process they want to simultaneously optimize and robustify. Hmm. I suppose robustify is not a word but it sounds good.

[Peter] What do you mean exactly? Remember that my native language is Flemish.

[Brad] Sorry. They just bought a machine that hermetically seals a package. There is a tab over the opening that the consumer pulls. The tab has to adhere to the package well enough to make sure the package stays airtight under normal conditions. It also has to pull off easily, so that the customers don't get frustrated with failed attempts to open the package. So, they have to get the amount of adhesion just right.

[Peter] This does not sound like a robustness study to me.

[Brad] I have not gotten to that part yet. Lone Star Snack Foods gets their packaging material from three different suppliers. All the suppliers are working to the same specifications but still the packages from different suppliers are not exactly alike. Twenty years ago, they would have just developed three different processes, one for each supplier. Nowadays, though, they want to have one process that works for all three. So, the process has to be relatively insensitive to the change in packaging material from one supplier to another.

[Peter] I see where the robustness comes in now.

[Brad] Anyway, you will hear all of this again in our meeting with the Packaging Development Department. I just wanted to give you a little head start.

They pull into the parking lot of the Plano research facility and sign in at the reception desk. In a few minutes, their contact, Dr. Pablo Andrés, arrives and escorts them to the meeting. The meeting starts with Dr. Andrés introducing the problem.

[Dr. Andrés] Engineers in Lone Star Snack Foods's Packaging Development Department want to find workable settings for the major factors involved in a new sealing process. The process involves closing a package using a high-speed hermetic sealing apparatus. The response variable is peel strength, which measures the amount of force required to open the package. If the peel strength is low, then there is a good chance that the package will leak. However, if the peel strength is very high, then consumers will complain that opening the package is too difficult. We plan to study the effects of temperature, pressure, and speed on the peel strength. Our first goal is to find factor settings to match a peel strength target value of four and a half pounds.

Table 4.1   Factor-level ranges for the robustness experiment.

Factor	Range	Unit
Temperature	193–230	°C
Pressure	2.2–3.2	bar
Speed	32–50	cpm
Supplier	1–3	–

cpm = cycles per minute

Different suppliers furnish the packaging material used in the sealing process. Supplier 1 is our main supplier and accounts for 50% of our orders. Suppliers 2 and 3 each supply 25% of the material. We do not want to change the manufacturing settings depending on which supplier is providing the packaging material for the current processing run. So, our second goal is to find settings for temperature, pressure, and speed that make the peel strength robust to the variation from one supplier to another. To be completely successful, we have to achieve both goals using only one set of processing conditions. Questions?

[Brad] What operating ranges are you considering for temperature, pressure, and speed in your study?

Dr. Andrés smiles and displays his next slide showing the factor ranges and measurement units in Table 4.1.

[Dr. Andrés] Good timing. Here you go. What else?

[Peter] Do any of these factors have levels that are hard to change from one processing run to the next?

[Dr. Andrés] No. Why do you ask?

[Brad] Peter's PhD work was partly about what to do in that case. I agree with him that this is always a good question to ask.

[Dr. Andrés] Why is it a good question to ask?

[Brad] When some of the experimental factors have levels that are hard to change, then completely randomizing the experiment is tedious at best. Experimenters then end up not resetting all the factor levels independently from one run to the next. This leads to correlated observations and requires a more complicated statistical analysis.

[Dr. Andrés] I see, but we don't have problems of that sort here. Next question?

[Brad] Do you have any idea what the package-to-package variation in peel strength is?

[Dr. Andrés] Well, currently our variation in peel strength is too high. We are seeing a standard deviation in peel strength of a little less than half a pound.

[Brad] What are your specification limits on peel strength?

[Dr. Andrés] We target peel strength at four and a half pounds, but anything between three and six pounds is acceptable.

[Brad] Have you given any thought to the model and design for this study?

[Dr. Andrés] We are only considering three continuous factors so we would like to fit a quadratic response surface model. Normally, we would just run a Box–Behnken or a central composite design, but our design also has to accommodate our three suppliers. The Box–Behnken design for three factors requires 15 runs if you use three center runs and the central composite design requires even more. If we ran the entire Box–Behnken design for each supplier, it would take 45 runs at a minimum. That is more than the budget for this study will allow.

[Peter] How many runs can you afford to perform?

[Dr. Andrés] I doubt that we can afford more than 30 or so. We will have to justify any choice we make on economic and statistical grounds.

[Brad] That budget could be more than adequate or tight depending on how many terms you add to your proposed model for the effect of the categorical factor, supplier.

[Dr. Andrés] We know what you mean. We have already thought about the different alternatives. The full quadratic model in three continuous factors has ten terms. There is the intercept, three main effects, three two-factor interaction effects, and three quadratic effects.

Dr. Andrés shows a slide containing Equation (4.1)—the model for the three continuous factors only:

$$Y = \beta_0 + \sum_{i=1}^{3} \beta_i x_i + \sum_{i=1}^{2} \sum_{j=i+1}^{3} \beta_{ij} x_i x_j + \sum_{i=1}^{3} \beta_{ii} x_i^2 + \varepsilon. \qquad (4.1)$$

[Dr. Andrés] If we assume that the effect of changing from one supplier to another only shifts the peel strength by a constant, then we get this model.

Dr. Andrés adds Equation (4.2) to the slide:

$$Y = \beta_0 + \sum_{i=1}^{3} \beta_i x_i + \sum_{i=1}^{2} \sum_{j=i+1}^{3} \beta_{ij} x_i x_j + \sum_{i=1}^{3} \beta_{ii} x_i^2 + \sum_{i=1}^{2} \delta_i s_i + \varepsilon. \qquad (4.2)$$

[Dr. Andrés] The supplier effect adds two more unknowns to estimate.

[Brad] But, if this equation describes reality, then you cannot make your process robust to supplier.

Dr. Andrés adds Equation (4.3) to the slide:

$$Y = \beta_0 + \sum_{i=1}^{3} \beta_i x_i + \sum_{i=1}^{2} \sum_{j=i+1}^{3} \beta_{ij} x_i x_j + \sum_{i=1}^{3} \beta_{ii} x_i^2 + \sum_{i=1}^{2} \delta_i s_i + \sum_{i=1}^{2} \sum_{j=1}^{3} \delta_{ij} s_i x_j + \varepsilon. \qquad (4.3)$$

[Dr. Andrés] Exactly! That is why we want to include the two-factor interactions involving supplier and all three continuous factors in our proposed model. Of course, this results in six additional unknown parameters, i.e., six two-factor interaction effects, to estimate. That brings us up to an 18-parameter model.

Dr. Andrés adds a fourth equation to the slide:

$$
\begin{aligned}
Y = {} & \beta_0 + \sum_{i=1}^{3} \beta_i x_i + \sum_{i=1}^{2} \sum_{j=i+1}^{3} \beta_{ij} x_i x_j + \sum_{i=1}^{3} \beta_{ii} x_i^2 \\
& + \sum_{i=1}^{2} \delta_i s_i + \sum_{i=1}^{2} \sum_{j=1}^{3} \delta_{ij} s_i x_j + \sum_{i=1}^{2} \sum_{j=1}^{2} \sum_{k=j+1}^{3} \delta_{ijk} s_i x_j x_k \qquad (4.4) \\
& + \sum_{i=1}^{2} \sum_{j=1}^{3} \alpha_{ijj} s_i x_j^2 + \varepsilon.
\end{aligned}
$$

[Dr. Andrés] If we add all the third-order effects involving supplier, we get this equation. This model allows for the possibility that each supplier requires a completely different response surface model in the three continuous factors.

[Peter] This last model is equivalent to fitting your first model, the second-order model in the three continuous factors, independently for each of the three suppliers. This would require 30 parameters, 10 for each of the three suppliers. Equivalently, there are 30 parameters in your last model. Since you need at least one experimental run per parameter, the model in your last equation will completely exhaust your run budget. Also, it will leave you no internal way to estimate your error variance.

[Dr. Andrés] You mean that we would not be able to compute the mean squared error to estimate the variance of the εs if we use that model?

Brad nods in consent.

[Dr. Andrés, pointing at Equation (4.3)] So, perhaps, we should use the third equation, which has 18 parameters, as our a priori model.

[Peter] That seems reasonable. Brad, what do you think?

[Brad] That is the model I would choose too, given your run budget and given the fact that third-order effects are usually negligible.

[Dr. Andrés] Great! Now, all we have to do is decide on a design approach and the number of runs required. It seems obvious to me that some kind of optimal design is our only choice here. What about using a D-optimal design?

[Brad] Certainly, a D-optimal design is a reasonable choice. But I would also suggest considering I-optimal designs.We could compare D-optimal designs with I-optimal designs for a couple of alternative numbers of runs.

[Dr. Andrés] Why would you suggest I-optimal designs? How are they different from the D-optimal ones?

[Brad] Since you have already narrowed your list of major sources of process variation to only a few factors and you have second-order terms in your model, I think you probably want to use your model to make predictions about the peel strength for various settings of temperature, pressure, and speed.

[Dr. Andrés] Of course.

[Brad] The main benefit of I-optimal designs is that they minimize the average variance of prediction using the a priori model over whatever region you specify. If one of your main goals is prediction, then I imagine that you would want to consider a design based on a criterion that emphasizes prediction.

[Dr. Andrés] That makes sense. How long does it take to compute these designs?

[Peter] That is one of the embarrassing things about commercial software for design these days. We can generate D-optimal and I-optimal designs for your situation in a few seconds. It almost seems too easy.

[Dr. Andrés] Well, let us get going then. How many runs shall we consider?

[Peter] How about looking at 24-run designs? Since you have 18 parameters, the 24-run design will give you six degrees of freedom for estimating your error variance. Also, 24 is divisible by 3 so we can devote the same number of runs to each supplier.

[Brad] Not so fast Peter. I just computed the D-optimal design for 24 runs. Look at the table here.

He turns his laptop so that Peter and Dr. Andrés can see the design shown in Table 4.2. There are differing numbers of runs for each supplier: seven for supplier 1, nine for supplier 2, and eight for supplier 3.

Table 4.2   D-optimal 24-run design for the robustness experiment.

Run number	Temperature	Pressure	Speed	Supplier
1	−1	0	−1	1
2	−1	1	−1	1
3	1	1	−1	1
4	1	−1	0	1
5	−1	−1	1	1
6	1	0	1	1
7	−1	1	1	1
8	−1	−1	−1	2
9	1	−1	−1	2
10	−1	1	−1	2
11	1	1	−1	2
12	0	0	0	2
13	−1	−1	1	2
14	1	−1	1	2
15	−1	1	1	2
16	1	1	1	2
17	0	−1	−1	3
18	0	1	1	3
19	1	0	−1	3
20	−1	0	1	3
21	−1	1	−1	3
22	1	−1	1	3
23	−1	−1	0	3
24	1	1	0	3

Table 4.3   I-optimal 24-run design for the robustness experiment.

Run number	Temperature	Pressure	Speed	Supplier
1	0	−1	−1	1
2	−1	0	−1	1
3	1	1	−1	1
4	1	−1	0	1
5	−1	1	0	1
6	−1	−1	1	1
7	1	0	1	1
8	0	1	1	1
9	−1	−1	−1	2
10	1	0	−1	2
11	−1	1	−1	2
12	0	0	0	2
13	0	0	0	2
14	1	−1	1	2
15	−1	0	1	2
16	1	1	1	2
17	1	−1	−1	3
18	−1	0	−1	3
19	0	1	−1	3
20	−1	−1	0	3
21	0	0	0	3
22	1	1	0	3
23	0	−1	1	3
24	−1	1	1	3

[Dr. Andrés] That does not seem right! How can the D-optimal design fail to balance the number of runs allocated to each supplier? Common sense would seem to dictate that balanced designs are better.

[Brad] I'll address that in just a minute. But first, let's take a look at the I-optimal design, which I just computed.

Brad shows Dr. Andrés Table 4.3 on his laptop.

[Dr. Andrés] Hmph! As far as I am concerned, we already have a good argument for using the I-optimal design.

[Brad] Generally, I prefer I-optimal designs for this kind of problem but I feel compelled to defend the D-optimal design just a little. Notice that the runs for supplier 2 in Table 4.2 form a 2^3 factorial with a center point. Also, considering the runs for supplier 3, each factor has 3 runs at its low setting, 2 runs at its middle setting, and 3 runs at its high setting. So, there is substantial embedded symmetry in the D-optimal design. That may be part of the reason that it is allocating different numbers of runs to the various suppliers.

Table 4.4   Relative variances of the factor-effect estimates, given by the diagonal elements of $(\mathbf{X}'\mathbf{X})^{-1}$, for the D-optimal design in Table 4.2 and the I-optimal design in Table 4.3.

Model term	I-optimal	D-optimal
Intercept	**0.265**	0.691
Temperature	0.066	0.050
Pressure	0.065	0.063
Speed	0.066	0.057
Supplier 1	0.086	0.103
Supplier 2	0.099	0.087
Temperature × Pressure	0.083	0.065
Temperature × Speed	0.086	0.062
Pressure × Speed	0.083	0.079
Temperature × Temperature	**0.220**	0.487
Pressure × Pressure	**0.227**	0.269
Speed × Speed	**0.220**	0.370
Temperature × Supplier 1	0.122	0.102
Temperature × Supplier 2	0.129	0.092
Pressure × Supplier 1	0.120	0.155
Pressure × Supplier 2	0.148	0.105
Speed × Supplier 1	0.122	0.129
Speed × Supplier 2	0.129	0.098

[Peter] Pretty geometry may convince managers, but I would rather compare things that really matter, like the variances of predictions and factor-effect estimates.

[Brad] Ouch! That sounds more like something I would say. Anyway, here is a table comparing the relative variances of the factor-effect estimates.

Brad shows Dr. Andrés and Peter Table 4.4 on his laptop.

[Brad] The I-optimal design is better than the D-optimal design at estimating the intercept and the three quadratic effects. The relative variances of these estimates are the bolded numbers in the middle column of this table. These relative variances are smaller because the I-optimal design has more settings at the middle level of each factor than does the D-optimal design.

[Dr. Andrés] What do you mean by the relative variances of the factor-effect estimates? How are they relative to anything?

[Peter] The actual variance of any one of the factor-effect estimates is given by multiplying the value shown in the table by the error variance. But, before we actually run the experiment, we do not have an estimate of the error variance. It is in this sense that the values displayed in the table are relative to the error variance. Once we run the experiment, we can estimate the error variance. Then, we can use the estimate of the error variance to calculate estimates of the absolute variances of all the factor-effect estimates. In the meantime, these relative variances provide good diagnostic information for comparing two designs.

[Dr. Andrés] Interesting . . .

[Brad] I also computed the relative D-efficiency of the I-optimal design compared to the D-optimal design. It is 93%. This is due to the fact that the D-optimal design is estimating most of the factor effects slightly more precisely than the I-optimal design. The average variance of prediction for the I-optimal design is 0.48 versus 0.66 for the D-optimal design. So, the I-efficiency of the D-optimal design is only 73%.

[Dr. Andrés] What do you mean by these D- and I-efficiencies?

[Peter] For D-efficiency the explanation is a little complicated. The D stands for the determinant of the so-called information matrix of the design, which is a measure of the information about the factor effects that is contained in the design. The relative D-efficiency of one design compared to another is the pth root of the ratio of the determinants of the two information matrices, where p is the number of unknown coefficients in your model. In this case, the determinant of the information matrix for the D-optimal design is 5.3467×10^{20}, and for the I-optimal design, it is 1.5121×10^{20}. Since you have 18 unknown coefficients in your model, the relative D-efficiency of the I-optimal design relative to the D-optimal design is, therefore, the 18th root of $1.5121/5.3467$, which is 0.93.

[Brad] The I-efficiency of one design with respect to another is the average prediction variance of that other design divided by the average prediction variance of the first design.

[Dr. Andrés, referring to Table 4.4] These efficiency measures make for easy comparisons, perhaps too easy. I would rather compare all the relative variances of the factor-effect estimates side-by-side as your table does. Is there some similar way to compare prediction variances?

[Peter] We certainly agree with you. Choosing a design solely on the basis of a summary measure is an inflexible approach. Since computers can calculate many diagnostic measures for any design, we can be more sophisticated about making design trade-offs.

[Brad] About your question concerning the comparison of prediction variances, take a look at this FDS plot.

Brad shows Dr. Andrés and Peter Figure 4.1 on his laptop.

[Dr. Andrés] How do I read this plot? What are the two solid curves in the plot?

[Brad] Actually, what you see here are two FDS plots in one, since I overlaid the curves for both the D-optimal and I-optimal designs on the same plot. So, one of the solid curves, the upper one, is for the D-optimal design, while the other is for the I-optimal design. Now, for a point with a given prediction variance value on the vertical axis, that point's horizontal coordinate gives the proportion of points in the region covered by the corresponding design that have a relative prediction variance less than or equal to the given prediction variance value. Note that I use the term relative prediction variance, because we don't know the error variance. So, we can only compute the prediction variance relative to the variance of the εs.

[Dr. Andrés] Let me see if I understand. The D-optimal design never has a relative prediction variance below .40, while the I-optimal design achieves relative variances of .40 or less over about 35% of the experimental region. Is that correct?

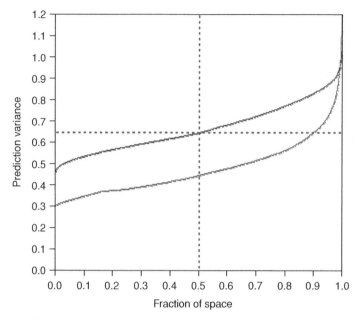

Figure 4.1   Fraction of Design Space plots: the D-optimal design in Table 4.2 is represented by the upper curve in the plot, and the I-optimal design in Table 4.3 is represented by the lower curve.

[Brad] You got it. To help you with the interpretation of the plot, I have added two dashed lines to the plot. These two dashed lines show that 50% of all the points in the experimental region have a relative prediction variance below 0.65 when the D-optimal design is used. In other words, the median relative prediction variance is 0.65 for the D-optimal design.

[Dr. Andrés] And the median prediction variance for the I-optimal design is only about 0.42, isn't it?

[Brad] Right. The fact that the curve for the I-optimal design is beneath that for the D-optimal design over most of the experimental region, or most of the design space, shows its excellent performance when it comes to making predictions. You can see that, except at the extreme right side of the plot, the I-optimal curve is substantially below the D-optimal curve. So, the I-optimal design gives better predictions over 98% of the region.

[Dr. Andrés, nodding] I like this plot. That is a very clever graphical idea!

[Peter] It surely is. No wonder it is getting popular as a diagnostic tool for designed experiments.

[Dr. Andrés] After having seen the FDS plot, I am still comfortable with my original assessment that we should go with the I-optimal design for this experiment.

[Peter] I am fine with that choice.

[Brad] Works for me too.

[Dr. Andrés] Good. I will contact you two once we have collected the experimental data.

4.2.2   Data analysis

A couple of weeks later, Dr. Andrés sends Brad the data in Table 4.5 via e-mail. Brad does an analysis of the data and shares his results with Peter.

[Brad] I started by fitting the a priori model we used to generate the design. It turns out that most of the estimated factor effects were negligible both statistically and practically, so I removed them.

[Peter] Were there any large two-factor interaction effects involving the categorical factor supplier? That is what Lone Star Snack Foods needs to make their process robust to differences in suppliers.

[Brad, showing Peter a page of paper with three hand-written equations] As it happens, the effect of speed is different from one supplier to another. Here are the

Table 4.5   Data for the robustness experiment.

Run number	Temperature	Pressure	Speed	Supplier	Peel strength
1	211.5	2.2	32	1	4.36
2	193.0	2.7	32	1	5.20
3	230.0	3.2	32	1	4.75
4	230.0	2.2	41	1	5.73
5	193.0	3.2	41	1	4.49
6	193.0	2.2	50	1	6.38
7	230.0	2.7	50	1	5.59
8	211.5	3.2	50	1	5.40
9	193.0	2.2	32	2	5.78
10	230.0	2.7	32	2	4.80
11	193.0	3.2	32	2	4.93
12	211.5	2.7	41	2	5.96
13	211.5	2.7	41	2	6.55
14	230.0	2.2	50	2	6.92
15	193.0	2.7	50	2	6.18
16	230.0	3.2	50	2	6.55
17	230.0	2.2	32	3	5.44
18	193.0	2.7	32	3	4.57
19	211.5	3.2	32	3	4.48
20	193.0	2.2	41	3	4.78
21	211.5	2.7	41	3	5.03
22	230.0	3.2	41	3	3.98
23	211.5	2.2	50	3	4.73
24	193.0	3.2	50	3	4.70

three prediction equations, in engineering units:

Peel strength (supplier 1) = 4.55 − 0.605 Pressure + 0.0567 Speed

Peel strength (supplier 2) = 4.45 − 0.605 Pressure + 0.0767 Speed

Peel strength (supplier 3) = 6.66 − 0.605 Pressure − 0.0080 Speed

[Peter] What does this tell us about how to run the process?

[Brad] Good question. Take a look at this plot.

Brad points to the screen of his laptop, where Figure 4.2 has appeared. The three curves in the plot show the differences across the suppliers for the low, medium, and high speeds when the pressure is set to 3.2 bar.

[Brad] I set the pressure to 3.2 bar to bring the peel strength down by 3.2 × 0.605 = 1.936 toward the target of 4.5.

[Peter] It seems pretty clear that lowering the speed decreases the supplier-to-supplier differences at a pressure setting of 3.2 bar.

[Brad] That is true. Now if we lower the speed to 26.15 cpm and raise the pressure to 3.23 bar, we get this picture.

Now, it is Figure 4.3 that is on the screen.

[Peter] I see that you have managed to find settings of the factors that get the predicted peel strength right on target for suppliers 2 and 3. Unfortunately, these settings are outside the experimental ranges, so you are extrapolating.

[Brad] You are right. I use too high a pressure and too low a speed here, compared to the settings used in the actual experiment. If Dr. Andrés is interested in the

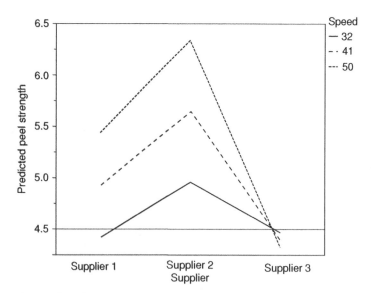

Figure 4.2   Predicted peel strength for three different speeds (32, 41, and 50 cpm) and the three suppliers when pressure is set to 3.2 bar.

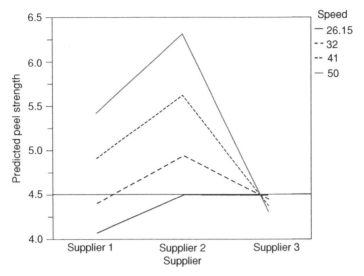

Figure 4.3   Predicted peel strength for four different speeds and the three suppliers when pressure is set to 3.23 bar.

possibility that our predictions at these settings are correct, he will have to run some extra trials to verify that the model still holds outside the original region of the factor levels.

[Peter] I remember Dr. Andrés saying that his budget allowed for 30 runs and he has only used 24 runs so far. So, conducting some extra runs might not be problematic.

[Brad] There is another problem too. Though the predictions for suppliers 2 and 3 are exactly on target, supplier 1 is the main supplier. Lone Star Snack Foods wants to use supplier 1 for 50% of its packaging material. For this solution to be ideal, Lone Star Snack Foods would have to take away the contract for supplier 1 and split their material acquisitions between the other two suppliers.

[Peter] I see that the effect of pressure does not change from one supplier to the next. Can't we change the pressure to split the difference between supplier 1 and the other two?

[Brad, displaying Figure 4.4] My thinking precisely.... Here is a plot of the predicted values where the pressure has been lowered to 2.86 bar. Note that I expanded the scale of the vertical axis. Now the predicted peel strength for supplier 1 is 0.2 pound below target, while the predicted peel strengths for suppliers 2 and 3 are 0.2 pound above target.

[Peter] This reminds me of the joke about the statistician who went hunting. He shoots above the duck and he shoots below the duck, but, on the average, the duck is dead.

[Brad] Yeah, yeah, very funny. Anyway I think we have dealt pretty effectively with this particular duck. Let's present the two solution strategies to Dr. Andrés.

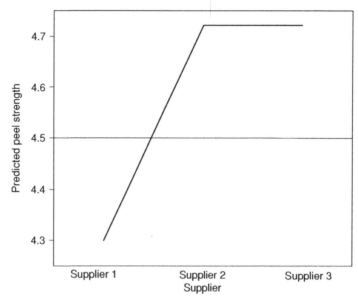

Figure 4.4   Predicted peel strength for the three suppliers when speed is set to 26.15 cpm and pressure is set to 2.86 bar.

4.3   Peek into the black box

4.3.1   Quadratic effects

After the vital few factors have been identified using a screening experiment, it is often important to refine the knowledge concerning the product or the process. A common objective is to find optimal settings for the important factors. For that purpose, it is important to study whether or not the continuous experimental factors have quadratic effects. To detect quadratic effects, we need to include terms of the form $\beta_{ii}x_i^2$ in the model. If all the experimental factors are continuous, then it is common to use the full quadratic model as the a priori model. If there are k such factors, then the full quadratic model is

$$Y = \beta_0 + \sum_{i=1}^{k} \beta_i x_i + \sum_{i=1}^{k-1} \sum_{j=i+1}^{k} \beta_{ij} x_i x_j + \sum_{i=1}^{k} \beta_{ii} x_i^2 + \varepsilon. \qquad (4.5)$$

This model includes an intercept, k main effects β_i, $k(k-1)/2$ two-factor interaction effects β_{ij}, and k quadratic effects β_{ii}. Hence, the model involves $2k + k(k-1)/2 + 1$ unknown parameters, meaning that at least $2k + k(k-1)/2 + 1$ experimental runs are required to estimate it. Depending on the magnitudes and signs of the effects, a full quadratic model may represent a variety of relationships between the experimental factors and the response variable. For example, consider a quadratic model involving a single experimental factor x. Figure 4.5 shows four possible quadratic relationships over a given operating range.

From the graphs in Figure 4.5, one can see why two design points would be insufficient to estimate quadratic models. It is clear that, for one factor, at least three settings of that factor are required in order to detect the curvature described by the quadratic model. More generally, in order to estimate quadratic effects in the a priori model, the design requires at least three levels per factor. This explains why the optimal designs in this chapter all have three levels, -1, 0, and $+1$, for every continuous factor.

4.3.2 Dummy variables for multilevel categorical factors

A key feature of the robustness experiment at Lone Star Snack Foods is that it involves a three-level categorical experimental factor, the supplier. One of the models considered in the discussion is one in which the effect of changing supplier only shifts the response, peel strength, by a constant.

A natural way to write down such a model involves one dummy variable for each of the three suppliers. The first dummy variable, s_1, has the value one for runs where the first supplier provides the packaging material and has the value zero for runs where one of the other two suppliers provides the material. The second dummy variable, s_2, is one for runs where the second supplier provides the packaging material and zero otherwise, and the third, s_3, is one for runs where the third supplier provides the packaging material and zero otherwise. The model that allows a shift in the response by supplier can then be written as

$$Y = \beta_0 + \sum_{i=1}^{3} \beta_i x_i + \sum_{i=1}^{2} \sum_{j=i+1}^{3} \beta_{ij} x_i x_j + \sum_{i=1}^{3} \beta_{ii} x_i^2 + \sum_{i=1}^{3} \delta_i s_i + \varepsilon. \qquad (4.6)$$

But, while specifying the model in this way is intuitive, we cannot estimate the parameters in this model. This is because the three dummy variables s_1, s_2, and s_3 and the intercept term are collinear. To see this, consider the model matrix $\mathbf{X}$ corresponding to the model in Equation (4.6):

$$\mathbf{X} = \begin{bmatrix} 1 & x_{11} & x_{21} & x_{31} & x_{11}x_{21} & \cdots & x_{21}x_{31} & x_{11}^2 & x_{21}^2 & x_{31}^2 & 1 & 0 & 0 \\ 1 & x_{12} & x_{22} & x_{32} & x_{12}x_{22} & \cdots & x_{22}x_{32} & x_{12}^2 & x_{22}^2 & x_{32}^2 & 1 & 0 & 0 \\ \vdots & \vdots & \vdots & \vdots & \vdots & \ddots & \vdots & \vdots & \vdots & \vdots & \vdots & \vdots & \vdots \\ 1 & x_{1i} & x_{2i} & x_{3i} & x_{1i}x_{2i} & \cdots & x_{2i}x_{3i} & x_{1i}^2 & x_{2i}^2 & x_{3i}^2 & 0 & 1 & 0 \\ \vdots & \vdots & \vdots & \vdots & \vdots & \ddots & \vdots & \vdots & \vdots & \vdots & \vdots & \vdots & \vdots \\ 1 & x_{1n} & x_{2n} & x_{3n} & x_{1n}x_{2n} & \cdots & x_{2n}x_{3n} & x_{1n}^2 & x_{2n}^2 & x_{3n}^2 & 0 & 0 & 1 \end{bmatrix}.$$

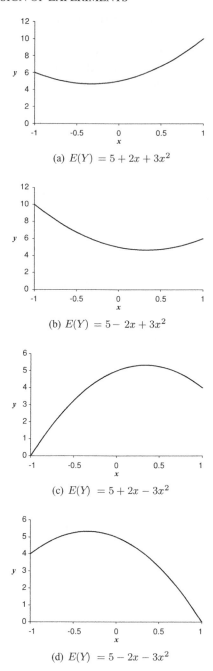

(a) $E(Y) = 5 + 2x + 3x^2$

(b) $E(Y) = 5 - 2x + 3x^2$

(c) $E(Y) = 5 + 2x - 3x^2$

(d) $E(Y) = 5 - 2x - 3x^2$

Figure 4.5   Four different quadratic functions in one experimental factor.

Here, the last three columns represent the values of s_1, s_2, and s_3. Note that the row-wise sum of the last three columns is always equal to the first column, so

$$s_1 + s_2 + s_3 = 1.$$

This perfect collinearity causes $\mathbf{X'X}$ to be singular, so that the inverse $(\mathbf{X'X})^{-1}$, and consequently the least squares estimator $(\mathbf{X'X})^{-1}\mathbf{X'Y}$ for the model parameters, do not exist.

To circumvent this technical problem, we can use one of the following remedies:

1. Drop the intercept from the model, in which case each parameter δ_i can be interpreted as the intercept for the model restricted to the ith supplier.

2. Drop one of the parameters δ_i from the model, in which case β_0 is the intercept for the model for the ith supplier, and, for each remaining supplier j, the term $\beta_0 + \delta_j$ is the intercept for the model for supplier j.

3. Constrain the estimates of the parameters δ_i to sum to zero, so that the intercept of the model restricted to supplier j is $\beta_0 + \delta_j$; this means that δ_j represents the shift in intercept that is required in order to fit supplier j.

Another remedy for the collinearity problem is to use effects-type coding instead of dummy variable coding for the suppliers. That approach is illustrated in the data analysis in Section 8.3.2.

There is no qualitative difference among these remedies. From a design perspective, it is important to stress that neither the D-optimal nor the I-optimal design is affected by the choice of remedy. In the model in Equation (4.2), Dr. Andrés used dummy variable coding and dropped the term involving δ_3 so that β_0 represents the intercept for the model restricted to supplier 3.

The model ultimately selected for the robustness experiment, Equation (4.3), is more involved than the model in Equation (4.2) since it involves interactions between the dummy variables s_1 and s_2 and the other experimental factors. Including the interactions involving s_3 would cause perfect collinearity involving (i) s_1x_1, s_2x_1, s_3x_1, and x_1 (because $s_1x_1 + s_2x_1 + s_3x_1$ sums to x_1), (ii) s_1x_2, s_2x_2, s_3x_2, and x_2 (because $s_1x_2 + s_2x_2 + s_3x_2$ sums to x_2), and (iii) s_1x_3, s_2x_3, s_3x_3, and x_3 (because $s_1x_3 + s_2x_3 + s_3x_3$ sums to x_3). For this reason, Dr. Andrés did not include any interaction effect involving the dummy variable s_3 in the model. This remedy for the collinearity problem is similar to that discussed above. Its impact is that the intercept β_0 and the main effect β_i should be interpreted as the factor effects in the situation where supplier 3 delivers the packaging material. The parameters δ_1 and δ_2 indicate how the intercepts for suppliers 1 and 2, respectively, differ from the intercept for supplier 3. The parameters δ_{11}, δ_{12}, and δ_{13} indicate how the main effects of the three quantitative factors x_1, x_2, and x_3 for supplier 1 differ from the main effects of these factors for supplier 3. Similarly, δ_{21}, δ_{22}, and δ_{23} define this difference for supplier 2.

4.3.3   Computing D-efficiencies

The model in matrix form is

$$Y = X\beta + \varepsilon.$$

The least squares estimator for the parameter vector, β, is

$$\hat{\beta} = (X'X)^{-1}X'Y$$

and the variance-covariance matrix of that estimator is $\sigma_\varepsilon^2(X'X)^{-1}$. The information matrix is the inverse of the variance-covariance matrix, that is, $\sigma_\varepsilon^{-2}X'X$. The variance σ_ε^2 in these expressions is an unknown proportionality constant that is irrelevant for the purpose of comparing experimental designs. It is, therefore, common to ignore σ_ε^2 (or, more precisely, to set σ_ε^2 equal to one) in the context of the optimal design of experiments, for instance, when comparing design in terms of D-optimality.

Recall that the D in the terms D-efficiency and D-optimal stands for determinant. More specifically, it refers to the determinant of the information matrix. A D-efficiency of a given design compares the determinant of the information matrix of that design to an ideal determinant corresponding to an orthogonal design. We use the term "ideal determinant" here because an orthogonal design for a given situation may not exist. If we denote the number of runs in a design by n and the number of parameters in β by p, then the ideal determinant is n^p. The D-efficiency is, therefore, computed as

$$\text{D-efficiency} = \left(\frac{|X'X|}{n^p}\right)^{1/p} = \frac{|X'X|^{1/p}}{n}. \tag{4.7}$$

Here, the pth root is taken so as to provide a measure that can be interpreted as a per-parameter measure. One problem with this definition of D-efficiency, which is reported in many software packages, is that it depends on the scale or coding used for the experimental variables. Another problem is that there are many practical experimental design problems for which the ideal determinant cannot be achieved because there is no design having a determinant as large as n^p. For these situations, D-efficiency, as defined in Equation (4.7), is not useful as an absolute measure. This definition of D-efficiency is only useful for comparing two designs that have the same scale or coding for the experimental factors as well as the same number of runs.

So, we do not recommend focusing on D-efficiency as defined in Equation (4.7). We do, however, find it useful to compute relative D-efficiencies to compare two competing designs. If D_1 is the determinant of the information matrix of one design and D_2 is the determinant of the information matrix of a second design, then the relative D-efficiency of the former design compared to the latter is defined as

$$\text{Relative D-efficiency of Design 1 versus Design 2} = (D_1/D_2)^{1/p}.$$

A relative D-efficiency larger than one indicates that Design 1 is better than Design 2 in terms of D-optimality. As in the definition of D-efficiency, we take the pth root of the ratio of the determinants of the two competing designs to obtain a per-parameter measure of relative efficiency.

The way we deal with multilevel categorical experimental variables (discussed in Section 4.3.2) has an impact on the exact value of the determinant of the information matrix, but it does not affect the ordering in terms of D-optimality or the relative D-efficiencies of alternative design options.

4.3.4 Constructing Fraction of Design Space plots

FDS plots display the performance of one or more designs in terms of precision of prediction. Let $\mathbf{f}(\mathbf{x})$ be a function that takes a vector of factor settings and expands that vector to its corresponding model terms. Thus, if $\mathbf{x}$ is a vector of factor levels that corresponds to one of the runs of the design, then $\mathbf{f}'(\mathbf{x})$ is the corresponding row of the matrix $\mathbf{X}$. For instance, suppose that the ith run of an experiment designed to estimate the model in Equation (4.5) is conducted at the vector of factor-level settings $\mathbf{x}_i = (x_{1i}, x_{2i}, x_{3i})$. Then,

$$\mathbf{f}'(\mathbf{x}_i) = \begin{bmatrix} 1 & x_{1i} & x_{2i} & x_{3i} & x_{1i}x_{2i} & x_{1i}x_{3i} & x_{2i}x_{3i} & x_{1i}^2 & x_{2i}^2 & x_{3i}^2 \end{bmatrix}.$$

The variance of the prediction's expectation at the setting $\mathbf{x}$ is

$$\mathrm{var}(\hat{Y} \mid \mathbf{x}) = \sigma_\varepsilon^2 \mathbf{f}'(\mathbf{x})(\mathbf{X}'\mathbf{X})^{-1}\mathbf{f}(\mathbf{x}).$$

The variance of the predicted response relative to the error variance σ_ε^2, therefore, is

$$\text{Relative variance of prediction} = \frac{\mathrm{var}(\hat{Y} \mid \mathbf{x})}{\sigma_\varepsilon^2} = \mathbf{f}'(\mathbf{x})(\mathbf{X}'\mathbf{X})^{-1}\mathbf{f}(\mathbf{x}).$$

To construct an FDS plot for a given design, we take a random sample of a large number of points (say 10,000 points) inside the experimental region. For each point, we compute the relative variance of the predicted value. Then, we sort these values from smallest to largest. Let v_i be the ith of these sorted variances. If there are N points in the sample, we plot the ordered pairs $(i/(N+1), v_i)$. The FDS plot then is the nondecreasing curve joining these N points. Each point on the horizontal axis of the plot, which is scaled from 0 to 1, corresponds to a fraction of the design space or the experimental region. The vertical axis covers the range from the minimum relative prediction variance to the maximum relative prediction variance at the sampled points. Suppose, for example, that the point $(0.65, 2.7)$ is on the FDS curve. Then, the relative variance of prediction is less than or equal to 2.7 over 65% of the experimental region.

4.3.5   Calculating the average relative variance of prediction

Section 4.3.4 introduced the relative variance of prediction at a given factor setting or factor-level combination $\mathbf{x}$. We can estimate the average relative variance of prediction for a given design by averaging the N relative variances obtained when constructing an FDS plot. However, we can calculate the average exactly by integrating the relative variance of prediction over the experimental region, which we call χ, and dividing by the volume of the region:

$$\text{Average variance } = \frac{\int_\chi \mathbf{f}'(\mathbf{x})(\mathbf{X}'\mathbf{X})^{-1}\mathbf{f}(\mathbf{x})d\mathbf{x}}{\int_\chi d\mathbf{x}}. \tag{4.8}$$

In this section, we show that it is relatively easy to calculate this expression for an arbitrary model. If there are k quantitative experimental factors and the experimental region is $[-1, 1]^k$, then the volume of the experimental region in the denominator is 2^k. To simplify the calculation of the numerator, we first observe that $\mathbf{f}'(\mathbf{x})(\mathbf{X}'\mathbf{X})^{-1}\mathbf{f}(\mathbf{x})$ is a scalar, so that

$$\mathbf{f}'(\mathbf{x})(\mathbf{X}'\mathbf{X})^{-1}\mathbf{f}(\mathbf{x}) = \text{tr}\left[\mathbf{f}'(\mathbf{x})(\mathbf{X}'\mathbf{X})^{-1}\mathbf{f}(\mathbf{x})\right].$$

We can now exploit the property that, when calculating the trace of a matrix product, we can cyclically permute the matrices. Therefore,

$$\text{tr}\left[\mathbf{f}'(\mathbf{x})(\mathbf{X}'\mathbf{X})^{-1}\mathbf{f}(\mathbf{x})\right] = \text{tr}\left[(\mathbf{X}'\mathbf{X})^{-1}\mathbf{f}(\mathbf{x})\mathbf{f}'(\mathbf{x})\right],$$

and

$$\int_\chi \mathbf{f}'(\mathbf{x})(\mathbf{X}'\mathbf{X})^{-1}\mathbf{f}(\mathbf{x})d\mathbf{x} = \int_\chi \text{tr}\left[(\mathbf{X}'\mathbf{X})^{-1}\mathbf{f}(\mathbf{x})\mathbf{f}'(\mathbf{x})\right]d\mathbf{x},$$

$$= \text{tr}\left[\int_\chi (\mathbf{X}'\mathbf{X})^{-1}\mathbf{f}(\mathbf{x})\mathbf{f}'(\mathbf{x})d\mathbf{x}\right].$$

Now, note that, since the factor-level settings of the design are fixed, the matrix $\mathbf{X}$, and hence $(\mathbf{X}'\mathbf{X})^{-1}$, is constant as far as the integration is concerned. Therefore,

$$\int_\chi \mathbf{f}'(\mathbf{x})(\mathbf{X}'\mathbf{X})^{-1}\mathbf{f}(\mathbf{x})d\mathbf{x} = \text{tr}\left[(\mathbf{X}'\mathbf{X})^{-1}\int_\chi \mathbf{f}(\mathbf{x})\mathbf{f}'(\mathbf{x})d\mathbf{x}\right],$$

so that we can rewrite the formula for the average relative prediction variance as

$$\text{Average variance } = 2^{-k}\,\text{tr}\left[(\mathbf{X}'\mathbf{X})^{-1}\int_\chi \mathbf{f}(\mathbf{x})\mathbf{f}'(\mathbf{x})d\mathbf{x}\right].$$

The integral in this expression is applied to a matrix of one-term polynomials (monomials). This notation is to be interpreted as the matrix of integrals of these monomials.

If the experimental region is $\chi = [-1, +1]^k$, then these integrals are quite simple. Let

$$\mathbf{M} = \int_{\mathbf{x} \in [-1,+1]^k} \mathbf{f}(\mathbf{x})\mathbf{f}'(\mathbf{x})d\mathbf{x}, \tag{4.9}$$

then,

$$\text{Average variance} = 2^{-k} \text{ tr} \left[(\mathbf{X}'\mathbf{X})^{-1}\mathbf{M} \right].$$

The matrix $\mathbf{M}$ is called the moments matrix. For a full quadratic model, $\mathbf{M}$ has a very specific structure, which we illustrate in Attachment 4.1 for a two-dimensional experimental region $\chi = [-1, +1]^2$, and then generalize for a k-dimensional cuboidal experimental region, $\chi = [-1, +1]^k$. From this, it is easy to see how the average variance in Equation (4.8) can be computed.

Attachment 4.1 Moments matrix $\mathbf{M}$ for a full quadratic model.

For a full quadratic model in two continuous factors x_1 and x_2 that can take values in the interval $[-1, +1]$, we have

$$\mathbf{f}'(\mathbf{x}) = \mathbf{f}'(x_1, x_2) = \begin{bmatrix} 1 & x_1 & x_2 & x_1x_2 & x_1^2 & x_2^2 \end{bmatrix},$$

so that the moments matrix is

$$
\begin{aligned}
\mathbf{M} &= \int_{\mathbf{x} \in [-1,+1]^2} \mathbf{f}(\mathbf{x})\mathbf{f}'(\mathbf{x})d\mathbf{x}, \\
&= \int_{-1}^{+1} \int_{-1}^{+1} \mathbf{f}(x_1, x_2)\mathbf{f}'(x_1, x_2)dx_1dx_2, \\
&= \int_{-1}^{+1} \int_{-1}^{+1}
\begin{bmatrix}
1 & x_1 & x_2 & x_1x_2 & x_1^2 & x_2^2 \\
x_1 & x_1^2 & x_1x_2 & x_1^2x_2 & x_1^3 & x_1x_2^2 \\
x_2 & x_1x_2 & x_2^2 & x_1x_2^2 & x_1^2x_2 & x_2^3 \\
x_1x_2 & x_1^2x_2 & x_1x_2^2 & x_1^2x_2^2 & x_1^3x_2 & x_1x_2^3 \\
x_1^2 & x_1^3 & x_1^2x_2 & x_1^3x_2 & x_1^4 & x_1^2x_2^2 \\
x_2^2 & x_1x_2^2 & x_2^3 & x_1x_2^3 & x_1^2x_2^2 & x_2^4
\end{bmatrix}
dx_1dx_2, \\
&= 2^2
\begin{bmatrix}
1 & 0 & 0 & 0 & 1/3 & 1/3 \\
0 & 1/3 & 0 & 0 & 0 & 0 \\
0 & 0 & 1/3 & 0 & 0 & 0 \\
0 & 0 & 0 & 1/9 & 0 & 0 \\
1/3 & 0 & 0 & 0 & 1/5 & 1/9 \\
1/3 & 0 & 0 & 0 & 1/9 & 1/5
\end{bmatrix}.
\end{aligned}
$$

This expression can be generalized to cases with $k > 2$ factors. All the constants corresponding to the main effects, the interaction effects and the quadratic effects in the moments matrix remain the same, except for the proportionality constant, which should be 2^k instead of 2^2 in the general case. So, in general, the moments matrix for a full quadratic model is

$$
\mathbf{M} = 2^k \begin{bmatrix}
1 & \mathbf{0}'_k & \mathbf{0}'_{k(k-1)/2} & \frac{1}{3}\mathbf{1}'_k \\
\mathbf{0}_k & \frac{1}{3}\mathbf{I}_k & \mathbf{0}_{k \times k(k-1)/2} & \mathbf{0}_{k \times k} \\
\mathbf{0}_{k(k-1)/2} & \mathbf{0}_{k(k-1)/2 \times k} & \frac{1}{9}\mathbf{I}_{k(k-1)/2} & \mathbf{0}_{k(k-1)/2 \times k} \\
\frac{1}{3}\mathbf{1}_k & \mathbf{0}_{k \times k} & \mathbf{0}_{k \times k(k-1)/2} & \frac{1}{5}\mathbf{I}_k + \frac{1}{9}(\mathbf{J}_k - \mathbf{I}_k)
\end{bmatrix},
$$

where $\mathbf{0}_k$ and $\mathbf{0}_{k(k-1)/2}$ are column vectors containing k and $k(k-1)/2$ zeros, $\mathbf{1}_k$ is a column vector containing k ones, $\mathbf{I}_k$ and $\mathbf{I}_{k(k-1)/2}$ are identity matrices of dimension k and $k(k-1)/2$, $\mathbf{0}_{k \times k}$, $\mathbf{0}_{k(k-1)/2 \times k}$, and $\mathbf{0}_{k \times k(k-1)/2}$ are $k \times k$, $k(k-1)/2 \times k$, and $k \times k(k-1)/2$ matrices of zeros, and $\mathbf{J}_k$ is a $k \times k$ matrix of ones.

We call a design that minimizes the average relative variance of prediction I-optimal. The I in I-optimal stresses the fact that an I-optimal design minimizes an integrated variance. Some authors prefer the terms V- or IV-optimal over I-optimal.

Another prediction-based criterion for selecting experimental designs is the G-optimality criterion, which seeks designs that minimize the maximum prediction variance over the experimental region. Recent work has shown that, in most cases, decreasing the maximum prediction variance comes at the expense of increasing the prediction variance in more than 90% of the experimental region. Therefore, we prefer I-optimal designs over G-optimal ones.

4.3.6 Computing I-efficiencies

If P_1 is the average relative variance of prediction of one design and P_2 is the average relative variance of prediction of a second design, then the relative I-efficiency of the former design compared to the latter is computed as

$$
\text{Relative I-efficiency of Design 1 versus Design 2} = P_2/P_1.
$$

A relative I-efficiency larger than one indicates that Design 1 is better than Design 2 in terms of the average prediction variance.

The way we deal with multilevel categorical experimental variables has no impact on the relative variances of prediction.

4.3.7 Ensuring the validity of inference based on ordinary least squares

In the models in Equations (2.3)–(2.5) in Chapter 2 and the full quadratic model in Equation (4.5) in this chapter, we have assumed that the random errors $\varepsilon_1, \varepsilon_2, \ldots, \varepsilon_n$ of the n runs of the experiment are statistically independent. Therefore, all the responses, $Y_1, Y_2, \ldots, Y_n$, are also independent. This independence is a key requirement for the ordinary least squares estimator to be the best possible estimator.

It is important that experimenters ensure that the random errors $\varepsilon_1, \varepsilon_2, \ldots, \varepsilon_n$ are truly independent. Failing to do so leads to correlated responses, which complicates model estimation and significance tests for the factor effects.

The presence of hard-to-change factors in the experiment is perhaps the most common reason why errors fail the independence assumption. If it is costly or time-consuming to change the levels of certain factors, then it is common to conduct the runs of the experiment in a systematic order rather than in a random order. The key feature of the systematic order is that the levels of the hard-to-change factors are changed as little as possible. The runs of the experiment are then performed in groups, where, within a group, the levels of the hard-to-change factors are not reset. This creates a dependence between all the runs in one group, and leads to clusters of correlated errors and responses. Ordinary least squares inference is then not appropriate.

This pitfall is quite well known now, and many researchers try to circumvent it by conducting the runs of the experiment in a random order. This is, however, not enough to ensure independence of the errors and the responses. Even when using a random run order, it is still necessary to set or reset the factor levels independently for each run. Otherwise, we again obtain groups of runs where one or more factor levels were not reset, resulting in clusters of correlated errors and responses.

The key lesson to draw from this is that, for ordinary least squares estimation, all the factor levels should be set independently for every run of an experiment, whether the levels are hard to change or not. Whatever the outcome of the randomization of the experimental runs is, an experiment with n runs requires n independent settings of the levels of each of the factors. If, from a practical or economic perspective, this is not feasible, split-plot designs and two-way-plot designs provide more appropriate alternative experimental designs. We refer to Chapters 10 and 11 for a detailed discussion of these designs.

Another common cause of dependent errors and responses is heterogeneity in experimental conditions. Whenever an experiment requires the use of several batches or is run over different days, for example, we often observe that the errors and responses obtained using the same batch or on the same day are more alike than errors and responses obtained from different batches or on different days. In such a case, it is important to capture the dependence, i.e., the batch-to-batch or day-to-day differences, using fixed or random block effects. How to do this is explained in Chapters 7 and 8.

4.3.8   Design regions

In Section 4.3.5, we mentioned the experimental region χ, which is the set of all possible factor-level combinations or design points that can be used in the experiment. Alternative terms for the experimental region are design space, design region and region of interest.

If all the experimental factors are continuous, then the coded experimental region is often $\chi = [-1, +1]^k$. This region is sensible if there are no constraints on the factor levels other than the lower and upper bounds. This experimental region is said to be cuboidal. Its key feature is that all the experimental factors can be set at their extreme levels simultaneously.

The most common alternative to a cuboidal region is a spherical region,

$$\chi = \{\mathbf{x} = (x_1, x_2, \ldots, x_k) | \sum_{i=1}^{k} x_i^2 \leq R\},$$

for some radius R that equals the maximum number of factors that can simultaneously take on the extreme levels -1 and $+1$. The spherical experimental region makes sense if it is infeasible to set all the factors at extreme levels at the same time.

In many experiments, however, the experimental region is neither cuboidal nor spherical. This may be due to the presence of categorical factors, or mixture variables, or due to constraints on the factor-level combinations used in the experiment. We refer to Chapters 5 and 6 for examples of experiments with constraints on the levels of the experimental factors and in the presence of mixture variables.

When choosing the best possible design for a given problem, it is of crucial importance to choose a design that matches the experimental region.

4.4   Background reading

FDS plots were introduced in Zahran et al. (2003) as an alternative to the Variance Dispersion graphs initially proposed by Giovannitti-Jensen and Myers (1989) and Myers et al. (1992). Both FDS plots and Variance Dispersion graphs visualize the performance of one or more designs in terms of prediction variance throughout the entire experimental region. Therefore, they provide more detailed information than the average or maximum prediction variances which are often used to compare different experimental designs. In recent years, FDS plots and Variance Dispersion graphs have been the subject of much research. A review of that work is given in Anderson-Cook et al. (2009) and the discussion of that paper.

The generation of I-optimal designs (which minimize the average prediction variance) is discussed in Meyer and Nachtsheim (1995), while the generation and the performance of G-optimal designs (which minimize the maximum prediction variance) is treated in Rodríguez et al. (2010). A key finding of the latter authors is that, to minimize the maximum variance of prediction, it is often necessary to allow worse prediction variances over most of the region of interest. In many practical cases, this is not an acceptable trade. For explicit formulas of the moments matrix $\mathbf{M}$ for cuboidal and spherical experimental regions, we refer to Wesley et al. (2009).

Along with the D-, I-, and G-optimality criteria, there exist numerous other criteria for selecting designs. Many of these are discussed and illustrated in Atkinson et al. (2007). For some, the existence of different design optimality criteria is seen as a weakness of the optimal experimental design approach. However, the existence of different design optimality criteria is completely in line with the philosophy of optimal experimental design: different experimental goals and problems require different experimental designs. The best way to achieve this is to adapt the design optimality criterion to the experimental goal. Obviously, in some experiments, there may be several goals. Rather than seeking a design that is best with respect to a single design optimality criterion, we then seek designs that perform well on multiple criteria simultaneously. Composite or compound optimal design criteria, which are usually weighted averages of the criteria defined above, are useful in such cases. We refer to Atkinson et al. (2007) for a discussion of these kinds of criteria.

The construction of D-optimal designs for experiments involving quantitative and categorical factors and the lack of balance in some of the optimal designs was discussed in Atkinson and Donev (1989). Some more details can also be found in Atkinson et al. (2007).

4.5 Summary

This chapter considered designed experiments for estimating models involving quadratic effects, in addition to main effects and two-factor interaction effects. The inclusion of quadratic effects in the a priori model requires a design to have three levels for each experimental factor. The case study in this chapter illustrates this, and also demonstrates that the optimal design of experiments approach makes it possible to set up experiments involving both continuous and categorical factors.

Optimal design algorithms generate experimental designs that are optimal with respect to a specified criterion. Though the generated design is, therefore, optimal in at least one sense, it is advisable to consider various design diagnostic tools to determine whether a given design is adequate to solve the problem at hand. One useful diagnostic is the relative variance of the factor-effect estimates. Better designs have lower values of these variances. A second diagnostic is the FDS plot. For a design to be good, its curve in an FDS plot should be as low as possible. This means that the design results in small prediction variances in large fractions of the experimental region.

This chapter introduced the I-optimality design criterion. Designs optimizing this criterion have the lowest average variance of prediction compared to any other design with the same number of runs, a priori model, and experimental region. Since the focus of I-optimal designs is on prediction, they are more appropriate than D-optimal designs for response surface experiments. D-optimal designs, with their focus on factor-effect estimation, are most appropriate for screening experiments.

5

A response surface design in an irregularly shaped design region

5.1   Key concepts

1. Quadratic effects are sometimes not sufficient to capture all the curvature in the factor–response relationship. In such cases, it is worth adding cubic effects to the model.

2. Using an algorithmic approach to design, it is possible to accommodate inequality constraints on the region of experimentation. This is useful in practical situations where certain factor-level combinations are known in advance to be infeasible.

In this chapter, we design an experiment for investigating the behavior of a chemical reaction using a full cubic model in two factors. An additional complication is that many factor-level combinations are known in advance to be infeasible. Consequently, the levels of the two factors cannot be varied completely independently. We discuss how to proceed with both the design and data analysis in such a situation.

5.2   Case: the yield maximization experiment

5.2.1   Problem and design

Peter and Brad are sitting in a Starbucks in Newton, Massachusetts, before an appointment at the Rohm and Haas division that was formerly Shipley Specialty Chemical Company.

Optimal Design of Experiments: A Case Study Approach, First Edition. Peter Goos and Bradley Jones.
© 2011 John Wiley & Sons, Ltd. Published 2011 by John Wiley & Sons, Ltd.

[Brad] We have to watch ourselves in this consulting session. Our client has already fired one design of experiments consultant.

[Peter] Wow! How do you know that?

[Brad] Our contact, Dr. Jane McKuen, told me when she called to ask me if we were interested.

[Peter] That is intriguing. Did she mention why?

[Brad] It seems that the other consultant ignored what the chemists were saying about the special nature of this problem.

[Peter] Our style is to always fit our design proposal to the unique aspects of the problem at hand. So, that makes me confident that we will be popular with the chemists.

[Brad, looking at his watch] Don't get too cocky. Anyway, it's time to get going. I'm really curious about the particulars of this problem.

A few minutes later, they meet Dr. McKuen at reception.

[Brad] Dr. McKuen, let me introduce my partner Peter Goos from Intrepid Stats. He is visiting the States this week and I invited him to join me on this project, if it is OK with you.

[Dr. McKuen, producing legal forms] It is no problem as long as you both sign our confidentiality agreement.

[Peter, signing his] Happy to oblige.

Peter and Brad hand over their signed forms and the three walk to the conference room, where a man in a lab coat is waiting.

[Dr. McKuen] Now, it is my turn to make introductions. Dr. Gray, this is Dr. Bradley Jones and Dr. Peter Goos from Intrepid Stats.

[Brad, shaking hands] Please just call us Brad and Peter.

[Dr. Gray] Pleased to meet you. I never use my first name but we can dispense with the titles if you want. Just call me Gray. I am anxious to get started. I feel that we have already lost too much time on this project.

[Brad] Yes, we heard that you were not happy with your first consultant.

[Gray] Hmph! That guy was supremely confident in his methodology and refused to pay much attention to our concerns. I guess that would have been OK if his design of experiments approach had worked.

[Brad] I hope you haven't given up on designed experiments!

[Gray] Let's just say that I am now quite skeptical, but I am willing to give it another try. What I am not prepared to do from now on is to blindly follow a consultant's advice.

[Peter] We would not want to have that kind of relationship with a client anyway. We don't think of design of experiments as if it were some magical potion that automatically makes processes better no matter how it is applied.

[Brad] Right, every process is different. We think that the designed experiment should accommodate the special constraints of the process it attempts to model.

[Gray] You guys already sound different.

[Brad] I hope so. Will you tell us about your previous experiment?

[Dr. McKuen] We want to start this investigation from scratch. We don't have any confidence in the earlier results.

Table 5.1   Data of the original yield maximization experiment.

Run	Time (s)	Temperature (K)	Yield (%)
1	430	500	10.0
2	430	525	44.1
3	430	550	51.6
4	540	500	12.3
5	540	525	49.6
6	540	550	42.9
7	650	500	14.5
8	650	525	53.7
9	650	550	35.1

[Peter] That is fine with us but perhaps we can still learn something from the data you have in hand.

[Gray] I have the data on this memory stick.

Brad takes the memory stick, fires up his computer and downloads the data file. He turns his computer, so Gray can see the data table shown in Table 5.1.

[Brad] Is this the data?

[Gray] Yes, the numbers are very familiar by now.

[Brad] It looks like your consultant ran a classical 3-by-3 full factorial design here.

[Gray] He called it a central composite design modified to put the axial points on the edges of a square.

[Peter, looking over Brad's shoulder] Yes, that is also true. It looks like the percent yield for the low temperature runs were pretty disappointing. Does your usual reaction run for 540 seconds at 525 K?

[Gray, looking a little surprised] Yes on both counts. How did you guess our standard settings?

[Peter] That setting is at the center of the design. It is a fairly common practice to put the current operating setting at the center of the design region. By comparing the yield at the center with the yields at low temperatures, I inferred that the latter were not satisfactory.

[Gray] We told our consultant that 500 K was too low and that not enough reactant would convert at that temperature. We also thought that the setting at the longest time and highest temperature would cause a secondary reaction to occur and also result in a lower yield. As you can see, running our reactor for 650 seconds at 550 K resulted in a lower yield.

[Brad] Can you tell us about this secondary reaction?

Gray goes to the white board and writes

$$A \rightarrow B \rightarrow C.$$

[Gray] This reaction is pretty simple really. When you heat up our initial reactant, A, it turns into our product, B. Unfortunately, once you start producing B, additional time at certain temperatures causes it to turn into C. So, the trick is to find the operating settings that convert the largest percentage of A into B without allowing enough time for B to further convert into C.

[Brad] So, you think that at low temperatures not enough A reacted, and at the long time and high temperature settings too much of B reacted and turned into C.

[Gray] Right, but we don't just think this. We know this because we also measured the percentages of A and C in the reaction vessel after we quenched the reaction.

[Brad] What I think I am hearing is that for short times and low temperatures you don't expect much conversion of A to B, and for long times and high temperatures you expect too much conversion of B to C.

[Gray] Precisely. That is what we told the other consultant too. He said that it was OK if the yield was not great in these regions. He said that he wanted his design to be orthogonal for the main effects and the two-factor interaction, and he also wanted to be bold in the exploration of the factor ranges.

[Peter] It is true that orthogonal designs have desirable mathematical properties.

[Brad] And if the effect of each factor on the response is linear or at least monotonic, varying the factors over a wider range will result in bigger changes in the response. That makes it easier to see which factors are having the largest effects even in the presence of substantial process variation.

[Gray] Hmph! It sounds like you are defending the other consultant.

[Peter] We would not try to defend what he did in your case. It just turns out that his advice, which is routinely good, failed in this specific application.

[Brad] For example, your consultant's 3-by-3 design often works very well. But this approach depends on the assumption that all possible factor-level combinations are feasible. In your case, four of the nine experimental runs were at impractical settings.

[Gray] You are right about the four settings: a temperature of 500 is too low for adequate conversion of A to B, and a temperature of 550 for 650 seconds allows too much time for the secondary reaction from B to C to occur. There is another matter I stressed that the other consultant ignored. I told him that the effects of time and temperature should be quite nonlinear. He said that his design would allow him to fit a full quadratic model and that is usually an adequate approximation.

[Peter] Well, that is often true but perhaps not in this case. Can we look at the analysis of this data?

Gray thumbs through his briefcase and pulls out a sheet of paper, showing Table 5.2.

[Brad, bending over his machine to run the analysis himself] Let me check these numbers.

Soon, Brad has reproduced Table 5.2.

[Brad] This analysis seems alright.

[Gray] Then, our consultant did at least one thing correctly. The factor effects in the column labeled "Estimate" in the table are based on transforming the data values for time and temperature, so that the low value corresponds to -1 and the high value

Table 5.2   Analysis of the data from the original yield maximization experiment.

Effect	Estimate	Standard error	t Ratio	p Value
Intercept	49.20	3.89	12.66	0.0011
Time	−0.40	2.13	−0.19	0.8629
Temperature	15.47	2.13	7.27	0.0054
Time × Time	−0.10	3.69	−0.03	0.9801
Time × Temperature	−5.25	2.61	−2.01	0.1375
Temperature × Temperature	−21.40	3.69	−5.80	0.0102

corresponds to +1. The first thing I see is that none of the effects involving time is statistically significant. I know that time is important in this reaction so this model cannot be any good.

[Brad, looking at the output on his screen] Let's have a look at the root mean squared error of the residuals of our analysis. Here it is: about 5.2%.

[Peter] When you run your reaction at your usual settings, how much does the yield vary from one run to the next?

[Gray] Over the last 2 weeks, we have not varied more than half a percent on either side of the 49.6% we got in the experiment. Why do you ask?

[Brad] That is a powerful indication that you are right about that model. If your residuals were only capturing run-to-run variation, they would be at least ten times smaller. The residuals are big because the model is missing at least one important term. I would venture a reasonable guess that your run-to-run standard deviation is about one-third of half a percent, say 0.2%. If that is correct, then the estimate of the time-by-temperature interaction effect would be highly statistically significant.

[Gray] So, how does this information help with what we are trying to do now?

[Brad] The a priori model that the previous consultant used was inadequate. So, it is not enough to include quadratic effects in the model.

[Peter] Another thing we have learned is that, in the next experiment, we want to avoid combinations of time and temperature where both are too low or too high.

[Gray] That is true, but I think that there is more to it than that. My knowledge of the chemistry suggests that the feasible region is shaped like this.

Gray goes to the white board and draws Figure 5.1.

[Brad] That is just the picture I wanted. Now, all we need are the end points of those diagonal line segments. I noticed that two of the runs in your original experiment had higher yields than your current process.

[Gray, getting animated] Yes, one was at a higher temperature than I expected to work. A further good feature of that run was that the reaction was 110 seconds shorter. The other was at the same temperature we normally use, but 110 seconds longer. That suggests that we might consider an even wider range of time.

[Brad] Your standard reaction takes 540 seconds or 9 minutes. How does an experimental range of 6–12 minutes sound? That is 360–720 seconds.

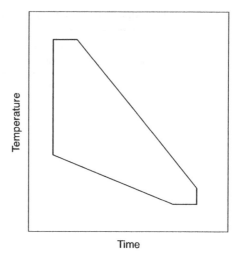

Figure 5.1   Region with feasible settings for the experimental factors time and temperature.

[Gray] It seems doable as long as a scatter plot of the runs you suggest falls inside a region that looks like the one I drew on the board. For the temperature, I suggest using a range from 520–550 K. Subject to the same constraints, of course.

[Peter] If you were going to do a run at 360 seconds, what is the minimum temperature you would consider?

[Gray] I don't know. Maybe 529 or 530 K.

[Peter] And if you were going to perform a run at 520 K, how long would you run your reaction?

[Gray] Probably 660 seconds. Eleven minutes.

[Brad] That gives us the end points of the lower diagonal line on your figure.

Brad walks up to the white board and adds the coordinates of these two end points, yielding Figure 5.2. Peter, in the meantime, has been scribbling on a piece of scratch paper.

[Peter, announcing his results] Here is the equation of Gray's lower inequality constraint:

$$0.03 \text{ Time} + \text{Temperature} \geq 539.8.$$

[Gray] I see what you guys are up to. Now you'll want to know the maximum time I want to run a reaction at 550 K, right?

Brad and Peter both nod.

[Gray] Given our previous results, it seems like 420 seconds would be a good guess.

[Brad] And what about the maximum temperature you would run at 720 seconds?

[Gray, scratching his head] That is a hard one. I would say 523 K or thereabouts.

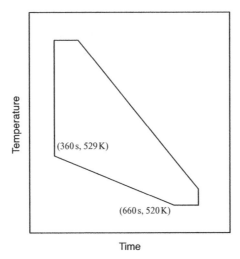

Figure 5.2   Region with feasible settings for the experimental factors time and temperature, along with the coordinates of two of its vertices.

Brad adds the new coordinates to the picture on the whiteboard, which results in Figure 5.3. Peter starts calculating again.

[Peter] If my quick calculation is correct, here is the equation of Gray's upper inequality constraint:

$$0.09 \text{ Time} + \text{Temperature} \leq 587.8.$$

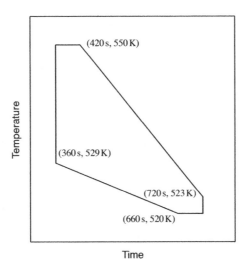

Figure 5.3   Region with feasible settings for the experimental factors time and temperature, along with the coordinates of four of its vertices.

Brad starts entering the factor ranges [520 K, 550 K] for temperature and [360 s, 720 s] for time, and the two inequality constraints in his design of experiments software application. He also indicates he wants to compute a D-optimal design.

[Peter, explaining to Gray] Brad is entering your factor ranges and constraints into his program. The program uses this information to construct a design that is optimal given these specifications.

[Brad, interrupting] I need to know which model we are going to fit and how many runs you are willing to perform.

[Gray] Earlier, Brad, you indicated that the quadratic model that we fit previously was not adequate. Could we fit a more flexible model?

[Peter] Good thinking! Suppose we add all the third-order terms to the quadratic model. For two factors, there are only four such terms.

[Gray] How do you get four terms? There are only two cubic terms, one for each factor.

[Peter] That is right but there are two other terms of the form $x_i x_j^2$.

[Gray] Oh, of course.

[Brad, impatiently] I have the full third-order model entered. How about a number of runs? At a very minimum we need ten. Peter and I usually recommend five extra runs to provide a reasonable internal estimate of the run-to-run variation.

[Gray] I suppose we can manage 15 runs.

Brad enters 15 runs into his program and presses the "Make Design" button in the user interface. Twenty seconds later, he turns his computer screen to display the design in Table 5.3.

Table 5.3 Optimal design for the new yield maximization experiment with the constrained experimental region. Runs 2 and 3 have the same settings as runs 13 and 7, respectively.

Run	Time (s)	Temperature (K)
1	580	528
2	360	529
3	480	525
4	720	523
5	360	545
6	660	520
7	480	525
8	420	550
9	360	536
10	454	539
11	720	520
12	630	531
13	360	529
14	510	542
15	360	550

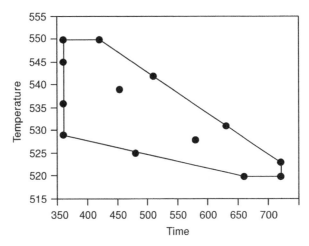

Figure 5.4   Graphical representation of the optimal design in Table 5.3 along with the region with feasible settings for the experimental factors time and temperature.

[Gray] I see 15 runs alright, but I can't tell if a scatter plot of the design points will fit in my figure.

[Brad] Just a second.

Brad makes a few graphics commands and produces the plot in Figure 5.4.

[Brad] You only see 13 points instead of 15 because two of the points are replicated.

[Gray] Which two?

[Brad] The point for a time of 360 seconds and a temperature of 529 K and the point for a time of 480 seconds and a temperature of 525 K.

[Gray] I like it. It is nice that there are several runs at shorter times and higher temperatures in your design. If we are lucky, we will find our optimal yield in that region. In that case, not only we would have higher yields but also we could run more reactions per day.

[Peter] So, are you willing to run this experiment?

[Gray] Absolutely. I will e-mail the yield data to you as soon as we perform the runs and assay the results.

[Brad] Outstanding! We look forward to hearing from you.

5.2.2   Data analysis

Three days later, Peter and Brad are in their US office. Brad gets an e-mail from Gray containing a table with the experimental results shown in Table 5.4 .

[Brad] Hey, Peter! Come over and look at these results I just got from Gray.

[Peter, looking at the table] Wow! Look at that 62.1% yield for 360 seconds and 545 K. That's a 20% improvement over the yields they were getting.

[Brad] I hope the cubic model fits well. If it does, the story is going to be fun to tell.

Table 5.4   Data for the new yield maximization experiment.

Run	Time (s)	Temperature (K)	Yield (%)
1	580	528	55.9
2	360	529	46.7
3	480	525	46.8
4	720	523	52.8
5	360	545	62.1
6	660	520	45.7
7	480	525	46.6
8	420	550	52.6
9	360	536	57.8
10	454	539	61.9
11	720	520	47.7
12	630	531	60.0
13	360	529	46.8
14	510	542	59.4
15	360	550	57.3

Brad fits the cubic model to the data and produces Table 5.5. His software automatically switches to coded units for the experimental factors.

[Peter] It looks as if you can remove the $Time^3$ term and the $Time^2 \times$ Temperature term to get a better model.

Brad removes both terms and shows Peter the new fitted model in Table 5.6.

[Brad] The root mean squared error is only 0.17%. That matches well with what Gray said their run-to-run variability was.

Table 5.5   Analysis of the data from the new yield maximization experiment: full model.

Effect	Estimate	Standard error	t Ratio	p Value
Intercept	61.08	0.10	598.66	< .0001
Time (360,720)	0.91	0.33	2.77	0.0391
Temperature (520,550)	5.06	0.34	14.97	< .0001
$Time^2$	−3.75	0.60	−6.21	0.0016
Time × Temperature	−16.50	1.08	−15.22	< .0001
$Temperature^2$	−21.32	0.56	−38.08	< .0001
Time × $Temperature^2$	−7.63	1.06	−7.19	0.0008
$Time^2$ × Temperature	−1.69	0.99	−1.71	0.1475
$Temperature^3$	−5.39	0.53	−10.11	0.0002
$Time^3$	−0.12	0.34	−0.36	0.7323

Table 5.6 Analysis of the data from the new yield maximization experiment: simplified model.

Effect	Estimate	Standard error	t Ratio	p Value
Intercept	60.99	0.10	610.21	< .0001
Time (360,720)	1.38	0.32	4.32	0.0035
Temperature (520,550)	5.22	0.40	13.19	< .0001
Time^2	−2.89	0.26	−10.92	< .0001
Time × Temperature	−14.69	0.42	−35.36	< .0001
Temperature^2	−20.32	0.27	−75.91	< .0001
Time × Temperature^2	−5.74	0.52	−11.08	< .0001
Temperature^3	−4.73	0.53	−8.90	< .0001

[Peter] Removing those two unimportant terms really dropped the standard errors of some of the remaining coefficients. With that cubic model and oddly shaped region, I was a little concerned that the standard errors of the factor effect estimates would be large due to variance inflation.

Brad raises an eyebrow and waits for Peter to continue....

[Peter] You know what I mean. Because of the very irregular shape of the region of experimentation and the cubic terms in the model, the D-optimal design you generated is not orthogonal. I thought that, therefore, there would be substantial correlation in our factor-effect estimates that would result in large standard errors or large variances.

[Brad] I also expected substantial variance inflation. But I did not want to worry Gray with this. Plus, I was not as worried about variance inflation once he told us how little variability their process has from one reaction to the next. I was more concerned about the possibility that even a third-order model would fail to approximate the true response surface. Here is the model's lack-of-fit test, using the two pure error degrees of freedom.

Brad scrolls down his screen to produce Table 5.7.

[Brad] The software reports this test because two of the combinations of time and temperature were replicated. The first replicated combination has a reaction time of 360 seconds and 529 K. The second combination has a reaction time of 480 seconds and a temperature of 525 K.

Table 5.7 Lack-of-fit test for the simplified model based on the data from the new yield maximization experiment.

Source	DF	Sum of squares	Mean square	F Ratio	p Value
Lack of fit	5	0.1827	0.0365	2.92	0.2743
Pure error	2	0.0250	0.0125		
Total error	7	0.2077			

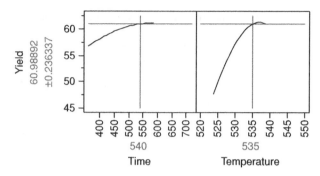

Figure 5.5   Graphical representation of the effect of the settings of time and temperature on the yield when time is fixed at 540 seconds or temperature is fixed at 535 K.

[Peter] There does not appear to be any strong reason to suspect that our cubic model is an inadequate approximation to the true response surface: the lack of fit is not close to being significant.

[Brad] Right. An interesting aspect of the model is that the impact of the factor time changes completely when the temperature is changed from low to high. Look at this plot.

Brad produces Figure 5.5, which shows what the impact of each of the experimental factors is when the other is fixed.

[Brad] This is a plot for a time of 540 seconds and a temperature of 535 K. On the left-hand side of the plot, you can read that the predicted yield for this setting is just under 61%. The increasing curve in the panel for the factor time shows that lowering the time will lead to a smaller yield and that you can improve the yield a tiny bit by increasing the time. Note that we cannot increase the time beyond 586 seconds when the temperature is 535 K, because that would violate the second constraint that Gray imposed on the factor settings.

[Peter] Right.

[Brad, modifying Figure 5.5 until he obtains Figure 5.6] Now, look at this. Here, I have maximized the predicted yield of Gray's process. The temperature is now fixed at 543 K instead of 535 K, and this has a massive effect on the picture for the time effect. At the new temperature, due to the interaction effect, lowering the time has a positive effect on the yield and the maximum yield is obtained for the lowest temperature that Gray wanted to study, 360 K. You can read what the maximum yield is right next to the vertical axis: about 62.5%.

[Peter] Plus or minus 0.3%. So, the 95% prediction interval for the mean yield is about [62.2%, 62.8%] at a time of 360 seconds and a temperature of 543 K.

[Brad, who continues playing with the graphical representation of the factor effects and eventually produces Figure 5.7] Here, I have a picture that shows the factor effects and the prediction interval together. Note that I had to use different scales for the axes to make the prediction interval visible. You can see that the

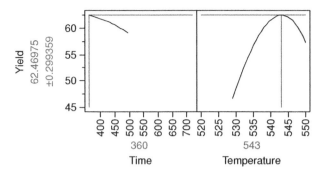

Figure 5.6   Graphical representation of the factor settings that produce the maximum predicted yield, along with the effects of the factors at the optimal settings.

prediction interval, or confidence band, is very narrow for each possible setting of the factors. So, there is little uncertainty about the predicted yield.

[Peter] Gray will be very pleased with these results.

Brad nods in agreement and continues to play with his computer for a minute or two. Then, he turns the screen to show the graph in Figure 5.8.

[Peter] I see you have drawn the hexagonal feasible region, defined by the constraints, as well as the contour lines for the predicted yield.

[Brad] This shows a couple of things you could not see in the other graph. First, you can clearly see the region of maximum yield, namely, the region enclosed by the contour for a yield of 62% and higher. Also, the graph suggests that there is still some opportunity for improved yields by lowering the reaction time still further.

[Peter] What do you think of the idea of broaching the subject of a follow-up experiment exploring even shorter reaction times in our presentation to Gray? He has indicated how important shorter reaction times are to Rohm and Haas.

[Peter] Let's see how it goes when we visit to present these results.

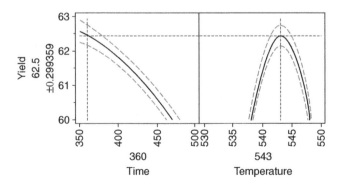

Figure 5.7   Graphical representation of the factor settings that produce the maximum predicted yield, the factor effects at these settings and the prediction interval.

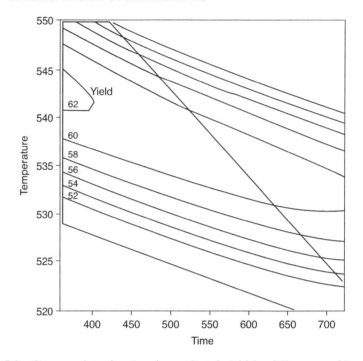

Figure 5.8   Contour plots showing the predicted yield for different combinations of time and temperature, with experimental region overlaid.

5.3   Peek into the black box

5.3.1   Cubic factor effects

Sometimes, the experimental data exhibit so much curvature that it is not enough just to include two-factor interaction effects and quadratic effects in the model. It is then worth considering the inclusion of higher-order terms. Beyond quadratic effects, we can consider cubic effects as well as other third-order effects. In the case study in this chapter, we saw the added value of including cubic effects in the context of a response surface experiment. Cubic effects also often turn out to be useful for modeling data from mixture experiments (see Chapter 6).

The most complicated model involving cubic effects is the full cubic model, which includes all possible third-order effects. If the number of factors, k, is greater than two, these terms have three different forms:

1. there are k terms of the form $\beta_{iii} x_i^3$,

2. there are $k(k-1)$ terms of the form $\beta_{iij} x_i^2 x_j$, and

3. there are $k(k-1)(k-2)/6$ terms of the form $\beta_{ijk} x_i x_j x_k$.

It is easy to see that, as the number of factors increases, the number of cubic effects increases rapidly. It is rare in industrial applications that the budget for runs can accommodate a full cubic model for more than four factors.

5.3.2   Lack-of-fit test

In the yield maximization experiment, there were 13 unique factor-level combinations or design points. Two of these factor-level combinations were run twice; we say that they were replicated. Replication is an important idea in design of experiments. In fact, it is one of the three main principles of experimental design, along with randomization and blocking.

The benefit of replicate observations is that they provide a model-independent estimate of the error variance. In experimentation without replication, the investigator must use the residuals from the fitted regression to estimate the error variance. If the model is missing some important effect, then it is probable that the error variance estimate will be inflated. This is due to variation in the data that comes from the real effect that is missing from the model. The consequence of an inflated error variance estimate is that fewer factor effects will appear as significant. This is an unavoidable risk of not having replicated factor-level combinations.

If there are both replicated factor-level combinations and more unique ones than parameters to be estimated, then it is possible to construct a hypothesis test to determine if it is likely that the model is missing an effect. This test is called a lack-of-fit test, because it identifies situations where the assumed model does not fit the data well. The null hypothesis for the lack-of-fit test is that the model is adequate. The alternative hypothesis is that one or more terms are missing from the model (typically terms involving higher-order effects, such as interaction effects, quadratic effects, or cubic effects).

The total residual sum of squares,

$$\text{SSE} = \sum_{i=1}^{n}(Y_i - \hat{Y}_i)^2,$$

measures the unexplained variation in the response of the regression model. This is nothing but the sum of squared residuals of the regression. This unexplained variation can be split into two parts. One part, called pure error, is due to the variability that is inherent to the system. This part is obtained from the replicated observations. The other part is due to the fact that the model is not perfect. That part is called the lack of fit.

Pure error can only be estimated if there are replicated factor-level combinations. What we need for the pure error calculation are the responses of the replicated points. If we denote the jth observed response at the ith replicated factor-level combination by Y_{ij}, the number of replicated factor-level combinations by m, and the number of replicates at the ith factor-level combination by n_i, then the pure error sum of squares is

$$\text{SSPE} = \sum_{i=1}^{m} \sum_{j=1}^{n_i} (Y_{ij} - \bar{Y}_i)^2,$$

where

$$\bar{Y}_i = \frac{1}{n_i} \sum_{j=1}^{n_i} Y_{ij}$$

is the average response at the ith factor-level combination. The lack-of-fit sum of squares then is

$$\text{SSLOF} = \text{SSE} - \text{SSPE}.$$

The two components of the residual sum of squares, i.e., SSPE and SSLOF, are statistically independent. The total degrees of freedom corresponding to the residual sum of squares is the total number of experimental runs minus the number of parameters in the model, $n - p$. The degrees of freedom value that corresponds to the pure error sum of squares SSPE is

$$\text{df}_{\text{PE}} = \sum_{i=1}^{m} (n_i - 1) = \sum_{i=1}^{m} n_i - m.$$

The difference in these degrees of freedom,

$$\text{df}_{\text{LOF}} = n - p - \text{df}_{\text{PE}},$$

is associated with the lack-of-fit sum of squares SSLOF. As can be seen in Table 5.4, there are 15 experimental runs in total in the yield maximization experiment and the simplified model in Table 5.6 involves eight parameters. Hence, $n = 15$ and $p = 8$. Therefore, the degrees of freedom associated with the residual sum of squares is $n - p = 7$. The design involves two duplicated points, so that $m = 2$ and $n_1 = n_2 = 2$. As a result, $\text{df}_{\text{PE}} = n_1 + n_2 - m = 2 + 2 - 2 = 2$ and $\text{df}_{\text{LOF}} = 7 - \text{df}_{\text{PE}} = 7 - 2 = 5$.

Under the lack-of-fit test's null hypothesis that the model is adequate, the ratio of the lack-of-fit mean square,

$$\frac{\text{SSLOF}}{\text{df}_{\text{LOF}}},$$

and the pure error mean square,

$$\frac{\text{SSPE}}{\text{df}_{\text{PE}}},$$

is distributed according to the F-distribution with numerator degrees of freedom equal to df_{LOF} and denominator degrees of freedom equal to df_{PE}. If this ratio were exceptionally large, it would call the validity of the null hypothesis into question. In our example, this ratio is a little less than three. A result of this magnitude is one that could easily occur by chance. This is reflected by the p value of 0.2743. So, there is no reason to suspect the presence of a lurking systematic effect inflating the error variance. In other words, there is no evidence of lack of fit.

5.3.3 Incorporating factor constraints in the design construction algorithm

Inequality constraints governing the feasibility of factor-level combinations restrict the values of the factor settings, and in turn restrict the values of the elements of the design matrix **D**, which we defined in Section 2.3.9. When replacing coordinates of the design points one by one, the coordinate-exchange algorithm must take into account the constraints. To make clear how this can be done, consider a small example.

Suppose that we have two continuous experimental factors, x_1 and x_2, and that each can take values on the interval $[-1, +1]$. Additionally, the sum of the two factors' settings must be less than or equal to one:

$$x_1 + x_2 \leq 1. \tag{5.1}$$

Finally, suppose that the coordinate-exchange algorithm is ready to replace the levels of x_1 and x_2 of a given row of the design matrix and that the current levels are -0.5 for the first factor, x_1, and 0.3 for the second factor, x_2.

For x_1, the coordinate-exchange algorithm can consider values as low as -1, without violating the inequality constraint in Equation (5.1). However, the maximum possible value for x_1, given that x_2 takes the value 0.3, is 0.7. We can see this by rewriting the constraint as

$$x_1 \leq 1 - x_2,$$

and observing that $1 - x_2 = 0.7$ when $x_1 = 0.3$. Therefore, for this specific element of the design matrix, the coordinate-exchange algorithm will replace the value of -0.5 with the best value it can find in the interval $[-1, 0.7]$. Depending on which optimality criterion you use, the best value is the one that maximizes the D-criterion value or minimizes the I-criterion value, depending on the optimality criterion in use.

For the second factor, x_2, assuming the value for x_1 is still -0.5, the coordinate-exchange algorithm will consider any value between its lower bound, -1, and its upper bound, $+1$. This is because neither the value of -1 for x_2 nor the value of $+1$ leads to a violation of the constraint.

In the coordinate-exchange algorithm described in Section 2.3.9, note that only one coordinate in the design matrix changes at each step. Thus, all the other entries remain fixed. We modify that algorithm in this way. For the factor level being considered for an exchange, we insert the known values of the other factor levels into the inequality constraints and solve each to give a range of allowable levels for the given factor. In some cases, there is only one solution to the set of inequality constraints, and that is the current factor level. If this happens, then we leave the current coordinate as it is and move to the next one. If the solution to the set of inequalities covers an interval, then we evaluate the optimality criterion for a number of factor levels along this interval. The level corresponding to the greatest improvement in the optimality criterion becomes the new coordinate in the design matrix.

5.4   Background reading

Most experiments involving constraints on the levels of the experimental factors in the literature have linear constraints only. Atkinson et al. (2007), however, describe a mixture experiment with nonlinear factor-level constraints. Mixture experiments often involve several constraints on the experimental factors. We discuss these types of experiments in the next chapter.

5.5   Summary

While most of the literature on response surface experiments considers design and analysis for full quadratic models involving main effects, two-factor interaction effects, and quadratic effects, there are some instances where models involving higher order terms are necessary.

Standard screening and response surface designs implicitly assume that all the factors can be varied independently. In other words, they assume that the allowable region of experimentation is either spherical or cubic. It is, however, quite often the case that certain combinations of the factor levels are infeasible, so that the experimental region is not spherical or cubic.

Using the optimal design of experiments approach allows the flexible specification of both the polynomial terms in the model and the shape of the experimental region.

6

A "mixture" experiment with process variables

6.1 Key concepts

1. A key advantage of designed experiments over observational studies is that observational studies can establish correlation, but not causation. Properly randomized experiments can demonstrate causation.

2. Mixture experiments require the ingredient proportions to sum to one (more generally, any constant). Therefore, the factor levels cannot vary independently in mixture experiments.

3. Models for mixture problems are different from standard models for factorial type experiments. For example, models for mixture experiments do not generally have an intercept or constant term, because such a term would be linearly dependent with the sum of the mixture ingredients.

4. Mixture experiments often include process variables in addition to the ingredients of the mixture. Process variables along with mixture ingredients have a further impact on standard models. For example, the main effects of process variables cannot appear in models including all the two-factor interactions of mixture factors and process factors.

The case study in this chapter begins with a cautionary tale about the nature of observational studies. In an observational study, the "independent" variables are merely observed without being explicitly varied. As a result, one can observe correlation between the independent variable and some response of interest, but this does not

Optimal Design of Experiments: A Case Study Approach, First Edition. Peter Goos and Bradley Jones.
© 2011 John Wiley & Sons, Ltd. Published 2011 by John Wiley & Sons, Ltd.

mean that variation in the dependent variable causes the observed variation in the response.

The case study involves a group of factors whose sum is a constant. Although the factors are not ingredients in a mixture, the nature of the problem is like a mixture experiment in this regard. In addition to the factors whose sum is constant, there are other factors in the experiment that may vary independently. Similarly, this feature makes the example analogous to a mixture-process variable experiment.

6.2   Case: the rolling mill experiment

6.2.1   Problem and design

Peter and Brad are driving through rural Minnesota, heading for a consultation with a supplier of sheet aluminum.

[Brad] The people who started this plant are smart. They built their facility just across a field from an aluminum recycling plant. That way they avoid much of the cost of raw material by just using the crushed cubes of aluminum they get from the recycling plant. Also, they got a bargain when they purchased their furnace and rolling mills from steel mills that are no longer producing.

Unfortunately, there are some quality problems at the plant. The aluminum sheets that come through their milling operation should shine like a mirror. Their customers will not buy dull or smudged looking sheets, so they have to dump the poor-quality stuff back into the furnace and try again.

[Peter] That does not sound so bad. After all, they can reuse the aluminum.

[Brad] There are two problems with that. First, the furnace requires substantial energy, so operating it twice for a batch of material raises the cost per sheet. Also, with the current amount of rework, they are having trouble keeping up with their orders.

[Peter] Who is our sponsor at the plant?

[Brad] It is Dr. Ferris Nighthome. He is their quality expert and right now he is pretty worried.

Later, inside the facility, Dr. Nighthome shows them the processing line, a schematic representation of which is given in Figure 6.1.

[Dr. Nighthome] This is where the molten aluminum comes out of the furnace. We have some control of the temperature of the metal at this point and we thought we had some evidence that higher melt temperatures resulted in better quality.

[Peter] You don't seem so sure now.

[Dr. Nighthome] Right. I'll show you the data after our tour of the processing line. Now, here is the milling operation. Three rollers mill the metal down to the thickness required. Each roller reduces the thickness by a particular amount and we can control that by changing the amount of force applied in that rolling operation.

[Brad] This operation is impressively loud even with our ear protection. And it is really hot here. It must be miserable in the summer.

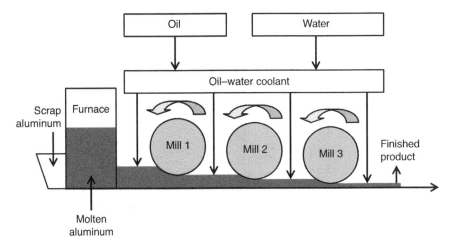

Figure 6.1   Schematic representation of the milling operation.

[Dr. Nighthome, grinning] Yeah, but, here in Minnesota, we really like the heat in the winter. Speaking of heat, we spray an oil–water coolant on the metal between milling operations. We can control the ratio of oil to water and the volume of coolant we apply.

After the tour, Peter and Brad have a good idea of the working of the milling operation. They get coffee and move to a conference room for a brainstorming session with the plant manager (PM).

[PM] Dr. Nighthome, why don't you show Drs. Goos and Jones what you have already done on this problem?

[Brad] Please just call us Peter and Brad. Drs. Goos and Jones sounds so formal.

[PM] That's fine with me, Brad.

[Dr. Nighthome] Let me start by giving you some more details on our goal here. Our response is the measured reflectivity of the sheet as it finishes processing. To be acceptable, the reflectivity has to be 5.5 or higher. Any sheets with reflectivity less than 5.5 go back into the furnace. The problem is that, lately, around 50% of the sheets are unacceptable.

He puts up a slide that shows Figure 6.2.

[Dr. Nighthome, pointing at the figure] Now, look at this scatter plot. We measure the temperature of the aluminum as it comes through the final milling operation. This plot seems to show that there is a positive linear relationship between the finish temperature and the reflectivity. We thought at first that all we needed to do was to raise the melt temperature of the metal as it comes from the furnace. That costs a little more in energy, but compared to doing everything twice, it is well worth it.

[Peter] The historical correlation between finish temperature and reflectivity is very strong. It is tempting to think that eliminating low temperature melts could solve the problem.

[Dr. Nighthome] Yes, we were very excited by the prospect of resolving our quality problem quickly. Unfortunately, things did not go as planned.

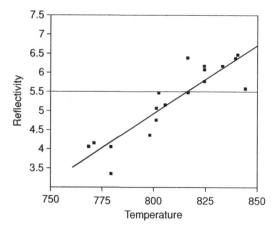

Figure 6.2   Scatter plot of reflectivity versus melt temperature.

[Brad] How so?

[Dr. Nighthome] We did a simple study where we raised the melt temperature to the point that the finish temperature corresponded to the good results in the scatter plot. The reflectivity did not improve. In fact, our scrap rate actually increased for the sheets in the study.

[Brad] Wow! That's a puzzler.

[Peter] Sounds like there must be a lurking variable somewhere.

[Dr. Nighthome, drily] Yeah, we know from our own hard experience now that correlation does not imply causation.

[PM] That was a valuable lesson for all of us even though it was an expensive and disappointing one. So, Ferris, what is your new plan?

[Dr. Nighthome] I brought Peter and Brad in because I want to do a more extensive experiment. We may have eliminated the melt temperature as a factor but we have a number of others to study.

[PM] I suppose you want to look at each of your factors one at a time while holding the others fixed to nail down the culprit. That's the scientific method according to one of my textbooks.

[Dr. Nighthome, embarrassed] Well, ah . . .

[Brad] I think that what Ferris wants to say is that the factorial approach is more common these days.

[PM] How is the factorial approach different?

[Brad] In a factorial study, you may vary all the potentially relevant factors from one processing run to the next.

[PM] Woah! That sounds dangerously like trial and error. You can waste a lot of time and money following your nose with that approach.

[Peter] We agree completely.

[PM, puzzled] Then what gives?

[Brad] The key to the factorial approach is that the group of processing runs you do has a very systematic structure. That structure enables a statistical analysis that can tease out the individual effects of each factor in a more cost-efficient way than the one-factor-at-a-time approach you mentioned.

There is another benefit too. Sometimes the behavior of a system is the result of synergistic or antagonistic effects involving more than one factor. With the right kind of study you can also quantify any such effects.

[PM] Ferris, I suppose you knew about this already?

[Dr. Nighthome] Yes, I am a strong advocate of the factorial approach. That was one of my courses in industrial engineering. Still, this problem has an aspect that does not jibe with my textbook approach. I know that Peter and Brad are experts in this area, so I brought them in to help.

[Peter] I don't know exactly what jibe means since my native language is Flemish. But from context, I can guess that your textbook design has something about it that does not work in this practical situation.

[Dr. Nighthome] You can say that again. So here is the problem.

We have five factors that my team agrees are potentially important. Three of the factors have to do with the three rollers. We also thought about changing the oil–water ratio in the coolant mixture as well as changing the volume of coolant that we spray between steps.

If we only consider high and low settings of each of these factors, we can use a two-level factorial design right out of my textbook. I was thinking about a half-fraction of a 2^5 factorial design because then we can estimate all the two-factor interaction effects—the synergistic or antagonistic effects Brad mentioned before. The good thing is that only 16 processing runs are necessary. We can do that in a couple of days.

[PM] The stuff about the factorial design sounded like a lot of jargon to me but studying five different factors with only 16 runs is attractive. Where is the problem?

[Dr. Nighthome] The problem comes with the rolling mill factors. We start with a melt that is, say, 1.1 units thick. Our finished sheet is 0.1 unit thick. Each roller can reduce the thickness between 0.1 and 0.8 unit.

He puts up a slide with the factorial design shown in Table 6.1.

[Dr. Nighthome] Notice the first run in the experiment. I am supposed to remove 0.8 with the first rolling mill, 0.8 with the second, and 0.8 with the third. All together I am supposed to remove 2.4 units of thickness, but there is only 1.1 to start with. Any operator would laugh at me if I were silly enough to ask them to do this run.

[Peter] I see your problem. The total reduction in the thickness of the material is a fixed quantity. So the sum of your three factor settings also needs to be a constant no matter what the individual settings are. If you consider the setting for a given rolling mill to be the proportion of the total reduction in thickness that mill accomplishes, then this has the same structure as a mixture experiment.

[Dr. Nighthome] I am not quite there yet. Could you be more explicit?

[Brad] Let me give it a try. Suppose the first milling operation brings the thickness down 20% of the way, the second brings it down another 20%, and the last operation drops the thickness the remaining 60%. This is like mixing three ingredients that have to sum to 100% of a mixture.

[Dr. Nighthome] Even though we are not mixing anything, I can see the similarity involving the constraint that the sum of the three factor settings has to be one. But

Table 6.1   Half fraction of a 2^5 factorial design for the five factors in the rolling mill experiment.

Mill 1	Mill 2	Mill 3	Spray volume	Oil–water
0.8	0.8	0.8	High	High
0.8	0.8	0.8	Low	Low
0.8	0.8	0.1	High	Low
0.8	0.8	0.1	Low	High
0.8	0.1	0.8	High	Low
0.8	0.1	0.8	Low	High
0.8	0.1	0.1	High	High
0.8	0.1	0.1	Low	Low
0.1	0.8	0.8	High	Low
0.1	0.8	0.8	Low	High
0.1	0.8	0.1	High	High
0.1	0.8	0.1	Low	Low
0.1	0.1	0.8	High	High
0.1	0.1	0.8	Low	Low
0.1	0.1	0.1	High	Low
0.1	0.1	0.1	Low	High

doesn't the whole concept of factorial design depend on being able to set each factor independently of the others? Having this constraint has to introduce collinearity, right?

[PM] More jargon!

[Peter] Yes, jargon is sometimes hard to avoid, but what Ferris wants to know is whether we will know how each milling operation affects the response independently of the others if we cannot vary them independently.

[Brad] And despite the jargon, Ferris has nailed the problem. Because you cannot vary the ingredients in a mixture independently, it affects the design, model fitting and interpretation of the results of the experiment.

[Peter] There are whole books written about the special nature of mixture experiments. I am not surprised that the course on design of experiments that Ferris took did not deal with them. Generally, mixture experiments come up in a second course.

[PM] OK then. So, can you talk about the problem using plain language?

[Brad] There is bad news and good news. The bad news is that you cannot talk about the effect of each of the milling operations independently of the others. But the good news is that, by investing in a designed study, you can predict the results of various combinations of roller settings—even the ones that you do not include in the study.

[PM] Your "good news" sounds suspiciously like magic to me.

[Peter] The accuracy of these predictions depends on the validity of the model that we fit to the data. That is why we always check our predictions by doing follow-up runs at factor settings that look promising.

Table 6.2   D-optimal 12-run design for the five factors in the rolling mill experiment.

Mill 1	Mill 2	Mill 3	Spray volume	Oil–water
0.10	0.80	0.10	Low	Low
0.80	0.10	0.10	Low	Low
0.10	0.45	0.45	Low	Low
0.10	0.10	0.80	High	Low
0.45	0.45	0.10	High	Low
0.45	0.10	0.45	High	Low
0.45	0.10	0.45	Low	High
0.10	0.10	0.80	Low	High
0.45	0.45	0.10	Low	High
0.80	0.10	0.10	High	High
0.10	0.80	0.10	High	High
0.10	0.45	0.45	High	High

[Brad] Good point, Peter. In addition, even though we know the model we fit to your process will not perfectly describe its behavior, sometimes the results do seem almost magical.

[PM] I am skeptical but we need to solve this quality problem. So, let's suppose we do this mixture experiment. How long will it take to you generate a test plan?

Brad takes out his laptop and starts typing rapidly.

[Peter] That part is almost embarrassingly fast and easy. Brad is entering your five factors into his software for designing experiments.

A couple of minutes later, Brad turns his laptop toward the plant manager to display Table 6.2.

[Brad] I created a 12-run design for your five factors assuming that there might be two-factor interactions involving the milling operations. But I only considered main effects in the spray volume and oil–water factors.

[Dr. Nighthome, excitedly] I see that, if you sum the settings of the mill factors for every row, you always get one.

[Peter] Yes, I also notice that the number of high and low settings of the spray volume and oil–water factors is the same. In fact, the four possible combinations of levels of these factors each appears three times. That has a nice symmetry.

[Brad] The ternary plot is a popular graphic for displaying the settings of mixture components in studies like these. Take a look at this.

Brad shows the group Figure 6.3.

[Brad] The first thing that the plot clearly shows is that there are six different level combinations of the milling factors. Each of these appears twice. Another nice feature of the design is that each of the six milling factor-level combinations appears once at the low setting of the oil–water ratio and once at the high setting. To show this in the plot, I made the plotting symbol for the low setting of the oil–water ratio an open circle and the plotting symbol for the high setting a plus sign.

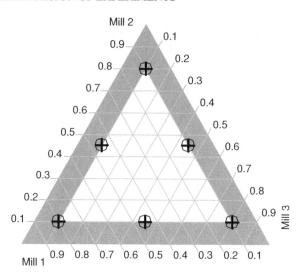

Figure 6.3   Ternary plot of the D-optimal 12-run design for the five factors in the rolling mill experiment.

[Dr. Nighthome] That is attractive indeed. Is the same thing true for the spray volume factor?

[Brad] Yes, it is.

[PM] You all are talking as if pretty geometry was the point here. I am more concerned about finding the source of the problem and fixing it.

[Peter] That is well put, and we feel the same way. Let's focus on the statistical properties then.

[PM] Good idea.

[Brad] I have checked and the worst relative variance of prediction is two-thirds. There is a theoretical result stating that for this number of runs and this model, two-thirds is the minimum value that the worst relative variance of prediction can have. So, in the sense of minimizing the worst prediction variance, this design is the best of all possible designs. The average relative variance of prediction is a little less than a half. That means that, when we make predictions using this model, the expected value of those predictions will have a variance that is less than half of the run-to-run variation of the process. That is, assuming our model is correct.

[Peter] If the experiment had more runs, we could drive these numbers even lower. You seemed to like the idea of a 16-run study. By starting with 12 runs, I was thinking in terms of reserving four runs to verify the results at any processing setting that looks promising.

[PM] Frankly, I am still skeptical. I will believe when I see the results.

[Peter] We appreciate your candor. How long do you think it will take to do the 12 processing runs?

[Dr. Nighthome] Given the urgency of our problem, I think we can do all the runs today.

[PM] Absolutely!

6.2.2   Data analysis

That night in his hotel room, Brad gets the following e-mail from Dr. Nighthome, with the data in Table 6.3 enclosed.

Brad,

Only two of the 12 runs yielded acceptable values for reflectivity. Both of the good runs resulted when Mill 1 and Mill 2 each did 10% of the thickness reduction and Mill 3 did the rest of the work. We also noticed that the finish temperature for these two runs was substantially higher than for the others.

Can you analyze these results and meet with us at 10:30 AM tomorrow morning. The plant manager is in a hurry!

Ferris

The next morning Brad and Peter meet with Dr. Nighthome and the plant manager in the same conference room as before.

[PM] I have looked at the data from yesterday's runs and things don't look very promising on the face of it. Only two of the 12 runs were acceptable.

[Brad] We actually think there may be some reason to celebrate. We do not run experiments expecting that the runs themselves will be good. The idea is to develop a predictive model that suggests a better way of operating.

[Peter] Yes, and we have three major findings to present:

1. The process variables, spray volume and oil–water, have no visible effect.

2. The relationship between the mills and reflectivity appears to be linear.

3. Taking off 10% with Mills 1 and 2 and 80% with Mill 3 results in a predicted reflectivity of 6.25. That is well above your quality specification of 5.5.

Table 6.3   Data for the rolling mill experiment.

Mill 1	Mill 2	Mill 3	Spray volume	Oil–water	Reflectivity
0.10	0.80	0.10	Low	Low	5.3
0.80	0.10	0.10	Low	Low	4.8
0.10	0.45	0.45	Low	Low	5.4
0.10	0.10	0.80	High	Low	6.3
0.45	0.45	0.10	High	Low	4.7
0.45	0.10	0.45	High	Low	5.4
0.45	0.10	0.45	Low	High	4.9
0.10	0.10	0.80	Low	High	6.8
0.45	0.45	0.10	Low	High	5.3
0.80	0.10	0.10	High	High	4.3
0.10	0.80	0.10	High	High	4.7
0.10	0.45	0.45	High	High	5.1

[Brad] We recommend that you do confirmation runs today by setting Mill 1 and 2 to each reduce thickness by 10% and Mill 3 to bring the thickness to nominal. You can use your standard settings for spray volume and oil–water ratio, since they do not appear to matter.

[Dr. Nighthome] Do you have a prediction equation for your model?

[Brad] Yes. Here it is:

$$\text{Reflectivity} = 4.55 \, \frac{\text{Mill } 1 - 0.1}{0.7} + 4.95 \, \frac{\text{Mill } 2 - 0.1}{0.7} + 6.25 \, \frac{\text{Mill } 3 - 0.1}{0.7}.$$

[Brad] The three ratios in this equation are called the pseudocomponents. Subtracting 0.1 from the factor setting of each mill and then dividing by 0.7 make the pseudocomponent values go from 0 to 1 when the actual components go from 0.1 to 0.8. That is a convenient scaling convention that makes it easy to make predictions for reflectivity at the extremes of each factor setting.

[Peter] When the Mill-1 pseudocomponent value is one, then the others are zero and the predicted reflectivity is 4.55. Similarly, when the Mill-2 pseudocomponent is one, then the predicted reflectivity is 4.95. Finally, and best, when the Mill-3 pseudocomponent is one, the predicted reflectivity is 6.25. This is the technical justification for Peter's last recommendation.

[Brad] The values 4.55, 4.95, and 6.25 are the estimates we got from a least squares fit of your data.

[PM] It has been a long time since I heard the term "least squares." Anyway, I am willing to go ahead with your confirmation runs and see what happens. Anything more to add?

[Brad] One side note you may find interesting is something that Ferris pointed out in his e-mail last night. The runs with high reflectivity also had high finish temperature. Do you think the last milling operation could have caused the temperature increase?

[PM] That is interesting. Maybe the extra work of milling off 80% in the last step is both improving the reflectivity and raising the finish temperature. Let's hope you are right.

One week later Brad gets the following e-mail from Dr. Nighthome.

Brad,

After getting your recommendations, we ran the three mills as you suggested. In the four confirmatory processing runs that day the reflectivity averaged 6.4 and all the runs were above 5.5. Since then, we have been operating round the clock to fill orders. Less than 5% of our runs have reflectivity less than 5.5. Of course, that is a huge improvement over the 50% defect rate we were experiencing.

The plant manager is very pleased. He asked me to send his congratulations and to tell you that he is now a believer. He also mentioned that the finish temperature has been high all week, so perhaps the mystery of the lurking variable is solved.

Regards,
Ferris

6.3   Peek into the black box

6.3.1   The mixture constraint

The unique feature of a mixture experiment is the mixture constraint. This is an equality constraint. If $x_1, x_2, \ldots, x_q$ are the proportions of q ingredients in a mixture, then

$$\sum_{i=1}^{q} x_i = x_1 + x_2 + \cdots + x_q = 1. \qquad (6.1)$$

More generally, the right-hand side of this equation can be 100, if you prefer percents to proportions. Or, if you are mixing by weight or volume, you can make the sum be any positive constant.

The most important assumption of mixture experiments is that the response is not a function of the amount of the mixture but only of the proportions of the various ingredients. The primary consequence of the equality constraint in Equation (6.1) is that, if you change the value of a particular ingredient proportion x_i, then at least one other ingredient proportion, say x_j, must also change to compensate. This means the ingredient proportions in a mixture experiment cannot be made orthogonal, and that their effects cannot be estimated independently.

6.3.2   The effect of the mixture constraint on the model

In all the experiments we have considered thus far, the a priori model included an intercept term. For instance, for a main-effects model, we assumed the following model for the jth response:

$$Y = \beta_0 + \sum_{i=1}^{q} \beta_i x_i + \varepsilon. \qquad (6.2)$$

The intercept is the β_0 term in the model. We can also write this model in matrix form:

$$Y = \mathbf{X}\boldsymbol{\beta} + \boldsymbol{\varepsilon}, \qquad (6.3)$$

where the first column of the model matrix $\mathbf{X}$ is a column of ones. For mixture experiments, the equality constraint in Equation (6.1) implies that the column of ones for the intercept is redundant because if there are q mixture ingredients and the model has a term for each component, then the sum of these mixture component columns in $\mathbf{X}$ is one for each row. As pointed out in Section 4.3.2 when discussing models involving categorical factors, if one column in the model matrix $\mathbf{X}$ is the sum (or any other linear combination) of other columns in $\mathbf{X}$, we cannot compute the ordinary least squares estimator

$$\hat{\beta} = (\mathbf{X}'\mathbf{X})^{-1}\mathbf{X}'\mathbf{Y}, \tag{6.4}$$

because $\mathbf{X}'\mathbf{X}$ cannot be inverted. One way to avoid this difficulty is to remove the unnecessary intercept term from the model. Whenever one column of the model matrix $\mathbf{X}$ is a linear combination of other columns, we say that there is linear dependence or perfect collinearity among the columns of $\mathbf{X}$.

There are two other similar issues that are worth mentioning. First, models having only the linear effect of each ingredient, such as the model in Equation (6.2), cannot capture curvature in the relationship between the ingredients and the response. It is natural to look for models that can deal with curvature. To achieve this end, we typically add two-factor interactions and pure quadratic effects to the model. However, if the model includes all the linear and two-factor interaction effects for each mixture component, then the pure quadratic effects are redundant. To see this, note that

$$\begin{aligned} x_i^2 &= x_i \left(1 - \sum_{\substack{j=1 \\ j \neq i}}^{q} x_j \right), \\ &= x_i - \sum_{\substack{j=1 \\ j \neq i}}^{q} x_i x_j, \end{aligned} \tag{6.5}$$

for every ingredient proportion x_i. As a result, the square of an ingredient proportion is a linear combination of that proportion and its cross-products with each one of the other ingredient proportions composing the mixture. As a specific example, suppose there are three ingredients with proportions x_1, x_2, and x_3. Then,

$$x_1^2 = x_1(1 - x_2 - x_3) = x_1 - x_1 x_2 - x_1 x_3. \tag{6.6}$$

Including quadratic terms in a model that already contains main effects and two-factor interaction effects, therefore, creates a model for which the least squares estimator does not exist. Therefore, to avoid problems with fitting the model, we should always leave pure quadratic effects of mixture ingredient proportions out of the model.

The second similar issue involves having a model with both the main effect of a process variable as well as all the two-factor interactions involving that process variable and all the mixture ingredients. In this case, the main effect of the process variable is redundant and should be removed from the model. To see this, suppose again that there are three ingredients with proportions x_1, x_2, and x_3, but this time there is an additional experimental factor, the process variable z. The sum of the cross-products between the value of the process variable, on the one hand, and the ingredient proportions, on the other hand, is

$$zx_1 + zx_2 + zx_3 = z(x_1 + x_2 + x_3) = z, \qquad (6.7)$$

because the sum of the ingredient proportions, $x_1 + x_2 + x_3$, is always one. Equality (6.7) implies that we should never include a process variable's main-effect term and its two-factor interactions with the ingredient proportions in the model together.

6.3.3 Commonly used models for data from mixture experiments

6.3.3.1 Mixture experiments without process variables

By far the most popular models for modeling the response Y of a mixture experiment are the Scheffé models. The first-order Scheffé model is given by

$$Y = \sum_{i=1}^{q} \beta_i x_i + \varepsilon, \qquad (6.8)$$

whereas the second-order Scheffé model is given by

$$Y = \sum_{i=1}^{q} \beta_i x_i + \sum_{i=1}^{q-1} \sum_{j=i+1}^{q} \beta_{ij} x_i x_j + \varepsilon. \qquad (6.9)$$

The special-cubic model can be written as

$$Y = \sum_{i=1}^{q} \beta_i x_i + \sum_{i=1}^{q-1} \sum_{j=i+1}^{q} \beta_{ij} x_i x_j + \sum_{i=1}^{q-2} \sum_{j=i+1}^{q-1} \sum_{k=j+1}^{q} \beta_{ijk} x_i x_j x_k + \varepsilon. \qquad (6.10)$$

6.3.3.2 Mixture experiments with process variables

We obtain models for analyzing data from experiments involving q ingredients proportions $x_1, x_2, \ldots, x_q$ and m process variables $z_1, z_2, \ldots, z_m$ (often referred to as mixture-process variable experiments) by combining Scheffé type models for the ingredient proportions with response surface models for the process variables. For example, a common mixture-process variable model is obtained by crossing the second-order Scheffé model

$$Y = \sum_{k=1}^{q} \beta_k x_k + \sum_{k=1}^{q-1} \sum_{l=k+1}^{q} \beta_{kl} x_k x_l + \varepsilon$$

for the mixture ingredients with a main-effects-plus-two-factor-interaction model in the process variables,

$$Y = \alpha_0 + \sum_{k=1}^{m} \alpha_k z_k + \sum_{k=1}^{m-1} \sum_{l=k+1}^{m} \alpha_{kl} z_k z_l + \varepsilon.$$

The combined model is

$$\begin{aligned}
Y = &\sum_{k=1}^{q} \gamma_k^0 x_k + \sum_{k=1}^{q-1} \sum_{l=k+1}^{q} \gamma_{kl}^0 x_k x_l \\
&+ \sum_{i=1}^{m} \left[\sum_{k=1}^{q} \gamma_k^i x_k + \sum_{k=1}^{q-1} \sum_{l=k+1}^{q} \gamma_{kl}^i x_k x_l \right] z_i \qquad (6.11) \\
&+ \sum_{i=1}^{m-1} \sum_{j=i+1}^{m} \left[\sum_{k=1}^{q} \gamma_k^{ij} x_k + \sum_{k=1}^{q-1} \sum_{l=k+1}^{q} \gamma_{kl}^{ij} x_k x_l \right] z_i z_j + \varepsilon.
\end{aligned}$$

In this expression, the terms

$$\sum_{k=1}^{q} \gamma_k^0 x_k + \sum_{k=1}^{q-1} \sum_{l=k+1}^{q} \gamma_{kl}^0 x_k x_l$$

correspond to the linear and nonlinear blending properties of the ingredients. Each term

$$\left[\sum_{k=1}^{q} \gamma_k^i x_k + \sum_{k=1}^{q-1} \sum_{l=k+1}^{q} \gamma_{kl}^i x_k x_l \right] z_i$$

contains the linear effect of the ith process variable z_i on the ingredients' blending properties, and terms of the form

$$\left[\sum_{k=1}^{q} \gamma_k^{ij} x_k + \sum_{k=1}^{q-1} \sum_{l=k+1}^{q} \gamma_{kl}^{ij} x_k x_l \right] z_i z_j$$

describe the interaction effect of process variables z_i and z_j on the blending properties. Note that the mixture-process variable model in Equation (6.11) does not have an intercept and that it does not have main-effect terms for the process variables. As explained in Section 6.3.2, the least squares estimator would not exist otherwise.

A major problem with the combined model in Equation (6.11) is that its number of parameters increases rapidly with the number of mixture components and the number of process variables. For q ingredients and m process variables, the number of parameters p in the model amounts to $[q + q(q - 1)/2] \times [1 + m + m(m - 1)/2]$. The large number of parameters in the model implies that at least $[q + q(q - 1)/2] \times [1 + m + m(m - 1)/2]$ runs are required in the experiment, and inspired Kowalski et al. (2000) to suggest a more parsimonious second-order model for mixture-process variable experiments. Their suggested model assumes that there are no interaction effects of the process variables on the blending properties (i.e., $\gamma_{kl}^{ij} = 0$ and $\gamma_k^{ij} = \alpha_{ij}$). Also, it is assumed that there is no linear effect of the process variables on the nonlinear blending properties either (i.e., $\gamma_{kl}^i = 0$). The nonlinear blending properties and the interactions between the process variables therefore enter the model only additively. Another difference between the model proposed by Kowalski et al. (2000) and the one in Equation (6.11) is that the former also involves quadratic effects in the process variables:

$$
Y = \sum_{k=1}^{q} \gamma_k^0 x_k + \sum_{k=1}^{q-1} \sum_{l=k+1}^{q} \gamma_{kl}^0 x_k x_l
$$
$$
+ \sum_{i=1}^{m} \left[\sum_{k=1}^{q} \gamma_k^i x_k \right] z_i + \sum_{i=1}^{m-1} \sum_{j=i+1}^{m} \alpha_{ij} z_i z_j + \sum_{i=1}^{m} \alpha_i z_i^2 + \varepsilon. \qquad (6.12)
$$

This model has $q + q(q - 1)/2 + qm + m(m - 1)/2 + m$ terms, so that this model is more parsimonious in terms of the number of parameters than the crossed model in Equation (6.11) in most practical situations. Along with a detailed strategy for building a suitable model, Prescott (2004) presents yet another mixture process-variable model, involving $q(q + 1)(q + 2)/6 + mq(m + q + 2)/2$ terms. That model is clearly not as parsimonious in terms of the number of parameters as the one in Equation (6.12).

6.3.3.3 Mixture-amount experiments

In some mixture experiments, the response does not just depend on the relative proportions x_i of the ingredients, but also on the total amount of each of the ingredients. For instance, the yield of an agricultural experiment depends on the composition of the fertilizers and herbicides used, but also on the total amounts of fertilizer and herbicides used. We call such experiments mixture-amount experiments.

To model the response from mixture-amount experiments, we use the same kind of model as for a mixture experiment with one process variable. The total amount of the mixture acts as a process variable z.

6.3.4 Optimal designs for mixture experiments

It is not difficult to construct optimal designs for ordinary mixture experiments when there are no constraints on the ingredient proportions other than the mixture constraint in Equation (6.1) and when there are no process variables. In that case, the experimental region is a regular simplex. For three ingredients, this is an equilateral

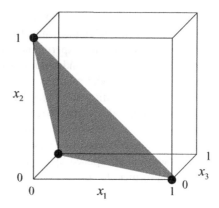

Figure 6.4 Graphical illustration of the experimental region in a mixture experiment.

triangle, while, for four ingredients, it is a tetrahedron. Figure 6.4 illustrates why the experimental region for three mixture ingredient proportions x_1, x_2, and x_3 is a triangle. Each of the three proportions can take values between 0 and 1. If they were factors that could be set independently, then the experimental region would be the unit cube shown in the figure. However, the sum of the three proportions has to be one, so that only the combinations of proportions on the gray triangle connecting the vertices $(1, 0, 0)$, $(0, 1, 0)$, and $(0, 0, 1)$ are feasible.

The optimal design for a first-order Scheffé model for q ingredients consists of one or more replicates of the q vertices of a regular simplex. A graphical representation of this design, which is called a $\{3, 1\}$ lattice design, for an experiment with three ingredients is shown in Figure 6.5. Every design point, i.e., every combination of ingredient proportions in the design, implies an experimental run with 100% of one ingredient and 0% of all other ingredients. When the available sample size is a multiple of three, then the optimal design has equal numbers of runs at each of the three design points. When the available sample size is not a multiple of three, the runs need to be distributed as evenly as possible across the three design points. Similarly, for q ingredients, the optimal design distributes the runs as evenly as possible across the q vertices of the simplex.

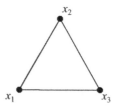

Figure 6.5 Optimal design for a first-order Scheffé model in three ingredients.

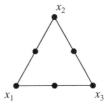

Figure 6.6   Optimal design for a second-order Scheffé model in three ingredients.

The optimal designs for a second-order Scheffé model for q mixture components consist of one or more replicates of the q vertices and the $q(q - 1)/2$ edge centroids of a regular simplex. The edge centroids correspond to binary blends, which involve 50% of each of two ingredients and 0% of all other ingredients. A graphical representation of this design, which is called a {3, 2} lattice design, for an experiment with three ingredients is displayed in Figure 6.6. When the available sample size is a multiple of $q + q(q - 1)/2$, then the optimal design has equal numbers of runs at each of the design points. When this is not the case, the runs are distributed as evenly as possible across the $q + q(q - 1)/2$ design points.

The optimal designs for a special-cubic Scheffé model for q mixture components consist of one or more replicates of the q vertices, the $q(q - 1)/2$ edge centroids, and the $q(q - 1)(q - 2)/3$ face centroids of the simplex. The face centroids correspond to ternary blends, which involve 33.3% of each of three ingredients and 0% of all other ingredients. A graphical representation of this design for an experiment with three ingredients is displayed in Figure 6.7. When the available sample size is a multiple of the total number of points in this design, then the optimal design has equal numbers of runs at each of the points. When this is not the case, the experimental runs are distributed as evenly as possible.

One odd thing about the optimal designs for mixture experiments is that, seemingly, they require performing runs with 100% of one ingredient and 0% of all other ingredients. Often, however, the ingredients $x_1, x_2, \ldots, x_q$ are pseudocomponents, which is jargon for saying that they are mixtures themselves. The reason for this is that, usually, there are lower bounds on the ingredient proportions. Sometimes, there are also upper bounds on the ingredient proportions. Suppose, for instance, that there

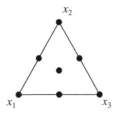

Figure 6.7   Optimal design for a special-cubic Scheffé model in three ingredients.

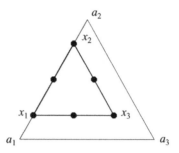

Figure 6.8   Optimal design for a second-order Scheffé model in three ingredients in the presence of lower bounds on the ingredient proportions.

are three ingredients in a mixture experiment, that we name their proportions a_1, a_2, and a_3, and that the following constraints apply:

$$0.2 \leq a_1 \leq 0.8,$$
$$0.2 \leq a_2 \leq 0.8,$$
$$0.0 \leq a_3 \leq 0.6.$$

The resulting experimental region is visualized in Figure 6.8, along with the optimal design points for estimating a second-order Scheffé model. It is clear that the constrained experimental region has the same shape as the unconstrained region. Therefore, the same kind of design is optimal for the constrained experimental region as for the unconstrained one. We can now define three new factors called pseudocomponents,

$$x_i = \frac{a_i - L_i}{1 - \sum_{i=1}^{q} L_i},$$

where L_i is the lower bound for proportion a_i. The key features of the pseudocomponents are that they take values between zero and one and that they satisfy the mixture constraint in Equation (6.1). Now, it should be clear that an experimental run with $x_1 = 100\%$ and $x_2 = x_3 = 0\%$ does not mean that only one of the three ingredients is used in the mixture. Instead, it means that $a_1 = 80\%$, $a_2 = 20\%$, and $a_3 = 0\%$.

6.3.5   Design construction algorithms for mixture experiments

There are two difficulties in developing a coordinate-exchange algorithm for constructing designs for mixture experiments. First, a starting design of feasible mixture points must be generated to begin the coordinate-exchange algorithm. A mixture point is feasible if it lies on or within all the constraint boundaries and satisfies the mixture constraint in Equation (6.1). Second, a coordinate value (i.e., a mixture ingredient proportion) cannot be changed independently of the other coordinates or ingredient proportions in a mixture. If a coordinate or proportion changes, then at least one other coordinate or proportion must also change to maintain the sum of the mixture ingredient proportions at one.

One way to find a starting design is by generating feasible design points one at a time until a starting design with the desired number of runs is obtained. For each design point, the process starts by generating a random point within the mixture simplex. If this point obeys all the constraints on the mixture ingredient proportions, then it is added to the starting design. Otherwise, it is projected onto the nearest constraint boundary. If all constraints are satisfied, the resulting point is added to the starting design. If not, the process continues by finding the next nearest constraint boundary and projecting the point onto that boundary. If the constraint set is consistent (i.e., if there are no unnecessary or incompatible constraints; see Cornell (2002) and Piepel (1983)), this process yields a feasible design point. The process is repeated until a starting design with the desired number of feasible points is obtained.

After a starting design has been constructed, a sequence of mixture ingredient coordinate exchanges is performed to generate an optimal design. Beginning with the first point of the starting design, the first mixture ingredient proportion can be varied. At the same time, the proportions of the other ingredients are changed so that their pairwise ratios remain fixed. Suppose, for instance, that the first design point in a three-component mixture experiment has coordinates

$$(x_1, x_2, x_3) = (0.30, 0.30, 0.40),$$

and that we consider increasing the first ingredient proportion, x_1, by 0.20, to 0.50. Because the ratio of the two other ingredient proportions, $x_2/x_3 = 0.30/0.40$, is 0.75, we reduce the proportions x_2 and x_3 of the other ingredients to 0.214 and 0.286, respectively. This leads to the modified design point

$$(x_1, x_2, x_3) = (0.50, 0.214, 0.286),$$

for which the ratio $x_2/x_3 = 0.214/0.286$ is still equal to 0.75, and the mixture constraint is still satisfied. Following this procedure means that we move the design points along the Cox-effect direction (Cornell 2002) to overcome the difficulty that the proportion of one mixture component cannot be changed independently. The Cox-effect direction for the point $(x_1, x_2, x_3) = (0.30, 0.30, 0.40)$ is shown using the dashed line in Figure 6.9. The general formula for recomputing the proportions $x_1, \ldots, x_{i-1}, x_{i+1}, \ldots, x_q$ after one proportion x_i is changed to $x_i + \delta$ is

$$x_j - \frac{\delta x_j}{1 - x_i}.$$

For the first coordinate of the first design point, the coordinate-exchange algorithm consider several points along the Cox-effect direction. The algorithm first checks the lower and upper bound constraints on the ingredient proportions, as well as any other linear inequality constraints on the proportions. That leads to a lower limit for the coordinate, as well as an upper limit. The algorithm then evaluates the optimality

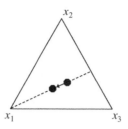

Figure 6.9 Moving along the Cox-effect direction for mixture ingredient 1 from
$(x_1, x_2, x_3) = (0.30, 0.30, 0.40)$ *to* $(x_1, x_2, x_3) = (0.50, 0.214, 0.286)$.

criterion for a certain number of different designs, say k, obtained by replacing the
original coordinate with k equidistant points on the Cox-effect direction line between
the lower and upper limit. The current design point is replaced by the point that
resulted in the best optimality criterion value, if that value is better than that for the
current design.

Next, the algorithm goes through the same procedure for the second coordi-
nate/proportion of the first design point. The process continues until all coordi-
nates/proportions of all the design points have been considered for exchange. If any
exchanges were made, then the entire process is repeated, starting with the first
coordinate/proportion of the first design point. The mixture coordinate-exchange algorithm ends when there have been no exchanges in a complete pass through all the
coordinates/proportions in the entire design.

6.4   Background reading

Key references when it comes to the design of mixture experiments and the analysis of data from mixture experiments are the books by Cornell (2002) and Smith
(2005). Besides a wide variety of examples, these books provide the reader with
a discussion of alternative models for mixture experiments data. They also include
a detailed discussion of analysis methods for mixture experiments, useful hypothesis tests and model reduction techniques. Mixture-amount experiments as well as
mixture experiments in the presence of process variables are discussed too.

For a detailed description of the mixture coordinate-exchange algorithm and an
application of it to a challenging 21-ingredient mixture experiment, we refer to Piepel
et al. (2005).

Often, it is impossible to run complete mixture experiments under homogeneous
circumstances. In these cases, it is useful to include block effects in the mixture model
and to generate an optimal block design for the mixture experiment. Essentially, this
is no different from the approach that we take in Chapters 7 and 8, where we show
how to use blocking for a screening experiment and a response surface experiment. A
recent reference on the subject of blocking mixture experiments is Goos and Donev
(2006b), who show how to compute optimal designs for blocked mixture experiments
and provide a review of the literature on this subject.

Useful references on the design and analysis of mixture experiments involving process variables are Cornell (1988), Kowalski et al. (2002), and Goos and Donev (2007), who stress that these types of experiments are often not completely randomized. Instead, mixture-process variable experiments are often (inadvertently) run using a split-plot design. Obviously, this has implications for the setup of the mixture-process variable experiment, as well as for the analysis of the resulting data. For the optimal design of split-plot experiments in general, we refer to Chapter 10. For the optimal design of split-plot mixture-process variable experiments in particular, we refer to Goos and Donev (2007).

In certain mixture experiments, the ingredients themselves are mixtures. These experiments have been referred to in the literature as categorized-components mixture experiment, multifactor mixture experiments or mixture-of-mixtures experiments. The ingredients that compose the final product are called the major components, whereas the subingredients are named the minor components. Examples of major components are cations and anions in experiments aiming to optimal fertilizer formulations, drug and excipient in drug formulation experiments, and chocolate, jam, and dough in the production of cookies. A review paper on this kind of mixture experiment is Piepel (1999).

6.5   Summary

An important advantage of experimental studies over observational studies is that the randomization built into designed experiments, along with the deliberate changes in the factor-level combinations from run to run, results in observed effects that are actually caused by the factors we are changing. Effects found in observational studies, by contrast, may be the result of an unobserved or lurking factor, rather than the factor that appears to be strongly related to a response of interest.

Mixture experiments have the property that at least some of the experimental factors are ingredients in a mixture. The proportions of these ingredients must add to one, which means that the levels of these factors are not independent. That is, if you know the proportions of all but one of the ingredients, you can calculate the proportion of the final ingredient. This property radically affects both the design and the analysis of such experiments.

In practice, it is rare that the only important factors are the ingredients in the mixture. Usually, there is interest in other factors as well. A general methodology for supporting mixture experiments, therefore, must allow for mixture factors as well as additional, continuous or categorical, factors in the same experiment. We name experiments involving mixture ingredients and additional factors mixture-process variable experiments.

7

A response surface design in blocks

7.1 Key concepts

1. When it is possible to group the runs of an experiment in such a way that runs in each group are more like each other than they are like runs in a different group, then it is statistically more efficient to make this grouping explicit. This procedure is called blocking the experiment.

2. It is important to identify the characteristic that makes the runs in a group alike. For example, runs done in one day of processing are generally more alike than those performed on separate days. In this case, the blocking factor is day, and runs performed on a given day constitute one block.

3. An important characteristic of a blocking factor is that such factors are not under the direct control of the investigator. For example, you can choose to include different lots of material from a supplier in your experiment, but you cannot reproduce any lot in the future.

4. It is ideal to make the blocking factor orthogonal to the other factors. Real-world constraints on model form, block sizes and the total number of runs can make this impossible. In such cases, optimal blocking is a useful alternative.

5. For making inferences about future blocks, it can be useful to model the characteristics that make one block different from another as random variation.

6. Characterizing the impact of the blocking factor as random variation requires the use of a mixed model (having both fixed and random components). Mixed models are better fit using generalized least squares (GLS) rather than ordinary least squares (OLS).

Optimal Design of Experiments: A Case Study Approach, First Edition. Peter Goos and Bradley Jones.
© 2011 John Wiley & Sons, Ltd. Published 2011 by John Wiley & Sons, Ltd.

When necessity dictates that you perform an experiment in groups of runs, we say that you have blocked the experiment. The blocking factor is the condition that defines the grouping of the runs. In our case study, the blocking factor is day, because the process mean shifts randomly from day to day and the investigator wants to explicitly account for these shifts in the statistical model.

There is more than one way to construct a blocked experiment. We discuss this, make comparisons, and recommend a general approach for the design and analysis of blocked experiments.

7.2   Case: the pastry dough experiment

7.2.1   Problem and design

At the reception on the evening of the opening day of the annual conference of the European Network for Business and Industrial Statistics, Peter and Brad are discussing the day's presentations on a rooftop terrace with a view on the Acropolis in Athens, Greece. One of the Greek conference attendees walks up to them.

[Maroussa] Hi. I am Maroussa Markianidou. I am a researcher in the Department of Food Science and Technology at the University of Reading in England.

[Peter] Nice to meet you, Maroussa. I am Peter, and this is Brad.

The two men and the lady shake hands and exchange business cards, after which Maroussa Markianidou continues the conversation.

[Maroussa, addressing Peter] I saw your presentation on the optimal design of experiments in the presence of covariates this afternoon. I liked the flexibility of the approach, and wondered whether it would be suitable for an experiment I am involved in.

[Peter] What kind of experiment are you doing?

[Maroussa] We're still in the planning phase. We decided to run an experiment at our lab in Reading about the baking of pastry dough to figure out the optimal settings for a pilot plant production process. It looks as if we are going to study three factors in the experiment, and we will measure two continuous responses that are related to the way in which the pastry rises.

[Peter] What kind of factors will you use?

[Maroussa] There are three continuous factors: the initial moisture content of the dough, the screw speed of the mixer, and the feed flow rate of water being added to the mix. We plan to study these factors at three levels because we expect substantial curvature. Therefore, we are considering a modified central composite design.

[Brad] Why a *modified* central composite design? What is wrong with the standard central composite design?

[Maroussa] Well, the experiment will require several days of work, and our experience with baking pastry is that there is quite a bit of day-to-day variation. Of course, we do not want the day-to-day variation to have an impact on our conclusions. So, we need to find a way to run the central composite design so that the effects of our three factors are not confounded with the day effects.

Table 7.1 Orthogonally blocked three-factor central composite design with two blocks of six runs and one block of eight runs.

Block	x_1	x_2	x_3
1	−1	−1	−1
	−1	1	1
	1	−1	1
	1	1	−1
	0	0	0
	0	0	0
2	−1	−1	1
	−1	1	−1
	1	−1	−1
	1	1	1
	0	0	0
	0	0	0
3	−1.633	0	0
	1.633	0	0
	0	−1.633	0
	0	1.633	0
	0	0	−1.633
	0	0	1.633
	0	0	0
	0	0	0

[Brad] I see. Please continue.

[Maroussa] I know that there are textbooks on experimental design and response surface methodology that deal with the issue of running central composite designs in groups, or blocks, so that the factor effects are not confounded with the block effects. The problem with the solutions in the textbooks is that they don't fit our problem. Look at this design.

Maroussa digs a few stapled sheets of paper out of her conference bag, and shows the design given in Table 7.1. In the table, the three factors are labeled x_1, x_2, and x_3, and the factor levels appear in coded form.

[Maroussa] According to the textbooks, this is *the* way to run a three-factor central composite design in blocks. The design has three blocks. The first two blocks contain a half fraction of the factorial portion of the central composite design plus two center runs. The third block contains the axial points plus two center runs.

[Peter] Why is it that this design doesn't fit your problem?

[Maroussa] There are several reasons. First, we can only do four runs a day. So, for our problem, we need blocks involving four runs only. Second, we have scheduled the laboratory for seven days in the next couple of weeks, so that we need a design with seven blocks. The third reason why I don't like the design is the fact that the third block involves the factor levels ± 1.633. Apparently, the value 1.633 is required to ensure that the design is what the textbooks call orthogonally blocked. You know better than me that this is jargon for saying that the factor effects are not confounded with the block effects. However, we would much rather use just three levels for every factor. The problem is that using ± 1 for the nonzero factor levels in the axial points violates the condition for having an orthogonally blocked design.

[Peter] Using ± 1 instead of ± 1.633 for the axial points would not destroy all of the orthogonality. The main effects and the two-factor interaction effects would still be unconfounded with the block effects, but the quadratic effects would not. So, you could then estimate the main effects and the two-factor interaction effects independently of the block effects, but not the quadratic effects.

[Maroussa, nodding] I see. Anyway, I was thinking that we needed a different approach. Since we can explore four factor-level combinations on each of seven days, we need a design with 28 runs. The standard central composite design for a three-factor experiment involves eight factorial points, six axial points, and a couple of center runs. I was thinking that a good start for our design problem would be to duplicate the factorial points and to use six replicates of the center point. This would give us 28 runs: the duplicated factorial portion would give us 16 runs, and we would have six center runs plus six axial runs.

[Brad] And how are you going to arrange these 28 runs in blocks of four?

[Maroussa] This is where I got stuck. I was unable to figure out a way to do it. It is easy to arrange the factorial points in four blocks of size four. That kind of thing is described in the textbooks too.

Maroussa flips a few pages, and shows the design in Table 7.2. The two replicates of the factorial portion of the central composite design are blocked by assigning the points with a $+1$ for the third-order interaction contrast column $x_1 x_2 x_3$ to one

Table 7.2   Duplicated two-level factorial design arranged in four orthogonal blocks of four runs using the contrast column $x_1 x_2 x_3$.

Replicate 1					Replicate 2				
Block	x_1	x_2	x_3	$x_1 x_2 x_3$	Block	x_1	x_2	x_3	$x_1 x_2 x_3$
1	1	1	-1	-1	3	1	1	-1	-1
	-1	-1	-1	-1		-1	-1	-1	-1
	-1	1	1	-1		-1	1	1	-1
	1	-1	1	-1		1	-1	1	-1
2	-1	-1	1	1	4	-1	-1	1	1
	1	-1	-1	1		1	-1	-1	1
	-1	1	-1	1		-1	1	-1	1
	1	1	1	1		1	1	1	1

block and those with a -1 for that column to another block. This ensures that the main effects and the two-factor interaction effects are not confounded with the block effects.

[Maroussa] The problem is that I don't find guidance anywhere as to how to arrange the axial points and the center runs in blocks. The only thing I found is that one sentence in my textbook on response surface methodology that says that it is crucial that all the axial points are in one block.

[Peter, grinning] That advice is pretty hard to follow if you have six axial points and you can only manage four runs in a block.

[Maroussa] Exactly. Do you have suggestions to help us out?

[Brad] My suggestion would be to forget about the central composite design. For the reasons you just mentioned, it is not a fit for your problem. My advice would be to match your design directly to your problem. So, you want 28 factor-level combinations arranged in seven blocks of size four that allow you to estimate the factor effects properly.

[Maroussa] What do you mean by *properly*?

[Brad] Ideally, independently of the day effects. If that is impossible, as precisely as possible.

[Maroussa] And how do you do that?

[Brad, taking his laptop out of his backpack] It is easy enough with the optimal design algorithms that are available in commercial software packages these days. The packages allow you to specify the blocking structure and then produce a statistically efficient design with that structure in the blink of an eye.

While Brad's laptop is starting up, Peter chimes in.

[Peter] Yeah, all you have to do is specify the number of factors, their nature, the model you would like to fit, the number of blocks, and the number of runs that can be done within a block. Brad will show you.

Brad, who is now ready for the demonstration, takes over.

[Brad] Can you remind me what the three factors in your experiment were? They were continuous, weren't they?

[Maroussa, flipping through the pages in front of her and finding one showing Table 7.3] Sure, here they are.

[Brad] Wow. You came prepared.

Brad inputs the data in the user interface of the software he is using.

[Peter] Is there a problem with randomizing the order of the runs within each day and resetting the factor levels for every run? Or would that be inconvenient or costly?

Table 7.3 Factors and factor levels used in the pastry dough mixing experiment.

Flow rate (kg/h)	Moisture content (%)	Screw speed (rpm)
30.0	18	300
37.5	21	350
45.0	24	400

[Maroussa] As far as I can see, any order of the runs is equally convenient or inconvenient.

[Brad, pointing to the screen of his laptop] This is where I specify that we want to fit a response surface model in the three factors. That means we would like to estimate the main effects of the factors, their two-factor interaction effects, and their quadratic effects. Next, I specify that we need 28 runs in seven blocks of four runs, and then all I have to do is put the computer to work.

Brad clicks on a button in his software, and, a few second later, the design in Table 7.4 pops up.

[Brad] I have a design for you.

[Maroussa] Gosh, I haven't even had time to sip my drink. How can I see now that this is the design we need back in Reading?

[Brad] The key feature of the design is that it matches your problem. We have seven blocks of four runs. Also, you can see that the design has three levels for each of the factors. For instance, the moisture content in the first run on the first day is 30%, which is the low level of that factor. In the second run, the moisture content is 45%, which is the high level. In the third run, the moisture content is right in the middle, at 37.5%. That we have three levels for every factor can be better seen by switching to coded factor levels.

Brad enters the formulas to transform the factor levels in the design in Table 7.4 into coded form:

$$x_1 = (\text{flow rate} - 37.5)/7.5,$$
$$x_2 = (\text{moisture content} - 21)/3,$$

and

$$x_3 = (\text{screw speed} - 350)/50.$$

Next to the design in uncoded form, the coded design in Table 7.5 appears, and Brad continues.

[Brad] Here you can easily see that we have three levels for every factor: a low, a high, and an intermediate level. This is what you wanted, isn't it?

[Maroussa, pointing to Table 7.1] Right. Now, how can I see that the design you generated is orthogonally blocked, just like the central composite design here.

[Peter] It isn't.

[Maroussa, looking skeptical] I beg your pardon?

[Peter] In many practical situations, it is impossible to create an orthogonally blocked response surface design. This is certainly so if the number of runs within each block is small, and when you have to fit a response surface model. Constructing an orthogonally blocked design is easier to do for two-level factorial and fractional factorial designs, when the number of runs within the blocks is a power of two and your model is not too complicated.

Table 7.4 D-optimal design for the pastry dough experiment in engineering units.

Day	Flow rate (kg/h)	Moisture content (%)	Screw speed (rpm)
1	30.0	18	300
	45.0	18	400
	37.5	21	350
	45.0	24	300
2	37.5	24	300
	45.0	24	400
	45.0	18	350
	30.0	21	400
3	37.5	18	300
	45.0	21	400
	45.0	24	300
	30.0	24	350
4	30.0	18	300
	45.0	18	400
	30.0	24	400
	37.5	21	350
5	45.0	24	350
	30.0	21	300
	37.5	18	400
	30.0	24	400
6	30.0	18	400
	45.0	18	300
	45.0	24	400
	30.0	24	300
7	30.0	18	350
	37.5	24	400
	30.0	24	300
	45.0	21	300

[Maroussa, nodding] But this means that the day-to-day variation in our process will influence the estimates of the factor effects. Doesn't that invalidate any conclusion I draw from the data?

[Peter] No, it's not that bad. The design Brad generated is a D-optimal block design. There is a link between the D-optimality of block designs and orthogonality. You

Table 7.5 D-optimal design for the pastry dough experiment in coded units.

Day	Flow rate x_1	Moisture content x_2	Screw speed x_3
1	−1	−1	−1
	1	−1	1
	0	0	0
	1	1	−1
2	0	1	−1
	1	1	1
	1	−1	0
	−1	0	1
3	0	−1	−1
	1	0	1
	1	1	−1
	−1	1	0
4	−1	−1	−1
	1	−1	1
	−1	1	1
	0	0	0
5	1	1	0
	−1	0	−1
	0	−1	1
	−1	1	1
6	−1	−1	1
	1	−1	−1
	1	1	1
	−1	1	−1
7	−1	−1	0
	0	1	1
	−1	1	−1
	1	0	−1

could say that, in a way, D-optimal block designs are as close to being orthogonally blocked as the number of blocks and the number of runs in each block allow.

[Maroussa, all ears] As close as possible to being orthogonally blocked might still not be close enough.

[Peter] My experience is that D-optimal block designs are close enough to being orthogonally blocked that you do not need to worry about the confounding between the factor effects and the block effects. There are several things you can do to quantify to what extent a block design is orthogonally blocked. One approach is to compute an efficiency factor for each factor effect in your model. The minimum and maximum values for the efficiency factors are 0% and 100%. An efficiency factor of 100% is, of course, desirable. Designs that are perfectly orthogonally blocked have efficiency factors of 100%. An efficiency factor of 0% results when a factor effect is completely confounded with the block effects. In that case, you can't estimate the factor effect. However, as soon as an efficiency factor is larger than 0%, you can estimate the corresponding factor effect. The larger the efficiency factor, the more precisely you can estimate it.

[Maroussa, pointing to Brad's laptop] And what would the efficiency factors for this design be?

[Peter] I trust that they are larger than 90%, which would indicate that the design is very nearly orthogonally blocked. So, I am pretty sure there is no reason to worry about confounding between factor effects and block effects when using this design. If you would like to be fully sure, I can compute the values of the efficiency factors later this evening and let you know what they are tomorrow.

[Maroussa, looking persuaded and taking out a memory stick] That would be great. Could you put that design on my memory stick?

[Brad, copying the design into a spreadsheet and saving it onto Maroussa's memory stick] Sure. Here you go.

That evening, in his hotel room, Peter confirms that the efficiency factors for the D-optimal design with seven blocks of four runs range from 93.4% to 98.3%, with an average of 96.1%. The efficiency factors for the three main effects, the three interaction effects, and the three quadratic effects are shown in Table 7.6, along with the corresponding variance inflation factors (VIFs). The VIFs show to what extent the variances of the parameter estimates get bigger due to the blocking of the experiment. The efficiency factors and the VIFs are the reciprocals of each other. Both are based on the variances of the factor-effect estimates in the presence and in the absence of blocking.

Table 7.6   Efficiency factors and variance inflation factors (VIFs) for the D-optimal design for the pastry dough mixing experiment, as well as the variances of the factor-effect estimates in the presence and in the absence of blocking.

Effect	Blocking	No blocking	Efficiency (%)	VIF
Flow rate	0.0505	0.0471	93.4	1.0705
Moisture content	0.0495	0.0465	94.1	1.0632
Screw speed	0.0505	0.0471	93.4	1.0705
Flow rate × Moisture content	0.0595	0.0579	97.3	1.0274
Flow rate × Screw speed	0.0580	0.0570	98.3	1.0178
Moisture content × Screw speed	0.0595	0.0579	97.3	1.0274
Flow rate × Flow rate	0.2282	0.2215	97.0	1.0305
Moisture content × Moisture content	0.2282	0.2215	97.0	1.0305
Screw speed × Screw speed	0.2282	0.2215	97.0	1.0305

7.2.2 Data analysis

A couple of weeks later, Brad enters Intrepid Stats's European office, where Peter is working on his desktop computer.

[Peter, handing Brad a sheet of paper containing Table 7.7]. You remember Maroussa Markianidou, the Greek lady from Reading in the UK? She ran the design

Table 7.7 Design and responses for the pastry dough experiment. The responses are the longitudinal (y_1) and cross-sectional (y_2) expansion indices.

Day	Flow rate (kg/h)	Moisture content (%)	Screw speed (rpm)	x_1	x_2	x_3	y_1	y_2
1	30.0	18	300	−1	−1	−1	15.6	5.5
	45.0	18	400	1	−1	1	15.7	6.4
	37.5	21	350	0	0	0	11.2	4.8
	45.0	24	300	1	1	−1	13.4	3.5
2	37.5	24	300	0	1	−1	13.6	4.7
	45.0	24	400	1	1	1	15.9	5.5
	45.0	18	350	1	−1	0	16.2	6.3
	30.0	21	400	−1	0	1	13.2	5.0
3	37.5	18	300	0	−1	−1	12.5	4.5
	45.0	21	400	1	0	1	14.0	4.1
	45.0	24	300	1	1	−1	11.9	4.3
	30.0	24	350	−1	1	0	9.2	4.3
4	30.0	18	300	−1	−1	−1	13.7	5.5
	45.0	18	400	1	−1	1	17.2	6.0
	30.0	24	400	−1	1	1	11.7	4.5
	37.5	21	350	0	0	0	11.9	4.9
5	45.0	24	350	1	1	0	10.9	4.1
	30.0	21	300	−1	0	−1	9.5	3.5
	37.5	18	400	0	−1	1	14.7	5.7
	30.0	24	400	−1	1	1	11.9	4.6
6	30.0	18	400	−1	−1	1	15.2	6.1
	45.0	18	300	1	−1	−1	16.0	5.1
	45.0	24	400	1	1	1	15.1	4.8
	30.0	24	300	−1	1	−1	9.1	4.4
7	30.0	18	350	−1	−1	0	15.6	6.6
	37.5	24	400	0	1	1	14.0	4.9
	30.0	24	300	−1	1	−1	11.6	4.8
	45.0	21	300	1	0	−1	14.2	4.5

that you generated for her in Athens. She e-mailed us this morning to ask whether we can analyze the data for her.

[Brad] Great. Please tell me she is going to pay us.

[Peter] Nope. I already replied to her that I was prepared to analyze her data if she agreed to put our names on her paper and if she shares the details about the application with us so that we can use her design problem as a case study in our book on modern experimental design. She skyped me soon after that to conclude the deal and to give me some of the background of the experiment.

[Brad] Fine. We need a chapter on blocking response surface designs.

Several days later, Peter sends Maroussa an e-mail report on the analysis of the data from her pastry dough experiment. Here is his e-mail:

Dear Maroussa: I analyzed the data from the pastry dough experiment you conducted to study the impact of the moisture content of the dough, the screw speed of the mixer and the feed flow rate on the two responses you measured, the longitudinal expansion index (y_1, expressed in centimeters per gram) and the cross-sectional expansion index (y_2, also expressed in centimeters per gram). I also determined factor settings that will give you the desired target values of 12 cm/g and 4 cm/g for the longitudinal and the cross-sectional expansion index, respectively.

The design that you ran was a D-optimal design with seven blocks of four runs, because you could do only four runs on each of seven days. Since you expected curvature, we generated the design assuming that the following second-order response surface model was adequate:

$$Y_{ij} = \beta_0 + \sum_{k=1}^{3} \beta_k x_{kij} + \sum_{k=1}^{2} \sum_{l=k+1}^{3} \beta_{kl} x_{kij} x_{lij} + \sum_{i=1}^{3} \beta_{kk} x_{kij}^2 + \gamma_i + \varepsilon_{ij}, \qquad (7.1)$$

where x_{kij} is the coded level of the kth factor at the jth run on day i (i goes from 1 to 7, and j goes from 1 to 4), β_0 denotes the intercept, β_k is the main effect of the kth factor, β_{kl} is the effect of the interaction involving the kth and the lth factor, and β_{kk} is the quadratic effect of the kth factor. The term γ_i represents the ith day (or block) effect, and ε_{ij} is the residual error associated with the jth run on day i. The γ_i in the model captures the day-to-day variation in the responses.

Fitting the second-order response surface model in Equation (7.1) to the two responses in Table 7.7 using generalized least squares (GLS) and restricted maximum likelihood (REML) estimation gave me the parameter estimates, standard errors and p values shown in Table 7.8. The left panel of the table shows the results for the longitudinal expansion index, while the right panel contains the results for the cross-sectional expansion index. For each of these responses, several p values are substantially larger than 5%, indicating that you can simplify the models. Using stepwise backward elimination, I got the models in Table 7.9.

Table 7.8 Factor-effect estimates, standard errors and p values obtained for the two responses, longitudinal expansion index and cross-sectional expansion index, in the pastry dough experiment when fitting the full second-order response surface model in Equation (7.1).

Effect	Longitudinal expansion index			Cross-sectional expansion index		
	Estimate	Standard error	p Value	Estimate	Standard error	p Value
β_0	11.38	0.66	<.0001	4.66	0.25	<.0001
β_1	1.01	0.21	0.0003	−0.05	0.08	0.5159
β_2	−1.47	0.20	<.0001	−0.61	0.08	<.0001
β_3	0.73	0.21	0.0040	0.35	0.08	0.0008
β_{12}	0.42	0.22	0.0822	0.03	0.09	0.7329
β_{13}	−0.07	0.22	0.7565	0.09	0.09	0.3391
β_{23}	0.21	0.22	0.3639	−0.11	0.09	0.2474
β_{11}	0.33	0.44	0.4692	0.04	0.17	0.8028
β_{22}	1.30	0.44	0.0114	0.72	0.17	0.0011
β_{33}	1.05	0.44	0.0331	−0.33	0.17	0.0738

The indices 1, 2 and 3 in the first column of the table refer to the flow rate, the moisture content and the screw speed, respectively.

For the longitudinal expansion index, the main effects of the three experimental factors, moisture content, screw speed and flow rate, have statistically significant effects. Also, the interaction effect involving moisture content and flow rate is significant, as well as the quadratic effects of the moisture content and the screw speed. For the cross-sectional

Table 7.9 Factor-effect estimates, standard errors, and p values obtained for the two responses, longitudinal expansion index and cross-sectional expansion index, in the pastry dough experiment when fitting simplified models.

Effect	Longitudinal expansion index			Cross-sectional expansion index		
	Estimate	Standard error	p Value	Estimate	Standard error	p Value
β_0	11.59	0.58	<.0001	4.46	0.19	<.0001
β_1	0.99	0.20	0.0001			
β_2	−1.46	0.19	<.0001	−0.62	0.08	<.0001
β_3	0.75	0.20	0.0015	0.33	0.08	0.0005
β_{12}	0.46	0.21	0.0463			
β_{22}	1.33	0.41	0.0058	0.69	0.17	0.0007
β_{33}	1.08	0.41	0.0197			

The indices 1, 2 and 3 in the first column of the table refer to the flow rate, the moisture content and the screw speed, respectively.

expansion index, only the main effects of the factors moisture content and screw speed are significant, as well as the quadratic effect of the moisture content.

I then used the simplified models in Table 7.9,

$$y_1 = 11.59 + 0.99x_1 - 1.46x_2 + 0.75x_3 + 0.46x_1x_2 + 1.33x_2^2 + 1.08x_3^2,$$

for the longitudinal expansion index, and

$$y_2 = 4.46 - 0.62x_2 + 0.33x_3 + 0.69x_2^2,$$

for the cross-sectional expansion index, to find factor settings that give you the desired target values, 12 cm/g and 4 cm/g, for these expansion indices. If you prefer to think in engineering units, then you might prefer this way of writing the two models:

$$\begin{aligned}
\frac{\text{Longitudinal}}{\text{expansion index}} &= 11.59 + 0.99 \, \frac{\text{Flow rate} - 37.5}{7.5} \\
&\quad - 1.46 \, \frac{\text{Moisture content} - 21}{3} + 0.75 \, \frac{\text{Screw speed} - 350}{50} \\
&\quad + 0.46 \, \frac{\text{Flow rate} - 37.5}{7.5} \, \frac{\text{Moisture content} - 21}{3} \\
&\quad + 1.33 \left(\frac{\text{Moisture content} - 21}{3} \right)^2 \\
&\quad + 1.08 \left(\frac{\text{Screw speed} - 350}{50} \right)^2,
\end{aligned}$$

and

$$\begin{aligned}
\frac{\text{Cross-sectional}}{\text{expansion index}} &= 4.46 - 0.62 \, \frac{\text{Moisture content} - 21}{3} \\
&\quad + 0.33 \, \frac{\text{Screw speed} - 350}{50} \\
&\quad + 0.69 \left(\frac{\text{Moisture content} - 21}{3} \right)^2.
\end{aligned}$$

Setting the flow rate to 40.6 kg/h, the moisture content to 22.3% and the screw speed of the mixer to 300 rpm gets you right on target. You can see this from Figure 7.1, which plots the dependence of your two responses on the three experimental factors. In the figure, the vertical dashed lines give the optimal factor levels for your process. The horizontal dashed lines indicate the corresponding predicted values for the expansion indices and for the overall desirability.

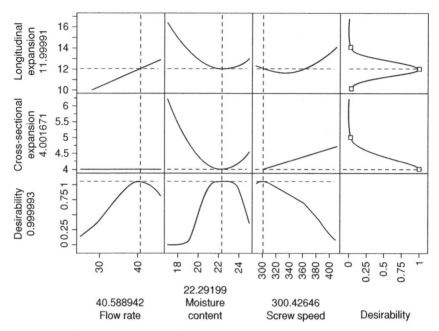

Figure 7.1   Plot of the factor-effect estimates, the desirability functions, and the optimal factor settings for the pastry dough experiment.

The upper panels of the figure show that the longitudinal expansion index linearly depends on the flow rate, and that it depends on the moisture content and the screw speed in a quadratic fashion. The middle panels show that the flow rate has no impact on the cross-sectional expansion index, that the screw speed has a linear effect, and that there is a quadratic relationship between the moisture content and the cross-sectional expansion index. The lower panels of the figure show how the overall desirability of the solution depends on the levels of the experimental factors. The overall desirability is a summary measure that tells us how close to target we are with our two responses. More specifically, it is the geometric mean of the individual desirabilities for your expansion indices, which are shown in the panels at the far right of Figure 7.1. You can see that the individual desirability function for the longitudinal expansion index is maximal when that index is 12, which is its target value. By contrast, the individual desirability function for the cross-sectional expansion index is maximal when that index is at its target value of 4. You can also see that the individual desirabilities drop as soon as we are no longer producing on target.

The curves at the bottom of the figure show that a flow rate of 40.6 kg/h, a moisture content of 22.3% and a screw speed of 300 rpm yield an overall desirability of 100%, which means that both of your responses are on target. The middle one of these curves shows that the overall desirability is not very sensitive to the moisture content, as long as it is somewhere between 21.5% and 23.5%.

Of course, I would advise you to carry out a couple of confirmatory runs of your process to verify that the factor settings I suggested do work.

That's about it I think. Perhaps, it is still worth saying that the day-to-day variation in the two responses was about as large as the variation in the residuals. So, you were definitely right to worry about the day-to-day variation.

Let me know if you have further questions.

Half an hour later, Peter finds Maroussa's reply in his mailbox:

Dear Peter:

Thanks for the swift reply. I do have a question though. You wrote me that you used generalized least squares and restricted maximum likelihood estimation to fit the model. Why generalized and not ordinary least squares? And what is restricted maximum likelihood estimation?

Maroussa

Peter responds:

Dear Maroussa:

When you use ordinary least squares or OLS estimation, you assume that the responses for all the runs are uncorrelated. In your study, this assumption is not valid because the responses obtained on any given day are more alike than responses from different days. So, the four responses measured within a given day are correlated. You have to take into account that correlation in order to get the most precise factor-effect estimates and proper significance tests. This is exactly what generalized least squares or GLS estimation does.

To account for that correlation, you first have to quantify it. This is where restricted maximum likelihood or REML estimation comes in.

To understand how REML estimates the correlation between responses collected on the same day, you have to know a bit more about the assumptions behind the model

$$Y_{ij} = \beta_0 + \sum_{k=1}^{3} \beta_k x_{kij} + \sum_{k=1}^{2} \sum_{l=k+1}^{3} \beta_{kl} x_{kij} x_{lij} + \sum_{i=1}^{3} \beta_{kk} x_{kij}^2 + \gamma_i + \varepsilon_{ij} \qquad (7.2)$$

that was in my earlier e-mail (and that we assumed when we generated the design for your study at the conference in Athens). The γ_i and ε_{ij} in the model are random terms, so the model contains two random terms:

1. The term ε_{ij} represents the residual error and captures the run-to-run variation in the responses, just as in the usual response surface models. By assumption, all ε_{ij} are independently normally distributed with zero mean and constant variance σ_ε^2. That variance is called the residual error variance.

2. The second random term in the model is the day or block effect, γ_i. The function of the term γ_i is to model the day-to-day differences between the responses. By assumption, the γ_i are also independently normally distributed with zero mean, but with a different variance, σ_γ^2. The variance σ_γ^2 is called the block variance.

A final assumption about the two random terms in the model is that the random day effects and the residual errors are all independent from each other.

Now, this is where it gets interesting. The interpretation of a small σ_γ^2 value (relative to the residual error variance σ_ε^2) is that the day-to-day variation of the responses is limited, whereas a large value for σ_γ^2 (again, relative to the residual error variance σ_ε^2) indicates a large day-to-day variation. That is easy to see. For example, if σ_γ^2 is small, all γ_i are pretty similar and so are the responses from different days, all other things being equal. Thus, a σ_γ^2 that is small relative to σ_ε^2 indicates that responses from different days do not vary much more than responses from within a given day. An alternative way to put that is to say that a small σ_γ^2 value indicates that responses from the same day have almost nothing in common, or, that the dependence is negligible. To cut a long story short, the magnitude of σ_γ^2 relative to σ_ε^2 measures the dependence in the responses.

The implication of that is that you have to estimate the variances σ_γ^2 and σ_ε^2 to quantify the dependence. Unfortunately, the estimation of variances is always a bit trickier than estimating means or factor effects, and there are a lot of ways to do it. However, REML estimation is a generally applicable procedure that gives you unbiased estimates of variances. Because it is not available in every commercial software package for industrial statisticians, it is still relatively unknown among experimenters in industry. I think that is a shame, since it is a technique that is more than 30 years old already. Also, it is standard practice to use REML in social science and medicine. I don't know how much more detail you want, but in case you want to understand a bit of the conceptual framework behind it, I have attached a short note I wrote to try to demystify REML a little bit (see Attachment 7.1).

Attachment 7.1   Restricted maximum likelihood (REML) estimation.

REML estimation is a general method for estimating variances of random effects in statistical models. It has become the default approach in many disciplines because it guarantees unbiased estimates. The benefits of REML estimation can best be illustrated by considering two situations where a single variance has to be estimated.

1. A well-known estimator of the variance of a population with mean μ and variance σ_ε^2 is the sample variance

$$S^2 = \frac{1}{n-1} \sum_{i=1}^{n} (X_i - \overline{X})^2, \qquad (7.3)$$

where X_i denotes the ith observation in a random sample of size n. The reason for the division by $n - 1$ rather than by n is to ensure that S^2 is an unbiased estimator of σ_ε^2. The division by $n - 1$ essentially corrects the sample variance for the fact that μ is unknown and needs to be estimated by the sample mean, $\overline{X}$. Now, the sample variance S^2 (involving the division by $n - 1$) can be viewed as the REML estimator of the population variance, σ_ε^2, if the population is assumed to be normally distributed. Dividing by n, which yields the maximum likelihood (ML) estimator, seems like a more intuitive thing to do. However, the ML estimator is biased.

2. The commonly accepted estimator for the residual error variance σ_ε^2 in a regression model with p parameters for analyzing a completely randomized design is the mean squared error (MSE)

$$\text{MSE} = \frac{1}{n - p} \sum_{i=1}^{n} (Y_i - \hat{Y}_i)^2 = \frac{1}{n - p} \sum_{i=1}^{n} R_i^2, \qquad (7.4)$$

where Y_i and $\hat{Y}_i$ are the observed and predicted responses, respectively, and R_i denotes the ith residual. Here, it is the division by $n-p$ rather than by n which guarantees the unbiasedness of the estimator for the residual error variance. It corrects the MSE for the fact that there are p fixed parameters in the regression model (intercept, main effects, ...) which need to be estimated. Just like the sample variance S^2, you can view the MSE as a REML estimator if the responses in the regression model are normally distributed. Dividing the sum of squared residuals by n yields the ML estimator, which is biased.

These examples illustrate REML's key characteristic: it takes into account that there are location parameters that need to be estimated in addition to variances. Roughly speaking, the only information in the data that REML uses for estimating variances is information that is not used for estimating location parameters, such as means and factor effects. The sample variance in Equation (7.3) takes into account that one unit of information is utilized to estimate the population mean. Similarly, the MSE in Equation (7.4) uses the fact that p information units are absorbed by the estimation of the p regression coefficients. The technical term for these units of information is degrees of freedom.

7.3   Peek into the black box

7.3.1   Model

The model for analyzing data from an experiment run in b blocks of k observations requires an additive block effect for every block. Equation (7.1) gives an example of the model if there are three continuous experimental factors and the interest is in estimating main effects, two-factor interaction effects, and quadratic effects. In general, the model becomes

$$Y_{ij} = \mathbf{f}'(\mathbf{x}_{ij})\boldsymbol{\beta} + \gamma_i + \varepsilon_{ij}, \tag{7.5}$$

where $\mathbf{x}_{ij}$ is a vector that contains the levels of all the experimental factors at the jth run in the ith block, $\mathbf{f}'(\mathbf{x}_{ij})$ is its model expansion, and $\boldsymbol{\beta}$ contains the intercept and all the factor effects that are in the model. For a full quadratic model with three continuous experimental factors,

$$\mathbf{x}_{ij} = \begin{bmatrix} x_{1ij} & x_{2ij} & x_{3ij} \end{bmatrix},$$

$$\mathbf{f}'(\mathbf{x}_{ij}) = \begin{bmatrix} 1 & x_{1ij} & x_{2ij} & x_{3ij} & x_{1ij}x_{2ij} & x_{1ij}x_{3ij} & x_{2ij}x_{3ij} & x_{1ij}^2 & x_{2ij}^2 & x_{3ij}^2 \end{bmatrix},$$

and

$$\boldsymbol{\beta}' = \begin{bmatrix} \beta_0 & \beta_1 & \beta_2 & \beta_3 & \beta_{12} & \beta_{13} & \beta_{23} & \beta_{11} & \beta_{22} & \beta_{33} \end{bmatrix}.$$

As in the model in Equation (7.1), γ_i represents the ith block effect and ε_{ij} is the residual error associated with the jth run on day i. If successful, including block effects in a model substantially reduces the unexplained variation in the responses. This results in higher power for the significance tests of the factor effects. So, including block effects in the model often results in higher chances for detecting active effects.

We assume that the block effects and the residual errors are random effects that are all independent and normally distributed with zero mean, that the block effects have variance σ_γ^2, and that the residual errors have variance σ_ε^2. The variance of a response Y_{ij} is

$$\mathrm{var}(Y_{ij}) = \mathrm{var}(\gamma_i + \varepsilon_{ij}) = \mathrm{var}(\gamma_i) + \mathrm{var}(\varepsilon_{ij}) = \sigma_\gamma^2 + \sigma_\varepsilon^2,$$

whereas the covariance between each pair of responses Y_{ij} and $Y_{ij'}$ from the same block i is

$$\mathrm{cov}(Y_{ij}, Y_{ij'}) = \mathrm{cov}(\gamma_i + \varepsilon_{ij}, \gamma_i + \varepsilon_{ij'}) = \mathrm{cov}(\gamma_i, \gamma_i) = \mathrm{var}(\gamma_i) = \sigma_\gamma^2,$$

and the covariance between each pair of responses Y_{ij} and $Y_{i'j'}$ from different blocks i and i' equals

$$\mathrm{cov}(Y_{ij}, Y_{i'j'}) = \mathrm{cov}(\gamma_i + \varepsilon_{ij}, \gamma_{i'} + \varepsilon_{i'j'}) = 0.$$

This is due to the independence of all the block effects and the residual errors, which implies that every pairwise covariance between two different random effects is zero. The correlation between a pair of responses from a given block i is

$$\mathrm{corr}(Y_{ij}, Y_{ij'}) = \frac{\mathrm{cov}(Y_{ij}, Y_{ij'})}{\sqrt{\mathrm{var}(Y_{ij})}\sqrt{\mathrm{var}(Y_{ij'})}} = \frac{\sigma_\gamma^2}{\sigma_\gamma^2 + \sigma_\varepsilon^2}.$$

This within-block correlation coefficient increases with σ_γ^2. So, the larger the differences between the blocks (measured by σ_γ^2), the larger the similarity between every pair of responses from a given block. The dependence between responses from the same block, $Y_{i1}, \ldots, Y_{ik}$, is given by the symmetric $k \times k$ matrix

$$\Lambda = \begin{bmatrix} \sigma_\gamma^2 + \sigma_\varepsilon^2 & \sigma_\gamma^2 & \cdots & \sigma_\gamma^2 \\ \sigma_\gamma^2 & \sigma_\gamma^2 + \sigma_\varepsilon^2 & \cdots & \sigma_\gamma^2 \\ \vdots & \vdots & \ddots & \vdots \\ \sigma_\gamma^2 & \sigma_\gamma^2 & \cdots & \sigma_\gamma^2 + \sigma_\varepsilon^2 \end{bmatrix}, \tag{7.6}$$

which is the variance-covariance matrix of all the responses in block i. The jth diagonal element of that matrix is the variance of Y_{ij}, and the off-diagonal element in row j and column j' is the covariance between Y_{ij} and $Y_{ij'}$.

A key assumption of the model in Equation (7.5) is that the block effects are additive, which implies that each block has a random intercept. Under this assumption, the factor effects do not vary from block to block. Obviously, this is a strong assumption, but it has been confirmed in many empirical studies. It is possible to allow the factor effects to vary from one block to another by including additional random effects in the model, but this falls outside the scope of this book.

Note also that we focus on scenarios where the number of runs is the same for every block. While we believe that this scenario is the most common one, the model can easily be adapted to handle experiments with blocks of different sizes.

The model in Equation (7.5) is called a mixed model, because it involves two sorts of effects. The factor effects are fixed effects, whereas the block effects and the residual errors are random effects. The distinction between the two types of effects is that the latter are assumed to be random variables, whereas the former are not. An interesting property of the mixed model for analyzing data from blocked experiments is that quantifying the differences between the blocks is done by estimating σ_γ^2. So, no matter how many blocks an experiment involves, it is only necessary to estimate one parameter to account for the block effects. An alternative model specification would treat the block effects as fixed effects, rather than random ones. However, the model specification involving fixed block effects requires the estimation of $b - 1$ parameters to quantify the differences between the blocks, and therefore necessitates a larger number of experimental runs. This is a major advantage of the mixed model. The next chapter deals explicitly with this issue.

7.3.2   Generalized least squares estimation

In Section 7.3.1, we showed that responses in the model in Equation (7.5) from the same block are correlated. This violates the independence assumption of the ordinary least squares or OLS estimator. While that estimator is still unbiased, it is better to use generalized least squares or GLS estimates, which account for the correlation and are generally more precise. The GLS estimates are calculated from

$$\hat{\beta} = (\mathbf{X}'\mathbf{V}^{-1}\mathbf{X})^{-1}\mathbf{X}'\mathbf{V}^{-1}\mathbf{Y}. \tag{7.7}$$

It is different from the OLS estimator,

$$\hat{\beta} = (\mathbf{X}'\mathbf{X})^{-1}\mathbf{X}'\mathbf{Y},$$

in that it is a function of the matrix $\mathbf{V}$ that describes the covariance between the responses.

To derive the exact nature of $\mathbf{V}$, we rewrite the model for analyzing data from blocked experiments in Equation (7.5) as

$$Y = \mathbf{X}\beta + \mathbf{Z}\gamma + \varepsilon, \tag{7.8}$$

where

$$Y = \begin{bmatrix} Y_{11} & \cdots & Y_{1k} & Y_{21} & \cdots & Y_{2k} & \cdots & Y_{b1} & \cdots & Y_{bk} \end{bmatrix}',$$

$$\mathbf{X} = \begin{bmatrix} \mathbf{X}_1 \\ \mathbf{X}_2 \\ \vdots \\ \mathbf{X}_b \end{bmatrix} = \begin{bmatrix} \mathbf{f}'(\mathbf{x}_{11}) \\ \vdots \\ \mathbf{f}'(\mathbf{x}_{1k}) \\ \mathbf{f}'(\mathbf{x}_{21}) \\ \vdots \\ \mathbf{f}'(\mathbf{x}_{2k}) \\ \vdots \\ \mathbf{f}'(\mathbf{x}_{b1}) \\ \vdots \\ \mathbf{f}'(\mathbf{x}_{bk}) \end{bmatrix}, \quad \mathbf{Z} = \begin{bmatrix} 1 & 0 & \cdots & 0 \\ \vdots & \vdots & \ddots & \vdots \\ 1 & 0 & \cdots & 0 \\ 0 & 1 & \cdots & 0 \\ \vdots & \vdots & \ddots & \vdots \\ 0 & 1 & \cdots & 0 \\ \vdots & \vdots & \ddots & \vdots \\ 0 & 0 & \cdots & 1 \\ \vdots & \vdots & \ddots & \vdots \\ 0 & 0 & \cdots & 1 \end{bmatrix},$$

$$\gamma = \begin{bmatrix} \gamma_1 & \gamma_2 & \cdots & \gamma_b \end{bmatrix}',$$

and

$$\varepsilon = \begin{bmatrix} \varepsilon_{11} & \cdots & \varepsilon_{1k} & \varepsilon_{21} & \cdots & \varepsilon_{2k} & \cdots & \varepsilon_{b1} & \cdots & \varepsilon_{bk} \end{bmatrix}'.$$

Note that the elements of Y and the rows of $\mathbf{X}$ are arranged per block, and that the model matrix is composed of one submatrix $\mathbf{X}_i$ for every block. The matrix $\mathbf{Z}$ is a matrix that has a one in the ith row and jth column if the ith run of the experiment belongs to the jth block, and a zero otherwise. For example, the $\mathbf{Z}$ matrix has ones in its first column for the first k rows, indicating that the first k runs belong to the first block. Because we assume that all the random effects are independent, the random-effects vectors γ and ε have variance–covariance matrices $\mathrm{var}(\gamma) = \sigma_\gamma^2 \mathbf{I}_b$

and $\text{var}(\boldsymbol{\varepsilon}) = \sigma_\varepsilon^2 \mathbf{I}_n$, respectively, where $\mathbf{I}_b$ and $\mathbf{I}_n$ are identity matrices of dimensions b and n, respectively.

The matrix $\mathbf{V}$ is the variance–covariance matrix of the response vector:

$$
\begin{aligned}
\mathbf{V} &= \text{var}(\boldsymbol{Y}), \\
&= \text{var}(\mathbf{X}\boldsymbol{\beta} + \mathbf{Z}\boldsymbol{\gamma} + \boldsymbol{\varepsilon}), \\
&= \text{var}(\mathbf{Z}\boldsymbol{\gamma} + \boldsymbol{\varepsilon}), \\
&= \text{var}(\mathbf{Z}\boldsymbol{\gamma}) + \text{var}(\boldsymbol{\varepsilon}), \\
&= \mathbf{Z}[\text{var}(\boldsymbol{\gamma})]\mathbf{Z}' + \text{var}(\boldsymbol{\varepsilon}), \\
&= \mathbf{Z}[\sigma_\gamma^2 \mathbf{I}_b]\mathbf{Z}' + \sigma_\varepsilon^2 \mathbf{I}_n, \\
&= \sigma_\gamma^2 \mathbf{Z}\mathbf{Z}' + \sigma_\varepsilon^2 \mathbf{I}_n,
\end{aligned}
$$

which is an $n \times n$ block diagonal matrix of the form

$$
\mathbf{V} = \begin{bmatrix}
\boldsymbol{\Lambda} & \mathbf{0}_{k \times k} & \cdots & \mathbf{0}_{k \times k} \\
\mathbf{0}_{k \times k} & \boldsymbol{\Lambda} & \cdots & \mathbf{0}_{k \times k} \\
\vdots & \vdots & \ddots & \vdots \\
\mathbf{0}_{k \times k} & \mathbf{0}_{k \times k} & \cdots & \boldsymbol{\Lambda}
\end{bmatrix},
$$

where $\boldsymbol{\Lambda}$ is the variance-covariance matrix for all the responses within a given block given in Equation (7.6) and $\mathbf{0}_{k \times k}$ is a $k \times k$ matrix of zeros. The interpretation of the elements of $\mathbf{V}$ is similar to that for the elements of $\boldsymbol{\Lambda}$. The diagonal elements of $\mathbf{V}$ are the variances of the responses Y_{ij}, and the off-diagonal elements are covariances between pairs of responses. The nonzero off-diagonal elements of $\mathbf{V}$ all correspond to pairs of responses from within a given block. All zero elements of $\mathbf{V}$ correspond to pairs of runs from two different blocks.

The GLS estimator in Equation (7.7) is a function of the matrix $\mathbf{V}$, and this complicates the estimation of the factor effects. This is because, even though the structure of $\mathbf{V}$ is known, we do not know the exact value of its nonzero elements. Therefore, these nonzero elements, which are functions of the block variance σ_γ^2 and the residual error variance σ_ε^2, have to be estimated from the data. So, to use the GLS estimator, we have to estimate σ_γ^2 and σ_ε^2, and insert the estimates into the covariance matrices $\boldsymbol{\Lambda}$ and $\mathbf{V}$. Denoting the estimators for σ_γ^2 and σ_ε^2 by $\hat{\sigma}_\gamma^2$ and $\hat{\sigma}_\varepsilon^2$, respectively, the estimated covariance matrices can be written as

$$
\hat{\boldsymbol{\Lambda}} = \begin{bmatrix}
\hat{\sigma}_\gamma^2 + \hat{\sigma}_\varepsilon^2 & \hat{\sigma}_\gamma^2 & \cdots & \hat{\sigma}_\gamma^2 \\
\hat{\sigma}_\gamma^2 & \hat{\sigma}_\gamma^2 + \hat{\sigma}_\varepsilon^2 & \cdots & \hat{\sigma}_\gamma^2 \\
\vdots & \vdots & \ddots & \vdots \\
\hat{\sigma}_\gamma^2 & \hat{\sigma}_\gamma^2 & \cdots & \hat{\sigma}_\gamma^2 + \hat{\sigma}_\varepsilon^2
\end{bmatrix} \tag{7.9}
$$

and

$$
\widehat{\mathbf{V}} = \begin{bmatrix} \hat{\mathbf{\Lambda}} & \mathbf{0}_{k \times k} & \cdots & \mathbf{0}_{k \times k} \\ \mathbf{0}_{k \times k} & \hat{\mathbf{\Lambda}} & \cdots & \mathbf{0}_{k \times k} \\ \vdots & \vdots & \ddots & \vdots \\ \mathbf{0}_{k \times k} & \mathbf{0}_{k \times k} & \cdots & \hat{\mathbf{\Lambda}} \end{bmatrix}.
$$

The GLS estimator of the factor effects then becomes

$$
\hat{\boldsymbol{\beta}} = (\mathbf{X}'\widehat{\mathbf{V}}^{-1}\mathbf{X})^{-1}\mathbf{X}'\widehat{\mathbf{V}}^{-1}\mathbf{Y}. \tag{7.10}
$$

That estimator is an unbiased estimator of $\boldsymbol{\beta}$, and is often named the feasible GLS estimator.

7.3.3   Estimation of variance components

There are different ways to estimate the variances σ_γ^2 and σ_ε^2 in an unbiased fashion, including an analysis of variance method called Henderson's method or Yates's method and REML (restricted or residual maximum likelihood) estimation. Henderson's method is computationally simple, and does not require the assumption of normality. It does, however, yield less precise estimates than REML if the fitted model is correctly specified. While REML estimation assumes that all the random effects are normally distributed, simulation results have indicated that it still produces unbiased estimates if that assumption is violated. Strong arguments in favor of REML estimates are that they have minimum variance in case the design is balanced and that, unlike other types of estimates, they are calculable for unbalanced designs too. For these reasons, and due to its availability in commercial software, REML estimation of variances in mixed models is the default method in many disciplines. For two familiar REML type estimators, we refer to Attachment 7.1 on page 150–151.

The REML estimates for σ_γ^2 and σ_ε^2 are obtained by maximizing the following restricted log likelihood function:

$$
l_R = -\frac{1}{2}\log|\mathbf{V}| - \frac{1}{2}\log|\mathbf{X}'\mathbf{V}^{-1}\mathbf{X}| - \frac{1}{2}\mathbf{r}'\mathbf{V}^{-1}\mathbf{r} - \frac{n-p}{2}\log(2\pi),
$$

where $\mathbf{r} = \mathbf{y} - \mathbf{X}(\mathbf{X}'\mathbf{V}^{-1}\mathbf{X})^{-1}\mathbf{X}'\mathbf{V}^{-1}\mathbf{y}$ and p is the number of parameters in $\boldsymbol{\beta}$. Like any other method, except for Bayesian methods, REML may provide a negative estimate of σ_γ^2 even in situations where the block-to-block variance is expected to be well above zero. The chances for getting a negative estimate for σ_γ^2 are larger when the number of blocks is small, the design is not orthogonally blocked and σ_γ^2 is small compared to σ_ε^2. In the event of such a negative estimate, we recommend setting

σ_γ^2 to a positive value when computing the GLS estimator to see how sensitive the factor-effect estimates, their standard errors and the p values are to its value, or to use a Bayesian analysis method.

7.3.4 Significance tests

To perform significance tests for each of the factor effects, three pieces of information are needed for each of the parameters in $\boldsymbol{\beta}$: the parameter estimate, its standard error, and the degrees of freedom for the t distribution to compute a p value. The parameter estimates are given by the feasible GLS estimator in Equation (7.10).

Finding the standard errors is more involved. If the variances σ_γ^2 and σ_ε^2 were known, then the standard errors would be the square roots of the diagonal elements of $(\mathbf{X}'\mathbf{V}^{-1}\mathbf{X})^{-1}$. However, σ_γ^2 and σ_ε^2 are unknown, and the logical thing to do is to substitute σ_γ^2 and σ_ε^2 by their estimates in $\mathbf{V}$. In this approach, the standard errors are the square roots of the diagonal elements of

$$\mathrm{var}(\hat{\boldsymbol{\beta}}) = (\mathbf{X}'\widehat{\mathbf{V}}^{-1}\mathbf{X})^{-1}. \tag{7.11}$$

In general, however, the standard errors obtained in this way underestimate the true standard errors a little bit. Slightly better standard errors can be obtained by using the diagonal elements from an adjusted covariance matrix

$$\mathrm{var}(\hat{\boldsymbol{\beta}}) = (\mathbf{X}'\widehat{\mathbf{V}}^{-1}\mathbf{X})^{-1} + \boldsymbol{\Omega} \tag{7.12}$$

instead of $(\mathbf{X}'\widehat{\mathbf{V}}^{-1}\mathbf{X})^{-1}$. The derivation of the adjustment, $\boldsymbol{\Omega}$, is too complex to give here, but it can be found in Kenward and Roger (1997), to whom the adjustment of the standard errors is due. The exact adjustment depends on the design, and, for designs which are orthogonally blocked or nearly orthogonally blocked (such as optimal designs), the adjustment is small (and sometimes even zero for some or all of the factor effects). Kenward and Roger (1997) also describe a method for determining the degrees of freedom for hypothesis tests based on the adjusted standard errors. The main software packages have already implemented the adjusted standard errors and degrees of freedom.

7.3.5 Optimal design of blocked experiments

To find an optimal design for a blocked experiment, we have to solve two decision problems simultaneously. First, we have to determine factor levels for each run, and, second, we have to arrange the selected factor-level combinations in b blocks of size k. The challenge is to do this in such a way that we obtain the most precise estimates possible. Therefore, we seek factor-level combinations and arrange them in blocks so that the covariance matrix of the GLS estimator, $(\mathbf{X}'\mathbf{V}^{-1}\mathbf{X})^{-1}$, is as small as possible. The way in which we do this is by minimizing the determinant of the covariance matrix $(\mathbf{X}'\mathbf{V}^{-1}\mathbf{X})^{-1}$, which yields the same result as maximizing the determinant of the information matrix $\mathbf{X}'\mathbf{V}^{-1}\mathbf{X}$. The block design that minimizes the determinant of the GLS estimator's covariance matrix or maximizes the determinant of the information matrix is the D-optimal block design. This results both in small variances of the parameter estimates and in small covariances between the parameter estimates. To find a D-optimal block design, we utilize the coordinate-exchange algorithm outlined in Section 2.3.9, but we use $|\mathbf{X}'\mathbf{V}^{-1}\mathbf{X}|$ instead of $|\mathbf{X}'\mathbf{X}|$ as the objective function to maximize.

A technical problem with finding the D-optimal design for a blocked experiment is that the matrix $\mathbf{V}$, and therefore also the D-optimality criterion $|\mathbf{X}'\mathbf{V}^{-1}\mathbf{X}|$, depends on the unknown variances σ_γ^2 and σ_ε^2. It turns out that the D-optimal block design does not depend on the absolute magnitude of these two variances, but only on their relative magnitude. As a result, software to generate optimal block designs requires input on the relative magnitude of σ_γ^2 and σ_ε^2. Our experience suggests that the block-to-block variance, σ_γ^2, is often substantially larger than the run-to-run variance, σ_ε^2. So, we suggest specifying that the variance ratio $\sigma_\gamma^2/\sigma_\varepsilon^2$ is at least one when generating a D-optimal block design. For the purpose of generating an excellent design, an educated guess of the variance ratio is good enough because the D-optimal design is not sensitive to the specified value. So, a design that is optimal for one variance ratio is also optimal for a broad range of variance ratios smaller and larger than the specified one. Moreover, in the rare cases where different variance ratios lead to different designs, the quality of these designs is almost indistinguishable. Goos (2002) recommends using a variance ratio of one for finding optimal designs for blocked experiments in the absence of detailed a priori information about it.

7.3.6   Orthogonal blocking

Textbooks on experimental design and response surface methodology emphasize the importance of orthogonal blocking. Orthogonally blocked designs guarantee factor-effect estimates that are independent of the estimates of the block effects. The general condition for orthogonal blocking is that

$$\frac{1}{k}\sum_{j=1}^{k}\mathbf{f}(\mathbf{x}_{ij}) = \frac{1}{k}\mathbf{X}_i'\mathbf{1}_k = \frac{1}{n}\mathbf{X}'\mathbf{1}_n = \frac{1}{n}\sum_{i=1}^{b}\sum_{j=1}^{k}\mathbf{f}(\mathbf{x}_{ij}) \tag{7.13}$$

for each of the b blocks. The interpretation of this condition is that, in an orthogonally blocked design, every $\mathbf{X}_i$ has the same average row. The condition for orthogonal blocking extends to designs with unequal numbers of runs in their blocks.

As an example, consider the orthogonally blocked central composite design in Table 7.1, meant for estimating the model in Equation (7.1). The design consists of three blocks, so the model matrix for that design consists of three submatrices, $\mathbf{X}_1$,

$\mathbf{X}_2$, and $\mathbf{X}_3$, one for each block:

$$
\mathbf{X} = \begin{bmatrix} \mathbf{X}_1 \\ \mathbf{X}_2 \\ \mathbf{X}_3 \end{bmatrix} =
\left[
\begin{array}{cccccccccc}
1 & -1 & -1 & -1 & 1 & 1 & 1 & 1 & 1 & 1 \\
1 & -1 & 1 & 1 & -1 & -1 & 1 & 1 & 1 & 1 \\
1 & 1 & -1 & 1 & -1 & 1 & -1 & 1 & 1 & 1 \\
1 & 1 & 1 & -1 & 1 & -1 & -1 & 1 & 1 & 1 \\
1 & 0 & 0 & 0 & 0 & 0 & 0 & 0 & 0 & 0 \\
1 & 0 & 0 & 0 & 0 & 0 & 0 & 0 & 0 & 0 \\
\hline
1 & -1 & -1 & 1 & 1 & -1 & -1 & 1 & 1 & 1 \\
1 & -1 & 1 & -1 & -1 & 1 & -1 & 1 & 1 & 1 \\
1 & 1 & -1 & -1 & -1 & -1 & 1 & 1 & 1 & 1 \\
1 & 1 & 1 & 1 & 1 & 1 & 1 & 1 & 1 & 1 \\
1 & 0 & 0 & 0 & 0 & 0 & 0 & 0 & 0 & 0 \\
1 & 0 & 0 & 0 & 0 & 0 & 0 & 0 & 0 & 0 \\
\hline
1 & -1.633 & 0 & 0 & 0 & 0 & 0 & 2.667 & 0 & 0 \\
1 & 1.633 & 0 & 0 & 0 & 0 & 0 & 2.667 & 0 & 0 \\
1 & 0 & -1.633 & 0 & 0 & 0 & 0 & 0 & 2.667 & 0 \\
1 & 0 & 1.633 & 0 & 0 & 0 & 0 & 0 & 2.667 & 0 \\
1 & 0 & 0 & -1.633 & 0 & 0 & 0 & 0 & 0 & 2.667 \\
1 & 0 & 0 & 1.633 & 0 & 0 & 0 & 0 & 0 & 2.667 \\
1 & 0 & 0 & 0 & 0 & 0 & 0 & 0 & 0 & 0 \\
1 & 0 & 0 & 0 & 0 & 0 & 0 & 0 & 0 & 0 \\
\end{array}
\right]
$$

For each of these three submatrices, the average row is

$$
\begin{bmatrix} 1 & 0 & 0 & 0 & 0 & 0 & 0 & 2/3 & 2/3 & 2/3 \end{bmatrix},
$$

where the first element corresponds to the intercept and the next three sets of three elements correspond to the main effects, the two-factor interaction effects, and the quadratic effects, respectively. The D-optimal design in Table 7.5 is not orthogonally blocked because the row averages of the submatrices $\mathbf{X}_i$ of the model matrix $\mathbf{X}$ are unequal. Table 7.10 shows this.

You can quantify the extent to which a design is orthogonally blocked in various ways. One approach is to compute an efficiency factor for each of the effects in the

Table 7.10 Average rows of the matrices $\mathbf{X}_i$ for each of the seven blocks of the D-optimal design in Table 7.5. The column for the intercept is not displayed.

	x_1	x_2	x_3	$x_1 x_2$	$x_1 x_3$	$x_2 x_3$	x_1^2	x_2^2	x_3^2
1	0.25	−0.25	−0.25	0.25	0.25	−0.25	0.75	0.75	0.75
2	0.25	0.25	0.25	0.00	0.00	0.00	0.75	0.75	0.75
3	0.25	0.25	−0.25	0.00	0.00	0.00	0.75	0.75	0.75
4	−0.25	−0.25	0.25	−0.25	0.25	0.25	0.75	0.75	0.75
5	−0.25	0.25	0.25	0.00	0.00	0.00	0.75	0.75	0.75
6	0.00	0.00	0.00	0.00	0.00	0.00	1.00	1.00	1.00
7	−0.25	0.25	−0.25	0.00	0.00	0.00	0.75	0.75	0.75

model. The efficiency factor is the ratio of the variance of the factor-effect estimate in the absence of block effects to the variance in the presence of block effects. If a design is orthogonally blocked, these two variances are identical because, in that case, the factor-effect estimates are independent of the block effects' estimates. This results in an efficiency factor of 100%. If a design is not orthogonally blocked, then the variance of some of the factor-effect estimates is inflated to some extent. This results in efficiency factors of less than 100%. The variances in the presence and absence of block effects required to compute the efficiency factors are the diagonal elements of the matrices $([\mathbf{X}\ \mathbf{Z}]'[\mathbf{X}\ \mathbf{Z}])^{-1}$ and $(\mathbf{X}'\mathbf{X})^{-1}$, respectively. The efficiency factors do not depend on the variances σ_γ^2 and σ_ε^2.

The inverse of the efficiency factors gives the variance inflation factors or VIFs due to the blocking. For the pastry dough experiment, the smallest efficiency factor is 93.4%, so the largest VIF is $1/0.934 = 1.0705$. This was shown in Table 7.6.

7.3.7   Optimal versus orthogonal blocking

The complexity of many experimental design problems is such that it is often difficult, if not impossible, to find an orthogonally blocked design. As a result, there is often no other possibility than to generate an optimal experimental design, that is not orthogonally blocked, for a practical experiment that has to be run in blocks.

However, there also exist scenarios where it is possible to construct an orthogonally blocked design, and where that design is in fact D-optimal. So, there is not necessarily a conflict between orthogonality and optimality of a design. As a matter of fact, Goos and Vandebroek (2001) provide theoretical evidence that D-optimal designs tend to be orthogonally blocked.

The reverse, however, is not generally true. It may be possible to construct orthogonally blocked designs for certain practical problems, which are not even close to being optimal. Goos and Donev (2006b) provide an example of an optimal block design which yields variances for the factor-effect estimates which are three times smaller than those produced by an orthogonally blocked alternative described in the literature.

It is our view that whether or not a design is orthogonally blocked is of secondary importance. What is of primary importance is that the factor-effect estimates can be estimated precisely, so that powerful statistical tests and precise predictions are possible. The results in Goos and Vandebroek (2001) indicate that a design that produces precise factor-effect estimates is automatically at least nearly orthogonally blocked.

7.4   Background reading

More details about the optimal design of experiments in blocks can be found in Chapters 4 and 5 of Goos (2002). These chapters review much of the literature on the subject, provide additional examples of optimal block designs, and discuss in detail the design of an optometry experiment. The optometry experiment involves one continuous factor with a quadratic effect. The blocks in the experiment are test subjects and have two observations each. Further discussion about the optimality and orthogonality of block designs is given in Goos (2006).

More technical details and discussion about the analysis of data from blocked experiments are given in Khuri (1992), Gilmour and Trinca (2000), and Goos and Vandebroek (2004). The REML method, which is recommended by Gilmour and Trinca (2000) for analyzing data from blocked experiments, was originally proposed by Patterson and Thompson (1971). The unbiasedness and variance of the feasible GLS estimator in Equation (7.10) are discussed in Kackar and Harville (1981, 1984), Harville and Jeske (1992), and Kenward and Roger (1997). Gilmour and Goos (2009) show how to use a Bayesian approach to overcome the problem of a negative estimate for σ_γ^2 for data from a split-plot experiment. Khuri (1996) shows how to handle data from blocked experiments when the block effects are not additive.

This chapter presented a case study with only one blocking factor, the day on which an experimental run was carried out. Gilmour and Trinca (2003) and Goos and Donev (2006a) discuss extensions to situations where there is more than one blocking factor. Goos and Donev (2006b) demonstrate the benefits of optimal design for setting up mixture experiments in blocks.

The case study in this chapter features two different responses. For the simultaneous optimization of these responses, we used a desirability function. For more details about desirability functions, we refer to Harrington (1965) and Derringer and Suich (1980).

7.5   Summary

Practical considerations often cause experimental runs to appear in groups. Common causes for this are that the experiment involves several days of work or that several batches of raw material are used in an experiment. Experiments in the semi-conductor industry, for example, often require the use of different silicon wafers. Other examples of experiments where the runs are grouped are agricultural experiments where multiple fields are used and medical, marketing, or taste experiments where each of different subjects tests and rates several factor-level combinations.

The groups of runs are called blocks, and the factor that causes the grouping is the blocking factor. The block-to-block differences often account for a large share of the overall variability in the responses, and necessitate a careful planning of the experiment and subsequent data analysis. Therefore, we recommend spending substantial effort on the identification of sources of variation such as blocking factors when planning an experimental study. This is especially important when the block-to-block variability is comparable to or larger than the run-to-run variability.

In this chapter, we showed how a mixed model, involving random block effects, can be used as a starting point for generating an optimal block design and for the analysis of the data from that design. In the next chapter, we discuss the design of another blocked experiment, and highlight the benefits of the random block effects model compared to the fixed block effects model.

8

A screening experiment in blocks

8.1 Key concepts

1. When the blocks in an experiment are groups of runs that are different from each other but not identifiable as random choices from a population of blocks, then we treat their effects as fixed rather than random.

2. Using fixed blocks does not allow the investigator to make predictions for individual runs from future blocks.

3. Ordinary least squares (OLS) regression is the standard analytical technique for experiments with fixed block effects.

4. Experiments assuming fixed blocks require more runs to fit a specified model than experiments treating the blocks as random. This is especially true when there are only two runs in each block. In such cases, the smallest fixed block design requires roughly twice as many runs as the smallest random block design.

Historically, most blocked industrial experiments have treated the effects of the blocking factor as fixed. This is partly due to the fact that, in general, the estimation of mixed models involving random block effects is more computationally demanding than the estimation of models with fixed block effects. Most textbook expositions of blocking deal with how to create designs with blocks that are orthogonal to the factor effects.

This chapter deals with a problem where the blocks cannot be orthogonal to the other factor effects in the model. We show how to create designs where the blocks are nearly orthogonal to the factor effects. We also discuss the advantages of assuming random blocks instead of fixed blocks, and show how to evaluate how

Optimal Design of Experiments: A Case Study Approach, First Edition. Peter Goos and Bradley Jones.
© 2011 John Wiley & Sons, Ltd. Published 2011 by John Wiley & Sons, Ltd.

much the variance of the factor effects' estimates is increased by the inability to block orthogonally.

8.2   Case: the stability improvement experiment

8.2.1   Problem and design

Peter and Brad are driving down the New Jersey Turnpike in route to a consulting visit with FRJ, a pharmaceutical manufacturer. Peter briefs Brad on the upcoming client call.

[Peter] Dr. Xu phoned me yesterday about a project they are considering at FRJ. They make vitamin preparations for over-the-counter sale at grocery stores. Unfortunately, many vitamins are photosensitive. They break down when exposed to light. Dr. Xu wants to run an experiment to see if combining the vitamin with certain fatty molecules can lower their sensitivity to light, thereby increasing their shelf life.

[Brad] That is vitally important work as far as I am concerned. I am a big believer in vitamin supplements. So, why is he calling us in? He is an expert in designed experiments already.

[Peter] Dr. Xu said something about day-to-day calibration problems with the photometric equipment they use. He did not want to talk much on the phone. These drug companies are very security conscious.

They arrive at the FRJ Research & Development site. Dr. Xu meets them and escorts them to a conference room.

[Dr. Xu] I am pleased you are here. Let me go through the details of the scenario with you.

[Peter] Sure!

[Dr. Xu] Many vitamins are sensitive to light and become inactive after being in sunlight for a very short time. For our study, we chose riboflavin since its behavior is typical. We have a theory and some evidence that, by embedding riboflavin in a complex with one or more fatty molecules that are light absorbing, we can dramatically decrease how fast it breaks down in direct sunlight.

Peter and Brad nod, and Dr. Xu continues.

[Dr. Xu] We can use riboflavin in its natural state or in a sugar complex. That is our first factor. We also have five different fatty molecules to test. Each fatty molecule can be present or absent in combination with the riboflavin. So, gives us six factors in all. We strongly suspect there are two-factor interactions, so we want to be able to estimate all 15 of them. Are you with me so far?

[Peter glancing at Brad for confirmation] We've got it.

[Dr. Xu] Here is where things get tough. Our measuring apparatus for calculating the decay rate due to exposure to light requires recalibration every day. We can evaluate four possible molecular complexes a day and we scheduled the lab for 8 days.

[Brad] You want to adjust the estimates of the effects of the factors and their two-factor interactions for any day-to-day differences in the calibration, right?

[Dr. Xu] Yes, and that is where I am having trouble. I consider the effect of changing the calibration each day to be a blocking factor with eight levels. It should be possible to estimate these eight block effects, six main effects, and 15 two-factor interaction effects with the 32 experimental runs. There are, after all, only 29 $(8 + 6 + 15)$ unknown parameters. Here is the best orthogonally blocked design I can generate.

Dr. Xu shows Peter and Brad the 2^{6-1} design in Table 8.1. It is the half fraction of the six-factor two-level or 2^6 factorial design for which $x_1x_2x_3x_4x_5x_6 = 1$. The 32 runs of that design are arranged in eight blocks of four using the contrast columns for the three-factor interactions $x_1x_3x_5$, $x_2x_3x_5$, and $x_1x_4x_5$ to split the runs in eight blocks of four. The three contrast columns $x_1x_3x_5$, $x_2x_3x_5$, and $x_1x_4x_5$ are shown in Table 8.1. They allow the 32-run design to be split in 2^3 or eight blocks of four runs.

[Brad] Don't tell me. Let me guess. You cannot estimate all the two-factor interaction effects.

[Dr. Xu] That's right. Three of the two-factor interaction effects, x_1x_2, x_3x_4, and x_5x_6, are completely confounded with block effects.

[Brad] This design does have some excellent qualities. The blocks are orthogonal to the model you can fit. And all the main effects and two-factor interaction effects except for those three have a variance inflation factor or VIF of one, which is perfect. A VIF of one means that you can estimate all these effects with maximum precision, or minimum variance. The block effects also have a variance inflation factor of one. The only problem is that you can't estimate the interactions involving x_1 and x_2, x_3 and x_4, and x_5 and x_6, which is the same as saying their variance is infinite.

[Dr. Xu] Well, that problem is a show-stopper. What is even more depressing is that, even if I double the number of runs to 64, I still cannot estimate all the main effects, two-factor interaction effects, and block effects independently.

[Peter] As you increase the number of runs from 32 to 64, you get a resulting increase in the number of blocks from 8 to 16. So, increasing the number of runs also increases the number of effects in your model.

[Dr. Xu] I had a difficult time getting management to approve the resources to fit all these two-factor interactions. If I use all these runs and still can't estimate them, I'll lose face big time.

[Brad] Not to worry. We can give you a design that will estimate all the effects you want. What were the names of your six factors again?

Brad pulls his laptop from his pack and fires it up.

[Dr. Xu] Hold on a minute! I think it is impossible to find an orthogonally blocked design that estimates all these effects in 32 runs. So, how do you think your computer is going to help?

[Brad] Let's suppose you are right that no orthogonally blocked design exists for your problem that allows you to estimate your desired model with the best possible precision. Since you can do 32 runs and only have 29 unknown factor effects, you could fit all these unknowns even if you used a totally random design.

[Dr. Xu] What do you mean by a totally random design?

[Brad] All of your factors are categorical and each has two levels. Let's use -1 and 1 for the two values. Now, for each factor, I randomly assign -1 or 1 to each of

Table 8.1 Dr. Xu's 2^{6-1} design arranged in eight blocks of four observations.

Block	x_1	x_2	x_3	x_4	x_5	x_6	$x_1x_2x_3x_4x_5x_6$	$x_1x_3x_5$	$x_2x_3x_5$	$x_1x_4x_5$
1	-1	-1	-1	-1	-1	-1	1	-1	-1	-1
1	-1	-1	1	1	1	1	1	-1	-1	-1
1	1	1	-1	-1	1	1	1	-1	-1	-1
1	1	1	1	1	-1	-1	1	-1	-1	-1
2	-1	-1	-1	1	-1	1	1	-1	-1	1
2	-1	-1	1	-1	1	-1	1	-1	-1	1
2	1	1	-1	1	1	-1	1	-1	-1	1
2	1	1	1	-1	-1	1	1	-1	-1	1
3	-1	1	-1	-1	-1	1	1	-1	1	-1
3	-1	1	1	1	1	-1	1	-1	1	-1
3	1	-1	-1	-1	1	-1	1	-1	1	-1
3	1	-1	1	1	-1	1	1	-1	1	-1
4	-1	1	-1	1	-1	-1	1	-1	1	1
4	-1	1	1	-1	1	1	1	-1	1	1
4	1	-1	-1	1	1	1	1	-1	1	1
4	1	-1	1	-1	-1	-1	1	-1	1	1
5	-1	1	-1	1	1	1	1	1	-1	-1
5	-1	1	1	-1	-1	-1	1	1	-1	-1
5	1	-1	-1	1	-1	-1	1	1	-1	-1
5	1	-1	1	-1	1	1	1	1	-1	-1
6	-1	1	-1	-1	1	-1	1	1	-1	1
6	-1	1	1	1	-1	1	1	1	-1	1
6	1	-1	-1	-1	-1	1	1	1	-1	1
6	1	-1	1	1	1	-1	1	1	-1	1
7	-1	-1	-1	1	1	-1	1	1	1	-1
7	-1	-1	1	-1	-1	1	1	1	1	-1
7	1	1	-1	1	-1	1	1	1	1	-1
7	1	1	1	-1	1	-1	1	1	1	-1
8	-1	-1	-1	-1	1	1	1	1	1	1
8	-1	-1	1	1	-1	-1	1	1	1	1
8	1	1	-1	-1	-1	-1	1	1	1	1
8	1	1	1	1	1	1	1	1	1	1

the 32 runs in the design. Here is the resulting design, along with a table showing VIFs for each coefficient.

Brad turns his computer toward Dr. Xu where he sees Tables 8.2 and 8.3.

[Dr. Xu, looking at the two tables and eventually focusing on Table 8.3] Is the random design any good? What do these two VIFs mean? I know that, in regression studies, we can compute VIFs, but then there is just one for each term in the regression model. Why do you have two VIFs per model term?

[Peter] The table shows two kinds of VIFs because there are two reasons why the random design is not very good. The first reason is that the factor settings, that is the values for x_1-x_6, in the random design are not orthogonal. The second reason is that the random design is not at all orthogonally blocked. The nonorthogonality of the factor settings inflates the variance of the estimates of the factor effects by a factor that ranges from 1.72 to 6.34. You can see this in the second column of the table. The nonorthogonality of the blocking leads to an additional inflation of the variance of the factor-effect estimates of the same order of magnitude. This is shown in the third column of the table. The last column gives you the total VIF for each effect. The smallest total VIF is 3.18 and the largest one is 23.45.

[Dr. Xu] These numbers are very large. Ideally, the VIFs are one for each factor effect. In that case, the estimation is fully efficient for all factor effects, and all the effects are estimated with maximum precision. Rules of thumb for VIFs say that we should start worrying about multicollinearity as soon as certain VIFs are larger than five. If I count correctly, then 16 of the 21 total VIFs are larger than five. So, that design is terrible!

[Peter] True, the random design Brad just built is worse than your design in every way except one.

[Dr. Xu] What's that?

[Brad] My design estimates all two-factor interaction effects and yours doesn't.

[Dr. Xu] OK, but so what? Surely, you are not suggesting that I actually run this design?

[Brad] No, of course not. My point in showing you the random design is that producing a design that gives you estimates of all 29 effects in your model is easy. Of course, the next step is to search among all the possible designs that fit your model for one that performs acceptably. That's when the computer becomes very useful. Let's see how we do when we search for a really good design.

Brad sets up Dr. Xu's design problem on his laptop and starts computing the design. In a few seconds, the computer displays the design in Table 8.4.

[Dr. Xu] Done already? That's amazing. Well, assuming this design is any good.

[Peter] Shall we look at the VIFs for each factor effect for this design?

[Dr. Xu] That seems reasonable.

[Brad] Here's the table with the VIFs due to blocking, which happen to be the total variance inflations for this design. That is because the design would allow independent estimation of all main effects and all two-factor interaction effects if there were no blocks. In other words, if there was no reason for the blocking, the variances of the estimates of the effects of x_1-x_6 would be minimal because the design is orthogonal.

Table 8.2 Random design for the stability improvement experiment.

Block	x_1	x_2	x_3	x_4	x_5	x_6
1	−1	1	1	−1	−1	1
1	−1	1	−1	−1	−1	1
1	1	1	−1	1	−1	1
1	−1	−1	−1	1	−1	1
2	−1	1	−1	1	−1	−1
2	1	−1	1	1	1	−1
2	−1	1	1	1	−1	1
2	1	−1	1	−1	1	1
3	1	1	−1	−1	−1	1
3	1	−1	−1	1	1	−1
3	−1	−1	1	1	1	1
3	−1	−1	1	−1	−1	−1
4	−1	1	1	1	−1	−1
4	−1	1	1	1	1	−1
4	−1	−1	1	−1	−1	1
4	1	−1	−1	−1	1	1
5	−1	1	−1	−1	1	1
5	−1	−1	−1	−1	1	−1
5	1	1	−1	1	−1	1
5	−1	1	1	−1	1	1
6	1	1	−1	−1	−1	−1
6	−1	−1	−1	1	−1	1
6	−1	1	−1	−1	1	−1
6	1	−1	−1	1	1	1
7	1	1	1	−1	−1	1
7	1	1	−1	1	1	−1
7	1	−1	−1	1	1	1
7	−1	−1	1	−1	1	−1
8	−1	1	1	1	−1	−1
8	1	−1	1	−1	1	1
8	1	−1	−1	−1	−1	1
8	1	1	1	−1	1	−1

Table 8.3 Variance inflation due to the nonorthogonality of the random design and due to the blocks.

Effect	Nonorthogonality	Blocks	Total
x_1	2.29	2.21	5.06
x_2	2.11	2.14	4.51
x_3	1.88	2.82	5.29
x_4	1.78	2.09	3.72
x_5	2.75	2.19	6.01
x_6	1.72	1.84	3.18
$x_1 x_2$	4.39	2.24	9.81
$x_1 x_3$	2.89	4.00	11.54
$x_1 x_4$	5.61	4.18	23.45
$x_1 x_5$	6.34	2.01	12.77
$x_1 x_6$	5.60	2.40	13.42
$x_2 x_3$	2.75	4.33	11.91
$x_2 x_4$	4.06	4.25	17.24
$x_2 x_5$	3.42	2.05	7.02
$x_2 x_6$	3.36	6.72	22.54
$x_3 x_4$	4.91	1.36	6.69
$x_3 x_5$	3.58	1.63	5.85
$x_3 x_6$	3.32	2.33	7.75
$x_4 x_5$	3.03	1.42	4.30
$x_4 x_6$	2.36	1.77	4.17
$x_5 x_6$	2.20	2.29	5.05

Brad turns his laptop so Dr. Xu can see Table 8.5.

[Dr. Xu] These VIFs are much better than those for the random design. Five of the coefficients have a VIF of one, so their estimation has maximum precision. None of the VIFs is larger than 2.57. I guess the trade-off here is that you can't estimate many of the coefficients with maximum precision but you can estimate them all?

[Brad] That's right.

[Dr. Xu] I'll buy that. Thank you, gentlemen. I've got to go to a meeting now. Mail me a copy of that design later today, please. I will get back to you in a few days with the data.

8.2.2 Afterthoughts about the design problem

Soon after the meeting, Brad and Peter are back in their car. While driving on the freeway, they discuss the meeting with Dr. Xu. When the debriefing is winding down, Peter remembers Dr. Xu mentioning the 64-run design with 16 blocks of four

Table 8.4   D-optimal design for estimating a model including main effects and two-factor interaction effects for the stability improvement experiment.

Block	x_1	x_2	x_3	x_4	x_5	x_6
1	1	1	1	1	−1	1
1	−1	−1	1	1	1	1
1	−1	−1	1	−1	−1	1
1	−1	1	−1	1	−1	−1
2	1	1	−1	−1	1	−1
2	−1	−1	−1	−1	−1	−1
2	1	1	−1	−1	−1	1
2	−1	1	−1	1	1	1
3	−1	1	1	1	1	−1
3	−1	−1	−1	1	−1	1
3	1	1	−1	1	1	1
3	1	−1	1	1	−1	−1
4	−1	−1	1	−1	1	−1
4	1	−1	−1	1	1	−1
4	1	1	1	−1	−1	−1
4	1	−1	1	1	1	1
5	−1	1	−1	−1	−1	1
5	−1	−1	−1	1	1	−1
5	1	−1	−1	−1	1	1
5	1	1	−1	1	−1	−1
6	1	−1	1	−1	1	−1
6	−1	1	1	−1	1	1
6	−1	1	−1	−1	1	−1
6	−1	−1	1	1	−1	−1
7	1	1	1	−1	1	1
7	1	−1	−1	1	−1	1
7	−1	1	1	1	−1	1
7	−1	−1	−1	−1	1	1
8	−1	1	1	−1	−1	−1
8	1	−1	1	−1	−1	1
8	1	−1	−1	−1	−1	−1
8	1	1	1	1	1	−1

Table 8.5 Variance inflation for the D-optimal design for the stability improvement experiment.

Effect	Nonorthogonality	Blocks	Total
x_1	1.00	1.29	1.29
x_2	1.00	1.16	1.16
x_3	1.00	1.85	1.85
x_4	1.00	1.41	1.41
x_5	1.00	1.27	1.27
x_6	1.00	1.60	1.60
$x_1 x_2$	1.00	1.25	1.25
$x_1 x_3$	1.00	1.00	1.00
$x_1 x_4$	1.00	1.16	1.16
$x_1 x_5$	1.00	1.00	1.00
$x_1 x_6$	1.00	1.00	1.00
$x_2 x_3$	1.00	1.43	1.43
$x_2 x_4$	1.00	1.27	1.27
$x_2 x_5$	1.00	2.57	2.57
$x_2 x_6$	1.00	1.32	1.32
$x_3 x_4$	1.00	1.16	1.16
$x_3 x_5$	1.00	1.00	1.00
$x_3 x_6$	1.00	1.61	1.61
$x_4 x_5$	1.00	2.20	2.20
$x_4 x_6$	1.00	1.32	1.32
$x_5 x_6$	1.00	1.00	1.00

experimental runs as an alternative to the 32-run design involving eight blocks of four observations.

[Peter] You remember Dr. Xu mentioning the 64-run design?

Brad, who is driving, nods.

[Peter] He said that doubling the number of runs from 32 to 64 doesn't solve his problem. That is not quite true. I can construct a 64-run design in blocks of four runs that allows estimation of all main effects and all two-factor interaction effects. I can even guarantee that all the factor-effect estimates are uncorrelated.

Brad casts a questioning look to his right.

[Peter, getting excited when he sees that Brad does not immediately find a solution] Dr. Xu only considered using the 2^6 full factorial design. That's not the best thing to do if you have 16 blocks and you want to estimate all two-factor interaction effects. Of course, you can't blame him because that is the only option tabulated in textbooks. It is better to replicate the half fraction he showed us, and to use different block generators for each replicate of the half fraction.

[Brad] Go on.

[Peter, who has opened his laptop] Well, the key feature of the 2^{6-1} half fraction Dr. Xu showed us is that it was obtained using the defining relation

$x_6 = +x_1x_2x_3x_4x_5$ and arranging the runs in blocks using the three-factor interaction contrast columns $x_1x_3x_5$, $x_2x_3x_5$, and $x_1x_4x_5$. As we discussed, this results in three interaction effects (those involving x_1 and x_2, x_3 and x_4, and x_5 and x_6) being fully confounded with the block effects. All other two-factor interaction effects, and the main effects, are estimated independently with that design. In other words, with maximum precision. Agree?

[Brad] Yep.

[Peter] Now, for the next 32 runs, we can use the same half fraction, but with a different block generator. If, for example, we use the three-factor interaction contrast columns $x_1x_2x_4$, $x_1x_3x_4$, and $x_1x_2x_5$ to generate the blocks in that second half fraction, then it is the interactions involving x_1 and x_6, x_2 and x_3, and x_4 and x_5 that are confounded with the blocks.

[Brad, nodding] That's cute. Combining the two half fractions then gives you a 64-run design that estimates the six main effects and the 15 two-factor interactions. Basically, this design splits the set of 21 factor effects in three. The first set of effects includes all the main effects and nine two-factor interaction effects, which can be estimated from each of the half fractions. The second set of effects includes the three interaction effects involving x_1 and x_2, x_3 and x_4, and x_5 and x_6, which can only be estimated from the second half fraction. The third set includes the three remaining interaction effects, involving x_1 and x_6, x_2 and x_3, and x_4 and x_5, which can be estimated from the first half fraction only.

[Peter] Right. Each of the 15 effects in the first set can therefore be estimated with maximum precision. The other effects, the interaction effects involving x_1 and x_2, x_3 and x_4, x_5 and x_6, x_1 and x_6, x_2 and x_3, and x_4 and x_5, are estimated only with half that precision. In other words, the VIF for 15 of the model effects is one, and that for the remaining effects is two.

[Brad] I agree that your design estimates all the effects but Dr. Xu originally wanted an orthogonally blocked design. Your design is clever but the blocks are not orthogonal to the effects. Still, your solution is elegant and efficient. Do you think it is D-optimal?

[Peter, toggling on his machine again] I doubt it. Let me compute an optimal design with 64 run in 16 blocks of four observations. Finding the most D-efficient 64-run design is quite a challenging problem. We will soon find out whether the algorithm in our software beats my cute 64-run design.

Peter parks his laptop on the back seat while it is computing the optimal design, and has a look at the newspaper. After a couple of minutes, he grabs his laptop again and, after doing a couple of extra computations, confirms the conjecture he made a few minutes before.

[Peter] The results are as I expected. The optimal design is about 3% better than my cute design.

Peter saves the two designs that he has constructed and which are displayed in Tables 8.6 and 8.7. He also creates Table 8.8 in which the variances of the factor-effect estimates from the two designs are compared side by side.

[Peter, looking at Table 8.8] Here we go. I now have the side-by-side comparison of the variances of the estimates for the two designs. The D-optimal design gives

Table 8.6   64-run design in 16 blocks of four runs obtained by combining two 2^{6-1} designs with eight blocks of four runs.

Block	x_1	x_2	x_3	x_4	x_5	x_6	Block	x_1	x_2	x_3	x_4	x_5	x_6
1	−1	−1	−1	−1	−1	−1	9	−1	1	1	1	1	−1
1	1	1	1	1	−1	−1	9	1	−1	−1	1	1	1
1	−1	−1	1	1	1	1	9	1	1	1	−1	−1	1
1	1	1	−1	−1	1	1	9	−1	−1	−1	−1	−1	−1
2	−1	1	1	−1	−1	−1	10	−1	−1	1	1	−1	−1
2	−1	1	−1	1	1	1	10	1	−1	1	−1	1	1
2	1	−1	−1	1	−1	−1	10	1	1	−1	1	−1	1
2	1	−1	1	−1	1	1	10	−1	1	−1	−1	1	−1
3	−1	1	1	1	1	−1	11	1	−1	1	1	1	−1
3	1	−1	1	1	−1	1	11	−1	−1	1	−1	−1	1
3	1	−1	−1	−1	1	−1	11	−1	1	−1	1	1	1
3	−1	1	−1	−1	−1	1	11	1	1	−1	−1	−1	−1
4	−1	−1	1	−1	−1	1	12	1	−1	−1	−1	1	−1
4	1	1	−1	1	−1	1	12	1	1	1	1	−1	−1
4	−1	−1	−1	1	1	−1	12	−1	1	1	−1	1	1
4	1	1	1	−1	1	−1	12	−1	−1	−1	1	−1	1
5	−1	−1	1	−1	1	−1	13	1	−1	−1	1	−1	−1
5	1	1	−1	1	1	−1	13	−1	−1	−1	−1	1	1
5	−1	−1	−1	1	−1	1	13	1	1	1	−1	1	−1
5	1	1	1	−1	−1	1	13	−1	1	1	1	−1	1
6	−1	1	−1	−1	1	−1	14	1	−1	1	−1	−1	−1
6	−1	1	1	1	−1	1	14	−1	−1	1	1	1	1
6	1	−1	−1	−1	−1	1	14	1	1	−1	1	1	−1
6	1	−1	1	1	1	−1	14	−1	1	−1	−1	−1	1
7	−1	1	1	−1	1	1	15	1	−1	1	1	−1	1
7	1	−1	1	−1	−1	−1	15	1	1	−1	−1	1	1
7	1	−1	−1	1	1	1	15	−1	−1	1	−1	1	−1
7	−1	1	−1	1	−1	−1	15	−1	1	−1	1	−1	−1
8	1	1	−1	−1	−1	−1	16	−1	1	1	−1	−1	−1
8	−1	−1	−1	−1	1	1	16	1	−1	−1	−1	−1	1
8	1	1	1	1	1	1	16	−1	−1	−1	1	1	−1
8	−1	−1	1	1	−1	−1	16	1	1	1	1	1	1

Table 8.7 D-optimal 64-run design with 16 blocks of four runs for estimating a model including main effects and two-factor interaction effects.

Block	x_1	x_2	x_3	x_4	x_5	x_6	Block	x_1	x_2	x_3	x_4	x_5	x_6
1	−1	−1	1	1	1	1	9	−1	−1	−1	−1	−1	−1
1	−1	−1	−1	−1	−1	−1	9	1	1	1	−1	−1	1
1	1	1	1	1	−1	−1	9	−1	1	1	1	1	−1
1	1	1	−1	−1	1	1	9	1	−1	−1	1	1	1
2	1	−1	−1	1	−1	−1	10	1	1	−1	1	−1	1
2	1	−1	1	−1	1	1	10	−1	1	−1	−1	1	−1
2	−1	1	1	−1	−1	−1	10	−1	−1	1	1	−1	−1
2	−1	1	−1	1	1	1	10	1	−1	1	−1	1	1
3	1	−1	1	1	−1	1	11	−1	1	−1	1	1	1
3	1	−1	−1	−1	1	−1	11	1	−1	1	1	1	−1
3	−1	1	−1	−1	−1	1	11	−1	−1	1	−1	−1	1
3	−1	1	1	1	1	−1	11	1	1	−1	−1	−1	−1
4	1	1	1	−1	1	−1	12	−1	−1	−1	1	−1	1
4	−1	−1	1	−1	−1	1	12	1	−1	−1	−1	1	−1
4	1	1	−1	1	−1	1	12	1	1	1	1	−1	−1
4	−1	−1	−1	1	1	−1	12	−1	1	1	−1	1	1
5	1	1	−1	1	1	−1	13	−1	−1	−1	−1	1	1
5	−1	−1	1	−1	1	−1	13	−1	1	1	1	−1	1
5	1	1	1	−1	−1	1	13	1	1	1	−1	1	−1
5	−1	−1	−1	1	−1	1	13	1	−1	−1	1	−1	−1
6	−1	1	−1	−1	1	−1	14	1	1	−1	1	1	−1
6	1	−1	1	1	1	−1	14	−1	−1	1	1	1	1
6	1	−1	−1	−1	−1	1	14	1	−1	1	−1	−1	−1
6	−1	1	1	1	−1	1	14	−1	1	−1	−1	−1	1
7	1	−1	−1	1	1	1	15	1	−1	1	1	−1	1
7	−1	1	1	−1	1	1	15	1	1	−1	−1	1	1
7	−1	1	−1	1	−1	−1	15	−1	−1	1	−1	1	−1
7	1	−1	1	−1	−1	−1	15	−1	1	−1	1	−1	−1
8	−1	−1	−1	−1	1	1	16	1	−1	−1	−1	−1	1
8	1	1	1	1	1	1	16	−1	1	1	−1	−1	−1
8	1	1	−1	−1	−1	−1	16	−1	−1	−1	1	1	−1
8	−1	−1	1	1	−1	−1	16	1	1	1	1	1	1

Table 8.8   Variances of factor-effect estimates for the replicated 2^{6-1} design in Table 8.6 and the D-optimal design in Table 8.7.

Effect	$2 \times 2^{6-1}$	Optimal	Effect	$2 \times 2^{6-1}$	Optimal	Effect	$2 \times 2^{6-1}$	Optimal
x_1	0.016	0.019	$x_1 x_3$	0.016	0.021	$x_2 x_6$	0.016	0.020
x_2	0.016	0.018	$x_1 x_4$	0.016	0.018	$x_3 x_4$	0.031	0.018
x_3	0.016	0.018	$x_1 x_5$	0.016	0.020	$x_3 x_5$	0.016	0.019
x_4	0.016	0.020	$x_1 x_6$	0.031	0.019	$x_3 x_6$	0.016	0.018
x_5	0.016	0.018	$x_2 x_3$	0.031	0.020	$x_4 x_5$	0.031	0.017
x_6	0.016	0.018	$x_2 x_4$	0.016	0.019	$x_4 x_6$	0.016	0.020
$x_1 x_2$	0.031	0.018	$x_2 x_5$	0.016	0.018	$x_5 x_6$	0.031	0.018

variances that are roughly the same for each of the effects: the smallest variance is 0.017 and the largest one is 0.021.

[Brad] The largest variance for your cleverly constructed design is 0.031, isn't it?

[Peter] Yeah. The optimal design doesn't have such large variances, but this comes at the expense of slightly larger variances for the other estimates. I'd say choosing between these two designs is not an easy matter.

[Brad] I would always recommend the optimal design. It is so easy to generate using commercial software nowadays, and doesn't require any knowledge at all about all these combinatorial construction methods.

[Peter] Where have I heard that before?

8.2.3   Data analysis

Two weeks later, Brad and Peter receive an e-mail from Dr. Xu containing the data from the stability improvement experiment. The data, obtained using the optimal design in Table 8.4, appear in uncoded form in Table 8.9. Five of the six factors in the experiment were the presence or absence of five light-absorbing fatty molecules. The sixth factor indicated whether the vitamin under study was free or complexed with γ-cyclodextrin, which is a kind of sugar. The measured response was the stability ratio. Dr. Xu's initial e-mail is followed by a short exchange of e-mails to fix a time for a new meeting at FRJ. Two days later, Brad has analyzed the data, and the two consultants are back in Dr. Xu's office. Brad's laptop is connected to a projector so that everyone present can see what is on his screen.

[Dr. Xu] Well, gentlemen. What can we learn from our experiment? Are all these fatty molecules and the sugar complex effective for improving the stability ratio?

Peter and Brad have no doubt that Dr. Xu, being an expert in design of experiments, has already studied the data to form his own opinion. However, Dr. Xu does not want to reveal any insights he has obtained at this point. Instead, he makes it clear that he wants to hear them talk first and Brad starts discussing the analysis of the data.

[Brad] We ran two slightly different regression analyses on your data. Even though the underlying factor settings are orthogonal, we needed regression methodology for your data because the design was not orthogonally blocked. This rules out classical analysis of variance, because that can only be used for completely orthogonal data sets.

Table 8.9 Data for the stability improvement experiment.

Block	Bounded-ness	Oil red O	Oxy-benzone	Beta carotene	Suliso-benzone	Deoxy-benzone	Stability ratio
1	Complex	In	In	In	Out	In	237
1	Free	Out	In	In	In	In	42
1	Free	Out	In	Out	Out	In	29
1	Free	In	Out	In	Out	Out	72
2	Complex	In	Out	Out	In	Out	229
2	Free	Out	Out	Out	Out	Out	11
2	Complex	In	Out	Out	Out	In	235
2	Free	In	Out	In	In	In	89
3	Free	In	In	In	In	Out	76
3	Free	Out	Out	In	Out	In	18
3	Complex	In	Out	In	In	In	246
3	Complex	Out	In	In	Out	Out	23
4	Free	Out	In	Out	In	Out	29
4	Complex	Out	Out	In	In	Out	44
4	Complex	In	In	Out	Out	Out	228
4	Complex	Out	In	In	In	In	46
5	Free	In	Out	Out	Out	In	83
5	Free	Out	Out	In	In	Out	30
5	Complex	Out	Out	Out	In	In	60
5	Complex	In	Out	In	Out	Out	228
6	Complex	Out	In	Out	In	Out	39
6	Free	In	In	Out	In	In	92
6	Free	In	Out	Out	In	Out	76
6	Free	Out	In	In	Out	Out	7
7	Complex	In	In	Out	In	In	240
7	Complex	Out	Out	In	Out	In	38
7	Free	In	In	In	Out	In	74
7	Free	Out	Out	Out	In	In	24
8	Free	In	In	Out	Out	Out	65
8	Complex	Out	In	Out	Out	In	34
8	Complex	Out	Out	Out	Out	Out	24
8	Complex	In	In	In	In	Out	237

Table 8.10   Results from the fixed block effects analysis.

Effect	Estimate	Standard error	p Value
Intercept	93.91	0.62	< .0001
Boundedness[Complex]	43.05	0.66	< .0001
Oil Red O[In]	63.06	0.64	< .0001
Sulisobenzone[In]	6.19	0.66	< .0001
Deoxybenzone[In]	5.55	0.72	< .0001
Boundedness[Complex]× Oil Red O[In]	35.02	0.67	< .0001
Block[1]	5.43	1.79	0.0069
Block[2]	−1.94	1.70	0.2665
Block[3]	−3.16	1.63	0.0682
Block[4]	2.53	1.79	0.1728
Block[5]	6.34	1.63	0.0010
Block[6]	−1.70	1.78	0.3516
Block[7]	−5.45	1.79	0.0065

Since Dr. Xu is nodding impatiently, Brad does not pause and displays the results from the first analysis on the screen. The results are shown in Table 8.10.

[Brad] These are the results we got using the traditional analysis with fixed block effects. This table shows the simplified model we got after removing the insignificant effects from the full model involving all the main effect and the two-factor interactions. As you can see, four of the six main effects show up as significant, and one of the two-factor interaction effects. I'd say that three of these effects are massive: the main effects of the factors Boundedness and Oil Red O and the interaction effect involving these two factors. The three effects are all positive, indicating that using the sugar complex and encapsulating the vitamin in Oil Red O yields a spectacular increase in the stability ratio.

[Dr. Xu, pointing at Table 8.9] The results match my own findings, which are not based on regression methodology but on a close examination of the data table. I immediately observed that eight of the 32 experimental runs we did yielded desirable stability ratios. That is stability ratios of 200 and more. These eight runs have only two-factor settings in common: the vitamins were bounded in the sugar complex and encapsulated in Oil Red O. So, the results from your regression confirm what I thought after having seen the data.

[Peter] Great. In any case, there is a very strong synergistic interaction effect involving the factors Boundedness and Oil Red O. The main effect of Boundedness and Oil Red O are hugely positive, and, on top of that, there is this strong positive interaction effect. It's always nice to see such overwhelmingly positive results from a screening study.

[Dr. Xu] Brad, you mentioned a second analysis that you ran on the data. What was that for? Given the clear conclusions we can draw from the analysis you just presented, I see no need for another analysis of the same data.

[Brad] I also ran an analysis using random instead of fixed block effects. The problem with the fixed block effects analysis is that it does not allow us to make predictions about future blocks, because that analysis assumes that the blocks are completely unrelated. That makes it impossible to generalize the results from a fixed block effects analysis to other blocks, for example by making predictions about future values of the stability ratio in your experiment. The second analysis I ran essentially assumes that the blocks are all related, or, in other words, that they share several characteristics. This is achieved technically by treating the blocks as if they come from a single population.

[Dr. Xu, scratching his chin while thinking aloud] And this makes it possible to make predictions? I have come across random effects in standard works about analysis of variance, but I cannot remember having seen them being using in textbooks on factorial and fractional factorial experiments, or response surface designs, in industry.

[Peter] I think the reason for this is purely technical. Many statisticians agree that using random instead of fixed block effects should be standard. However, if your design is not fully balanced as in those analysis of variance textbooks, then the estimation of the random block effects model and hypothesis testing becomes computationally more involved.

[Brad] Right. But these days, it is just as easy to use random blocks as it is to use fixed blocks. Modern statistical software is able to handle balanced as well as unbalanced data, which allows researchers to make predictions for future blocks.

[Peter] Plus, it offers researchers more flexibility when setting up experiments. This is because the random block effects model does not require the explicit estimation of all the block effects—just the variance of the block effects.

[Dr. Xu] Using random block effects requires thus only one parameter to be estimated for the blocks, the variance of the block effects, whereas the use of fixed block effects requires the estimation of as many block effects as you have blocks in your experiment. Because fewer model parameters have to be estimated when using random block effects, is it sometimes possible to reduce the number of runs and still estimate the factor effects you're interested in?

[Brad] That is another advantage of designing with a random block effects model in mind.

[Dr. Xu] That sounds attractive indeed. Now, because you mentioned predictions, the thought occurred to me that we will need prediction intervals for our stability ratio. Could you provide such intervals? The reason I ask is that we will have to estimate the shelf life of the riboflavin. Of course, we would like to have a conservative estimate. The lower bound of a prediction interval for the stability ratio then is a crucial piece of information, isn't it?

[Brad] Definitely. Let me see what I can do for you.

Brad digs up the random block effects analysis of the data from the stability improvement experiment, and Table 8.11 flashes up on his screen.

[Brad, putting Tables 8.11 and 8.10 side by side] Here we go. This is the simplified model I got from the random block effects analysis. The significant terms happened to be the exact same ones as for the fixed block effects analysis for your data. As you can see, the qualitative conclusions from the two analyses agree with each other.

Table 8.11   Results from the random block effects analysis.

Effect	Estimate	Standard error	DF	t Value	p Value
Intercept	93.91	1.51	6.93	62.07	< .0001
Boundedness[Complex]	43.01	0.66	20.21	65.34	< .0001
Oil Red O[In]	63.01	0.64	19.70	99.06	< .0001
Sulisobenzone[In]	6.16	0.66	20.21	9.36	< .0001
Deoxybenzone[In]	5.50	0.71	21.52	7.72	< .0001
Boundedness[Complex] × Oil Red O[In]	35.10	0.66	20.32	53.10	< .0001

[Peter] An interesting aspect of the random block effects analysis is that the degrees of freedom for the significance tests, which are displayed in the column labeled DF, are all larger than 19, which is the residual degrees of freedom that is used for the hypothesis tests in the fixed block effects analysis.

[Dr. Xu, checking the degrees of freedom for the fixed block analysis] 32 runs minus one for the intercept, four for the four significant main effects, one for the significant two-factor interaction effect, and seven for the block effects. Gives us 19 residual degrees of freedom alright. Now, where do these noninteger degrees of freedom in the random block effects analysis come from? And, more importantly, what do they mean?

[Peter] Where they come from is a very technical issue. Let me just say that they are degrees of freedom calculated using the method of two British researchers, Kenward and Roger. The fact that these Kenward–Roger degrees of freedom are larger than 19 essentially means that the random block effects analysis extracts more information from your data than the fixed block effects analysis. For the first main effect, for instance, the random block effects analysis reports 20.21 degrees of freedom. Roughly speaking, this means that the analysis has been able to find 20.21 units of information about the first main effect in the data. The fixed block effects analysis retrieved only 19 units of information about that main effect.

[Dr. Xu] Are you saying that the random block effects analysis gets more out of the data than the fixed block analysis?

[Peter] Yes, that's right. The fixed block effects analysis essentially only uses what statisticians call intra-block or within-block information, whereas the random block effects analysis also recovers inter-block or between-block information.

The conversation continues and ends with a prediction of shelf life.

8.3   Peek into the black box

8.3.1   Models involving block effects

As we explained in the previous chapter, the model for analyzing data from an experiment run in b blocks of k observations requires an additive block effect for every block. A general expression for a regression model involving additive block effects is

$$Y_{ij} = \mathbf{f}'(\mathbf{x}_{ij})\boldsymbol{\beta} + \gamma_i + \varepsilon_{ij}, \tag{8.1}$$

where $\mathbf{x}_{ij}$ is a vector that contains the levels of all the experimental factors at the jth run in the ith block, $\mathbf{f}'(\mathbf{x}_{ij})$ is its model expansion, and $\boldsymbol{\beta}$ contains the intercept and all the factor effects that are in the model. The parameter γ_i represents the ith block effect and ε_{ij} is the random error associated with the jth run on day i.

The function of the block effects is to capture the differences between the responses obtained from different blocks. The block effects reduce the unexplained variation in the responses. This results in more powerful hypothesis tests concerning the factor effects. The beneficial impact of including block effects in the model is well known and it is described in many textbooks on the design and analysis of experiments.

What is not so commonly known is that we can make two different assumptions regarding the block effects: we can assume that the block effects are fixed or random. The assumption of fixed block effects is the one used in currently available textbooks on industrial design and analysis of experiments.

The use of fixed block effects implies that the block effects are treated as unknown fixed constants. In the random block effects approach, the block effects are viewed as realizations of a random variable with a zero mean and a certain variance, σ_γ^2. We then assume that the block effects are randomly drawn from an underlying population, so that the block effects are not arbitrarily different from each other. If we model using fixed block effects, then we implicitly assume that the block effects are just numbers that have no relationship to each other. The logical implication of this is that there is no basis for comparing experimental results from different blocks. As a result, if fixed blocks are used, the factor effects have to be estimated by just comparing responses from observations within each of the blocks.

We prefer treating the block effects as random for three reasons:

1. Assuming random blocks is more often realistic than working with fixed block effects because, in many cases, the blocks actually used in the experiment have a lot in common and can be viewed as belonging to some larger population of blocks. The assumption of random blocks effects also offers a framework for making predictions about future observations. This is due to the fact that every future block is selected from the same population of blocks, and, hence, will share characteristics with the blocks that were used in the experiment. If we use fixed block effects, then there is no implied relatedness between one block effect and another. Therefore, under the fixed block effects assumption, we cannot use the experimental data to make predictions about blocks other than those used in the experiment. This is undesirable in many practical situations. To avoid this, it seems reasonable to view the blocks used in the experiment as representative of the population of blocks that we will observe in the future.

That allows us to generalize the experimental results to future blocks and thus to make predictions.

2. Assuming random block effects does not require the estimation of all block effects. Instead, only the variance of the block effects needs to be estimated. As a result, fewer degrees of freedom are sacrificed and more degrees of freedom become available for estimating factor effects. We consider this a very important advantage since, after all, experiments are run to acquire information about the factors' effects. Making degrees of freedom available for estimating these effects is thus a good thing.

3. A third reason for using random instead of fixed block effects is that, for a given design, it results in more precise estimates of the factor effects. This is because, with random block effects, the data are used more efficiently. More specifically, the random block effects approach uses inter-block information on top of intra-block information to estimate the factor effects. This is different from the fixed block effects approach which uses only intra-block information. Roughly speaking, assuming fixed blocks comes down to assuming that the blocks in the experiment are so different that nothing can be learned from comparing responses from different blocks. The factor effects are then estimated using the information obtained by comparing responses that belong to the same block only. That information is called intra-block or within-block information. The assumption of random block effects does not go as far as saying that intra-block information is the only information available. Instead, it uses inter-block information, or between-block information, as well to estimate the factor effects. Inter-block information is the information that is obtained from comparing responses across blocks. The usefulness of the inter-block information depends on how different the blocks in the experiment are. If the blocks are very different from each other, then comparing responses across blocks does not make much sense because any difference observed is almost entirely due to the block effect. In that case, the inter-block information does not add much to the analysis. However, if the blocks are not very different, then it does make sense to compare responses across blocks. In that case, the inter-block information offers substantial added value. An interesting aspect of assuming random block effects is that the resulting statistical analysis, which involves generalized least squares (GLS) estimation of the factor effects and restricted maximum likelihood (REML) estimation of the variance components, combines the intra-block and the inter-block information so that the resulting factor effects' estimates have maximum precision.

In our opinion, the first and the second advantage are the most important ones. The first advantage is of crucial importance to predict whether future output will be within specifications. The second advantage is important because it opens the prospect of estimating more factor effects at the same experimental cost, or to reduce the experimental cost while still being able to estimate the same set of factor effects. It is sometimes even possible to estimate a larger number of factor effects at a lower experimental cost. The third advantage associated with assuming random block effects, the gain in precision of the factor effects estimates, is less spectacular. This gain even completely evaporates when the differences between the blocks are large relative to the residual error variance, when the design is orthogonally blocked, or when the number of blocks is extremely small.

8.3.2   Fixed block effects

In matrix notation, the model for analyzing data from blocked experiments is

$$Y = X\beta + Z\gamma + \varepsilon, \tag{8.2}$$

where Y, X, β, Z, γ, and ε are defined as in Section 7.3.2. If we assume that the block effects are fixed parameters, then the only random effects in the model are the residual errors. We assume that the residual errors are independent and normally distributed with zero mean and variance σ_ε^2.

Under these assumptions, the responses, collected in the vector Y, are all independent so that their variance–covariance matrix, var(Y), equals $\sigma_\varepsilon^2 I_n$. We can, therefore, estimate the factor effects β and the block effects γ using OLS. There is, however, a small complication: the intercept and the block effects cannot be estimated independently because there is perfect collinearity between the columns of Z and the column of X that corresponds to the intercept. As a matter of fact, summing the rows of Z produces a column of ones which is identical to the intercept column of X. We already encountered that problem in Section 4.3.2 when dealing with categorical experimental factors, and in Section 6.3.2 when dealing with mixture experiments. To circumvent the problem in the context of blocked experiments here, we can use one of the following remedies:

1. Drop the intercept from the model, in which case any parameter γ_i represents the average response from the ith block.

2. Drop one of the parameters γ_i from the model, in which case the intercept can be interpreted as the average response from the ith block and every remaining γ_j is the difference between the average responses from blocks j and i.

3. Constrain the estimates of the parameters γ_i to sum to zero, so that the intercept can be interpreted as the overall average response across the blocks and every γ_i indicates the difference between the average response from block i and the overall average across blocks.

Although there is no qualitative difference between each of these three remedies (each of the remedies leads to the same estimates for the factor effects), we prefer the latter solution because it preserves the natural interpretation of the intercept as the average response. This solution to the collinearity problem can be obtained using a different kind of coding, named effects-type coding. The model then becomes

$$Y = X\beta + Z_{ET}\gamma_{ET} + \varepsilon, \tag{8.3}$$

where

$$Z_{ET} = \begin{bmatrix} 1 & 0 & \cdots & 0 \\ \vdots & \vdots & \ddots & \vdots \\ 1 & 0 & \cdots & 0 \\ 0 & 1 & \cdots & 0 \\ \vdots & \vdots & \ddots & \vdots \\ 0 & 1 & \cdots & 0 \\ \vdots & \vdots & \ddots & \vdots \\ -1 & -1 & \cdots & -1 \\ \vdots & \vdots & \ddots & \vdots \\ -1 & -1 & \cdots & -1 \end{bmatrix} \tag{8.4}$$

and

$$\gamma_{ET} = \begin{bmatrix} \gamma_1 & \gamma_2 & \cdots & \gamma_{b-1} \end{bmatrix}'.$$

This model involves an intercept and $b-1$ block effects. The key feature of effects-type coding is that we create $b-1$ new three-level factors to be included in the model. The ith three-level factor takes the value 1 for runs in block i, the value -1 for runs in the last block, and the value 0 for runs in any other block. This is why the matrix Z_{ET}, which collects the $b-1$ three-level factors, has elements equal to -1, 0, or 1. Its first column has ones in the first few rows because the first few runs belong to the first block. Its first column has negative ones in its last rows because the last few runs belong to the last block. The intercept and the factor effects, contained within the vector β, and the $b-1$ block effects in γ_{ET}, are estimated using the OLS estimator

$$\begin{bmatrix} \hat{\beta} \\ \hat{\gamma}_{ET} \end{bmatrix} = \left(\begin{bmatrix} X & Z_{ET} \end{bmatrix}' \begin{bmatrix} X & Z_{ET} \end{bmatrix} \right)^{-1} \begin{bmatrix} X & Z_{ET} \end{bmatrix}' Y,$$

$$= \begin{bmatrix} X'X & X'Z_{ET} \\ Z'_{ET}X & Z'_{ET}Z_{ET} \end{bmatrix}^{-1} \begin{bmatrix} X'Y \\ Z'_{ET}Y \end{bmatrix}, \tag{8.5}$$

which has variance–covariance matrix

$$\text{var} \begin{bmatrix} \hat{\beta} \\ \hat{\gamma}_{ET} \end{bmatrix} = \sigma_\varepsilon^2 \begin{bmatrix} X'X & X'Z_{ET} \\ Z'_{ET}X & Z'_{ET}Z_{ET} \end{bmatrix}^{-1}. \tag{8.6}$$

The most relevant part of that matrix is the one corresponding to the estimates of the factor effects, $\boldsymbol{\beta}$. It can be shown that part equals

$$\begin{aligned}\text{var}(\hat{\boldsymbol{\beta}}) &= \sigma_\varepsilon^2\{\mathbf{X}'\mathbf{X} - \mathbf{X}'\mathbf{Z}_{\text{ET}}(\mathbf{Z}'_{\text{ET}}\mathbf{Z}_{\text{ET}})^{-1}\mathbf{Z}'_{\text{ET}}\mathbf{X}\}^{-1}, \\ &= \sigma_\varepsilon^2\{(\mathbf{X}'\mathbf{X})^{-1} + \boldsymbol{\Delta}\}, \end{aligned} \tag{8.7}$$

where

$$\boldsymbol{\Delta} = (\mathbf{X}'\mathbf{X})^{-1}\mathbf{X}'\mathbf{Z}_{\text{ET}}(\mathbf{Z}'_{\text{ET}}\mathbf{Z}_{\text{ET}} - \mathbf{Z}'_{\text{ET}}\mathbf{X}(\mathbf{X}'\mathbf{X})^{-1}\mathbf{X}'\mathbf{Z}_{\text{ET}})^{-1}\mathbf{Z}'_{\text{ET}}\mathbf{X}(\mathbf{X}'\mathbf{X})^{-1}.$$

Because $\boldsymbol{\Delta}$ is a positive definite matrix, this shows that, in general, the variance of $\hat{\boldsymbol{\beta}}$ is larger than $\sigma_\varepsilon^2(\mathbf{X}'\mathbf{X})^{-1}$, which was the variance–covariance estimator for $\hat{\boldsymbol{\beta}}$ in the absence of blocks. The impact of running an experiment in blocks thus is that the variances of the factor-effect estimates are inflated.

There is, however, one exception to this rule. If the design is chosen such that $\mathbf{X}'\mathbf{Z}_{\text{ET}} = \mathbf{0}_p$, then the variance–covariance matrix of the factor-effect estimates $\hat{\boldsymbol{\beta}}$ reduces to

$$\text{var}(\hat{\boldsymbol{\beta}}) = \sigma_\varepsilon^2(\mathbf{X}'\mathbf{X})^{-1} \tag{8.8}$$

so that there is no variance inflation due to the blocking. It can be shown that the condition that $\mathbf{X}'\mathbf{Z}_{\text{ET}} = \mathbf{0}_p$ is equivalent to the condition for orthogonal blocking in Equation (7.13). In that case, we also have that

$$\hat{\boldsymbol{\beta}} = (\mathbf{X}'\mathbf{X})^{-1}\mathbf{X}'Y, \tag{8.9}$$

which means that the factor-effect estimates are independent of the block effects.

We can compute a VIF for every factor effect by taking the ratio of the diagonal elements of the variance–covariance matrix in Equation (8.7) and those of $\sigma_\varepsilon^2(\mathbf{X}'\mathbf{X})^{-1}$. This VIF quantifies the extent to which each individual variance is increased due to the presence of blocks in the experiment. The smaller the VIFs, the smaller the impact of the blocks on the factor-effect estimates. The inverse of the VIF is the efficiency factor, which we used in Chapter 7.

8.4   Background reading

More details about the optimal design of experiments in blocks can be found in Chapters 4 and 5 of Goos (2002). In these chapters, which review much of the literature on the subject, the focus is on random block effects. Goos (2006) gives some further discussion about the optimality and orthogonality of block designs.

Goos and Vandebroek (2001) study D-optimal designs for blocked experiments with random blocks, and show that, when it comes to finding a D-optimal design, fixed block effects are a special, limiting, case of random block effects. Atkinson and

Donev (1989) and Cook and Nachtsheim (1989) study optimal designs for the case of fixed block effects.

Wu and Hamada (2009) provide tables for blocking two-level designs and three-level designs, created while assuming fixed block effects. Myers et al. (2009) and Montgomery (2009) discuss how to arrange central composite designs and Box–Behnken designs in blocks. In these textbooks, the focus is on orthogonality.

Khuri (1992), Gilmour and Trinca (2000), and Goos and Vandebroek (2004) give more technical details and discussion about the analysis of data from blocked experiments. The restricted maximum likelihood or REML method, which is recommended by Gilmour and Trinca (2000) for analyzing data from blocked experiments, was originally proposed by Patterson and Thompson (1971).

8.5   Summary

Blocking is a key concept in designed experiments. It is especially valuable in experimenting with systems where the signal-to-noise ratio is small, i.e., when there is a lot of variation in the responses that cannot be explained by the effects of the experimental factors. In these situations, the reduction of the unexplained variation in the responses due to the inclusion of a blocking factor in the model makes it possible to detect smaller effects.

We recommend modeling the blocks using random effects. This allows us to make predictions about individual runs in future blocks as long as the blocks in the experiment come from the same population of blocks as the future blocks. Modeling the blocks using random effects also offers the advantage that only one parameter, the block effects' variance, has to be estimated to account for the block-to-block differences in the analysis. Therefore, as a result of the blocking and assuming random block effects, one fewer factor effect can be estimated.

Historically, investigators have modeled blocks using fixed effects. Doing this has two important drawbacks. First, it is not possible to predict the response for a future processing run. Second, using, say, eight blocks in an experiment means that $8 - 1 = 7$ block effects have to be estimated. This implies that seven fewer factor effects can be estimated than in a situation where there are no blocks.

In conclusion, experiments based on a model involving random block effects either allow the fitting of the same model with fewer runs, or the estimation of a model with more factor effects with the same number of runs, than experiments based on a model involving fixed block effects.

The trade-off involved in using random blocks is that ordinary least squares or OLS is no longer the best way to estimate the factor effects. The correct analysis requires the use of generalized least squares, which is more involved. However, modern computing power has obviated this problem.

9

Experimental design in the presence of covariates

9.1 Key concepts

1. Even though you cannot control some input variables, if you know their values in advance of experimentation, you can design your experiment maximizing the total information given the values of the known but uncontrolled inputs.

2. When processes are subject to linear drift over time, and experimental runs happen roughly uniformly over time, then treating the time when each run is done as a covariate factor can remove the effect of the drift. Choosing the other factors to be nearly orthogonal to the time covariate maximizes the information about all the effects.

3. Another useful example of the use of covariate information comes from making prior measurements of important characteristics of the experimental material. Where people are the subjects of experimentation, relevant demographic information (also known as concomitant variables) about each subject can also serve as a covariate factor or factors.

When setting up experiments, we would like to use all the available information. Sometimes, part of that information comes in the form of covariates, also known as concomitant variables. These variables provide information about the experimental units. In this chapter, we show how to take covariate information into account when setting up an experiment.

Optimal Design of Experiments: A Case Study Approach, First Edition. Peter Goos and Bradley Jones.
© 2011 John Wiley & Sons, Ltd. Published 2011 by John Wiley & Sons, Ltd.

9.2   Case: the polypropylene experiment

9.2.1   Problem and design

Peter and Brad are in a train approaching the town of Oudenaarde in Belgium. Peter is briefing Brad about Oudenaarde and the meeting they are about to have with Marc Desmet, a chemical engineer at BelgoPlasma.

[Peter] The most well-known landmark in Oudenaarde is definitely its late gothic town hall built in 1538. It is on the UNESCO's World Heritage list. Before we take the train back, we should have a drink on the grand market right in front of the town hall so you can enjoy the local color in Flanders.

[Brad] Sounds like a plan. Can you tell me more about the company we are visiting?

[Peter] Sure. BelgoPlasma is a company that specializes in surface treatments for a wide variety of applications. Whether it is for making textiles water-absorbing or water-resistant, activating polyamide films, coating insulation products for aeronautics, for pretreatments for printing, gluing or coating of plastics, or treatment of catheters and syringes for medical applications, BelgoPlasma has the right solution.

[Brad] Judging from the name BelgoPlasma, they use some kind of plasma treatment?

[Peter] Right. A gas plasma treatment to be precise. Plasma is the fourth state of matter. Adding energy to matter changes it from a solid to a liquid to a gas, and from gas to plasma. What BelgoPlasma uses is low pressure plasma equipment where plasma is created by an electromagnetic discharge into a gas at low pressure. To treat an object, you place it in a vacuum chamber for some time, add a special type of gas at a certain rate using the right power, and you get a chemical reaction at the surface of your object.

[Brad] Interesting. How do you know all this?

[Peter] I helped Marc Desmet, the engineer we will meet in half an hour, with a few studies before.

[Brad] Why does he need our help now?

[Peter, looking outside the window as the train arrives in Oudenaarde's train station] On the phone, he said something about determining tolerances for additives in the plastics they treat.

A short taxi ride later, Brad and Peter are welcomed by Marc Desmet in a small meeting room where many of the products BelgoPlasma fabricates are on display.

[Marc] Hi Peter. Good to see you again, and thanks for coming at such short notice.

[Peter] You're welcome. Good to see you too. I brought my partner from Intrepid Stats, Brad Jones, who is also a design of experiments expert.

Brad and Marc shake hands and exchange the usual introductory chitchat. After ordering coffee for everyone, Marc comes to the point.

[Marc] For one of our applications, we are currently processing polypropylene products having different formulations. Almost always, the polypropylene formulation involves various additives such as ethylene propylene diene monomer rubber (also known as EPDM), ethylene, or ethylene vinyl acetate (EVA). The polypropylene I am interested in right now involves between 0% and 10% of polypropylene and between 0% and 15% of ethylene. About half the time, it contains 0.5% of EVA. EVA is an additive you need if you want to give your polypropylene a certain color.

[Brad] So, we're talking about polypropylene with three additives here.

[Marc] Yes. The problem is that these additives have an impact on the adhesion of coatings and glues to the polypropylene, even after a gas plasma treatment. What we would now like to do is determine tolerances for these three additives, and use these tolerances as specifications for our suppliers.

[Peter] Would you like to run a designed experiment to determine the tolerances?

[Marc] Yes, what we would like to do is an experiment where we study the impact of the three additives as well as four process variables: the power, the reaction time, the gas flow rate, and the type of gas we use for the gas plasma treatment. The problem is that we cannot choose the values for the additives at liberty.

[Peter] Why not? I am sure you can order any polypropylene formulation you need for your designed experiment.

[Marc] That is undeniably so. But on our shelves, we have plenty of polypropylene samples that are representative of what we process here. We would like to use some of these samples to save time and expense. We save expense because we don't have to order any new samples, and we save time because we avoid the whole burden of seeking an appropriate supplier and ordering the exact polypropylene formulations that we want.

[Brad] That makes sense if the proportions of EPDM and ethylene in the samples you have in stock span the ranges you named before. What were they? From 0% to 15% for ethylene and from 0% to 10% for EPDM?

[Marc, turning the screen of his laptop in the direction of Brad and Peter so that they can see Table 9.1] You needn't worry about that. This table shows a list of the EPDM and ethylene proportions for each of the 40 samples we want to work with. I also created two scatter plots of the proportions of EPDM and ethylene, one for the 17 samples containing EVA and one for the 23 samples not containing EVA. Here they are.

A few mouse clicks later, the scatter plots in Figure 9.1 appear on Marc's screen.

[Peter] There is a nice spread of the proportions indeed, for samples with EVA as well as for samples without EVA. It is easy to see from the scatter plots that the proportions were rounded to two decimal places.

[Marc] That should not cause any concern. What we would like to do is to select a subset of these samples to use in our experiment. We believe we can afford 18 runs, so we need a subset of 18 samples.

[Brad] On top of that, you need 18 settings for the other factors you want to study, power, reaction time, flow rate, and gas type.

[Marc] Correct. I have two kinds of factors. The first kind of factors are the proportions of EPDM and ethylene, and whether or not EVA is present. For these factors, I have a limited number of choices that I can make. The second kind of factors, I'd call ordinary experimental factors because I can choose their levels myself.

Table 9.1 Proportions of additives contained within the polypropylene formulation of the 40 samples at BelgoPlasma.

Sample	EPDM	Ethylene	EVA	Sample	EPDM	Ethylene	EVA
1	0.04	0.00	Yes	21	0.04	0.07	Yes
2	0.02	0.07	Yes	22	0.04	0.08	No
3	0.01	0.14	Yes	23	0.04	0.08	Yes
4	0.01	0.06	No	24	0.02	0.07	No
5	0.09	0.15	Yes	25	0.08	0.10	No
6	0.09	0.12	No	26	0.03	0.07	Yes
7	0.04	0.00	No	27	0.03	0.11	No
8	0.01	0.11	Yes	28	0.04	0.06	No
9	0.09	0.02	No	29	0.08	0.09	Yes
10	0.10	0.13	No	30	0.09	0.09	Yes
11	0.05	0.14	No	31	0.04	0.02	No
12	0.01	0.09	No	32	0.09	0.10	No
13	0.09	0.02	Yes	33	0.07	0.05	No
14	0.10	0.09	Yes	34	0.08	0.12	No
15	0.07	0.02	No	35	0.03	0.08	Yes
16	0.10	0.12	Yes	36	0.07	0.03	Yes
17	0.01	0.02	No	37	0.07	0.04	Yes
18	0.10	0.04	Yes	38	0.07	0.07	No
19	0.04	0.06	No	39	0.06	0.09	No
20	0.04	0.09	No	40	0.04	0.03	No

[Peter] We call the first kind of factors you have covariates. The term covariate is jargon for a property of an experimental unit. Here your samples are the experimental units. In medical experiments or marketing studies, where the term covariate is more commonly used, the experimental unit is a patient or a respondent.

[Marc] That sounds as if you are familiar with my design problem?

[Peter] I have seen design problems that resemble yours, but your kind of problem, involving three covariates and 40 experimental units—or samples—to choose from, is certainly not a routine industrial design of experiments problem.

[Brad] Peter, when you say you have seen design problems that resemble Marc's, are you thinking about experiments involving time-trend effects?

[Peter] I am.

[Marc] What are these time-trend effects?

[Brad] In some experimental settings, there is a drift in the response due to the wear of the equipment or warming-up of machines. Ideally, this drift is accounted for when you set up the experiment. To do so, you have to include time as an explanatory variable in your model. You can view the time at which you perform a run of the experiment as a covariate of the experimental unit "time window."

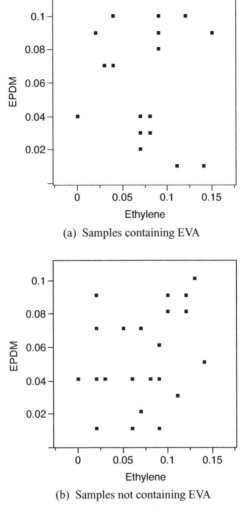

(a) Samples containing EVA

(b) Samples not containing EVA

Figure 9.1   Scatter plots of the proportion of EPDM versus the proportion of ethylene in Marc's samples.

[Marc, impatiently] Hmm. I can see the analogy, but I'd rather focus on my problem before talking about other people's problems.

[Peter] One thing we need to know before we can create a good design for your problem is the model you want to fit.

[Marc] Remember that in the previous experiments you helped us design, a main-effects model provided a good fit to the data every time, even for our newest vacuum chambers.

[Peter] Right. Much to our surprise, there was virtually never a practically important interaction effect involving the additives and the process variables.

[Brad, grabbing his laptop] There is one other thing we need to know. You mentioned gas type as an experimental factor. That is a categorical factor. How many levels does it have?

[Marc] We usually study two different activation gases plus an etching gas. So, three levels.

[Brad, toggling on his machine] Thanks. It is convenient that a main-effects model is good enough, because a small number of runs suffices to fit such a model. Hmm, we need a model that has an intercept, three main effects for the covariates, three main effects for the quantitative process variables time, power and flow rate, and two parameters to capture the main effect of the categorical factor gas type.

[Peter] So, we need to estimate nine model parameters in total: eight factor effects and an intercept. The 18 runs you can afford will be plenty.

By now, Brad is bent over his laptop and typing the names of the experimental factors. Marc and Peter follow that process on the screen of Brad's laptop, nodding in agreement.

[Brad, turning around as he hands over his memory stick to Marc] Could you put the table with the covariate values of your 40 samples on my memory stick? That is the last piece of information I need before I can produce a design for you.

Marc quickly copies the table onto Brad's memory stick and it does not take very long before Brad has opened the data table in his design of experiments software.

[Brad] Oops! We forgot to ask you what responses you will measure. I hope your response is not binary or categorical?

[Marc] No. Our response is quantitative. The surface tension, as usual.

[Peter] I forgot to tell you, Brad, that this is Marc's favorite response because the surface tension of the polypropylene is an excellent predictor of the quality of adhesion that can be achieved for coatings and glues.

[Brad] I see.

[Marc, looking at Brad] Why are you concerned about the response?

[Brad] Well, the type of response you have determines the kind of model you fit to the data. If your response were binary—for instance whether or not you have good adhesion—then a linear model would not be appropriate. Instead you'd need a logistic regression model, and the experimental design you need for such a model is different from the one that is required for the linear models we usually work with. If your response were an ordered categorical one—for instance, whether there is no, poor, good, or perfect adhesion—then you'd need an experimental design for yet another model, the cumulative logit model.

[Marc] But the usual techniques for linear regression models work for my response?

Peter and Brad nod, and Brad goes back to his laptop. It only takes him a few minutes to produce the design in Table 9.2.

[Marc] That was impressively fast, but I'd be even more impressed if you can convince me this is a good design. Can you visualize which of the 40 samples have been chosen? Do the low and high levels appear equally often? Are each of the gases used an equal number of times?

[Brad, creating the scatter plots in Figure 9.2] One question at a time. Look at these scatter plots. They are identical to the ones you showed us before, but in my

Table 9.2   D-optimal design for the polypropylene experiment involving the proportions of EPDM and ethylene and the presence of EVA as covariates.

EPDM	Ethylene	EVA	Flow rate	Power	Time	Gas type
0.01	0.14	Yes	High	High	Low	Activation gas 1
0.10	0.13	No	Low	High	Low	Etching gas
0.01	0.02	No	Low	High	Low	Activation gas 1
0.09	0.02	No	Low	High	Low	Etching gas
0.04	0.00	Yes	High	Low	Low	Etching gas
0.09	0.15	Yes	Low	Low	Low	Activation gas 2
0.01	0.11	Yes	Low	Low	High	Etching gas
0.09	0.02	Yes	Low	High	High	Activation gas 1
0.10	0.04	Yes	High	Low	High	Activation gas 2
0.04	0.00	No	Low	Low	Low	Activation gas 2
0.05	0.14	No	Low	Low	High	Activation gas 1
0.09	0.12	No	High	High	High	Activation gas 2
0.01	0.09	No	High	Low	High	Etching gas
0.10	0.12	Yes	High	Low	Low	Activation gas 1
0.02	0.07	Yes	Low	High	High	Activation gas 2
0.10	0.09	Yes	High	High	High	Etching gas
0.07	0.02	No	High	Low	High	Activation gas 1
0.01	0.06	No	High	High	Low	Activation gas 2

plots, the selected samples have larger point sizes than those that were not selected. You can immediately see that the selected samples are those near the boundaries of the square, both for samples containing EVA and samples without EVA.

[Marc] Is it correct to say that the samples selected are those that are closest to the boundaries?

[Peter] Yes. Because these samples contain most information when you are interested in main effects only. You will certainly remember from your basic design of experiments course that designs for main-effects models have all their runs at the lowest and highest possible factor levels.

[Brad, seeing that Marc is nodding in consent] Right, you can also see that nine of the selected samples do contain EVA, whereas the other nine don't.

[Marc, focusing on Table 9.2, showing the design] I can also see that each of the three gas types is used equally often. And I see that the low and high levels of the factors power, time and flow rate each occur nine times too. That is attractive. By the way, we will use 1000 sccm and 2000 sccm as the levels of the flow rate, 500 Watts (W) and 2000 W for the power, and 2 and 15 minutes for the reaction time.

Brad starts toggling again, and, moving his mouse up and down rapidly, clicks repeatedly on the interactive histograms he has been creating. He uses the histograms to visualize the points he is making.

[Brad, focusing on the histograms in Figure 9.3(a), where the darkly shaded areas correspond to runs involving activation gas 1] The design is really beautiful. Look at

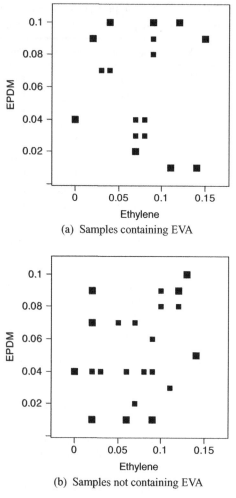

Figure 9.2   Scatter plots of the proportion of EPDM versus the proportion of ethylene showing the samples selected in the D-optimal design in Table 9.2. The larger squares represent the selected samples.

the six runs with the first activation gas. The shading in this graph shows that you need to do three of these runs at the low level of flow rate and the other three at the high level.

[Marc] So, you selected the six runs with activation gas 1 by clicking on the corresponding bar of the histogram for gas type? And then, I can see that I need to perform three of these runs at the low flow rate and three at the high flow rate. And also, these six runs show balance across the levels of the factors power and reaction time. Do I have it right?

[Brad, showing Figures 9.3(b) and (c)] Yes. Moreover, if we move on to the second activation gas and then to the etching gas, we can see exactly the same thing.

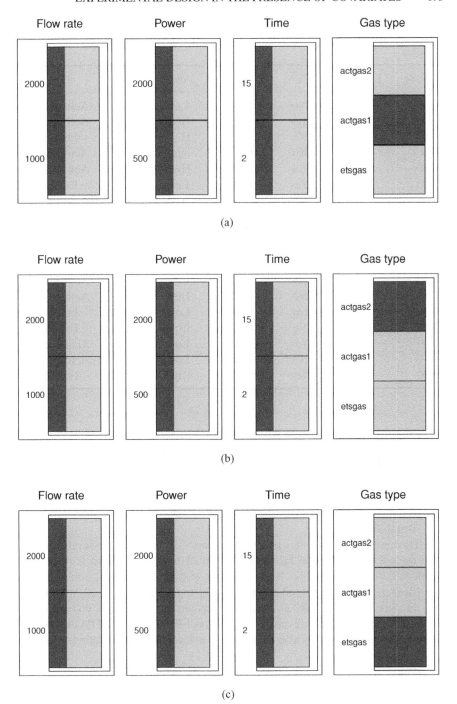

Figure 9.3   Histograms showing the frequencies of high and low levels for the factors flow rate, power, and time, for each gas type.

Table 9.3 D-optimal design for the polypropylene experiment treating the proportions of EPDM and ethylene and the presence of EVA as ordinary experimental factors.

EPDM	Ethylene	EVA	Flow rate	Power	Time	Gas type
0.00	0.15	No	Low	Low	Low	Activation gas 2
0.10	0.00	Yes	Low	High	High	Activation gas 1
0.10	0.00	No	Low	High	High	Etching gas
0.00	0.15	No	High	High	Low	Activation gas 1
0.10	0.00	Yes	High	High	High	Activation gas 2
0.00	0.15	Yes	Low	Low	High	Etching gas
0.00	0.15	No	Low	High	High	Activation gas 1
0.10	0.00	Yes	Low	High	Low	Etching gas
0.10	0.15	Yes	High	Low	High	Activation gas 1
0.10	0.15	Yes	Low	High	Low	Activation gas 2
0.00	0.00	No	High	High	High	Activation gas 2
0.00	0.00	Yes	High	Low	Low	Activation gas 1
0.10	0.15	No	High	Low	Low	Activation gas 2
0.10	0.15	No	High	Low	High	Etching gas
0.00	0.00	No	High	Low	Low	Etching gas
0.00	0.00	Yes	Low	Low	High	Activation gas 2
0.10	0.00	No	Low	Low	Low	Activation gas 1
0.00	0.15	Yes	High	High	Low	Etching gas

[Marc] I think this design will work fine. I have one final question though. We decided to use the polypropylene samples we have in stock. Would we have done much better had we ordered new samples with specifications dictated by an experimental design?

[Peter] Brad, why don't you generate a design in which you do not treat the proportions of EPDM and ethylene and the presence of EVA as given?

[Brad] Sure.

This time, Brad computes a new design with six ordinary two-level factors and one three-level categorical factor. The result of his computations is displayed in Table 9.3.

[Brad, using the same kind of interactive histogram as in Figure 9.3] This design has the same kind of balance as the one I computed before, but now you can see that, on top of that, we have nine runs at the low level of the factor EPDM and nine runs at its high level, as well as nine runs at the low level of the factor ethylene and nine runs at its high level.

[Peter] What does this additional balance give in terms of precision of the factor-effect estimates?

It takes a while before Brad has created Table 9.4. He enlarges the font size of the values in the table so that Marc and Peter can read them easily.

Table 9.4 Relative standard errors of the design in Table 9.2 treating the proportions of EPDM and ethylene and the presence of EVA as covariates and the design in Table 9.3 treating the proportions of EPDM and ethylene and the presence of EVA as ordinary experimental factors.

Coefficient	Table 9.2	Table 9.3
Intercept	0.2362	0.2358
EPDM	0.2893	0.2408
Ethylene	0.3491	0.2408
EVA	0.2435	0.2408
Flow rate	0.2394	0.2408
Power	0.2396	0.2408
Time	0.2400	0.2408
Gas type 1	0.3335	0.3333
Gas type 2	0.3338	0.3333

[Brad] This table compares the two designs I created. The column on the left shows the standard errors of the factor-effect estimates for the design that uses Marc's polypropylene samples, relative to σ_ε, the standard deviation of the errors. The column on the right shows the results for the design that treats the covariates as ordinary factors.

[Peter] The standard errors for the effects of the covariates EPDM and ethylene produced by the first design are substantially bigger than those produced by the second design, but the standard errors for the other factor effects are almost identical for the two designs. Note that, also for the categorical covariate EVA, there is almost no difference between the two designs. That's because the two designs balance the number of runs with and without EVA.

[Marc] The largest standard error is the one for ethylene, 0.3491. That seems rather large compared to the standard errors of the factor-effect estimates for the ordinary experimental factors.

[Brad] How large is your signal-to-noise ratio, usually? I mean, what is the ratio of your factor effects to σ_ε, the square root of the error variance?

[Peter] I had a look at some of the old reports I wrote for you, Marc, and the smallest signal-to-noise ratio I came across was larger than three. So, no worries.

The meeting then winds down, and Marc promises to run the design in Table 9.2 in the days following the meeting, and to send the data to Peter and Brad by e-mail.

9.2.2 Data analysis

A week later, Peter and Brad are back in Oudenaarde to present their analysis of the data from the polypropylene experiment. The data are displayed in Table 9.5. The flow rates are expressed in standard cubic centimeters per minute (sccm). The

Table 9.5   Design and responses for the polypropylene experiment.

EPDM	Ethylene	EVA	Flow rate	Power	Time	Gas type	Surface tension
0.01	0.14	Yes	2000	2000	2	Activation gas 1	39.02
0.10	0.13	No	1000	2000	2	Etching gas	56.81
0.01	0.02	No	1000	2000	2	Activation gas 1	43.29
0.09	0.02	No	1000	2000	2	Etching gas	62.69
0.04	0.00	Yes	2000	500	2	Etching gas	39.51
0.09	0.15	Yes	1000	500	2	Activation gas 2	26.18
0.01	0.11	Yes	1000	500	15	Etching gas	40.37
0.09	0.02	Yes	1000	2000	15	Activation gas 1	57.80
0.10	0.04	Yes	2000	500	15	Activation gas 2	32.61
0.04	0.00	No	1000	500	2	Activation gas 2	23.40
0.05	0.14	No	1000	500	15	Activation gas 1	27.26
0.09	0.12	No	2000	2000	15	Activation gas 2	45.34
0.01	0.09	No	2000	500	15	Etching gas	37.80
0.10	0.12	Yes	2000	500	2	Activation gas 1	23.93
0.02	0.07	Yes	1000	2000	15	Activation gas 2	50.37
0.10	0.09	Yes	2000	2000	15	Etching gas	66.56
0.07	0.02	No	2000	500	15	Activation gas 1	32.59
0.01	0.06	No	2000	2000	2	Activation gas 2	36.67

levels of the factor power are expressed in Watts, and the reaction times are expressed in minutes. The surface tension response was measured in milliNewton per meter (mN/m). When sending the data to Peter, Marc had written that he had played with the data himself to get a first impression but that he was unsure about the coding that was used for the categorical factors in the study. That made it difficult for him to interpret his software's output. Therefore, Peter uses more technical detail than usual during the presentation of his results.

[Peter] I have fitted a main-effects model to your data. The resulting model is on the next two slides. First, I show you what the model looks like when the product you want to process contains EVA.

Peter's next slide shows three different equations:

$$\text{Surface tension (EVA, etching gas)} = 41.06 + 1.50 + 9.27 + 2.98 \text{ EPDM}$$
$$- 3.27 \text{ Ethylene} - 1.37 \text{ Flow rate}$$
$$+ 10.05 \text{ Power} + 3.24 \text{ Time}$$
$$= 51.83 + 2.98 \text{ EPDM}$$
$$- 3.27 \text{ Ethylene} - 1.37 \text{ Flow rate}$$
$$+ 10.05 \text{ Power} + 3.24 \text{ Time}$$

$$\text{Surface tension (EVA, activation gas 1)} = 38.89 + 2.98 \text{ EPDM}$$
$$- 3.27 \text{ Ethylene} - 1.37 \text{ Flow rate}$$
$$+ 10.05 \text{ Power} + 3.24 \text{ Time}$$

Surface tension (EVA, activation gas 2) $= 36.97 + 2.98$ EPDM

$$-3.27 \text{ Ethylene} - 1.37 \text{ Flow rate}$$
$$+ 10.05 \text{ Power} + 3.24 \text{ Time}$$

[Peter] And here is what the model looks like when the product you want to process does not contain EVA.

Peter's next slide shows these three equations:

Surface tension (no EVA, etching gas) $= 41.06 - 1.50 + 9.27 + 2.98$ EPDM

$$-3.27 \text{ Ethylene} - 1.37 \text{ Flow rate}$$
$$+ 10.05 \text{ Power} + 3.24 \text{ Time}$$
$$= 48.83 + 2.98 \text{ EPDM}$$
$$-3.27 \text{ Ethylene} - 1.37 \text{ Flow rate}$$
$$+ 10.05 \text{ Power} + 3.24 \text{ Time}$$

Surface tension (no EVA, activation gas 1) $= 35.89 + 2.98$ EPDM

$$-3.27 \text{ Ethylene} - 1.37 \text{ Flow rate}$$
$$+ 10.05 \text{ Power} + 3.24 \text{ Time}$$

Surface tension (no EVA, activation gas 2) $= 33.97 + 2.98$ EPDM

$$-3.27 \text{ Ethylene} - 1.37 \text{ Flow rate}$$
$$+ 10.05 \text{ Power} + 3.24 \text{ Time}$$

[Peter] These equations show all we want to know about the factor effects. Each of the six equations shows that adding EPDM has a positive effect on the surface tension, while adding ethylene has a negative effect. The flow rate has a small negative effect. Finally, increasing the power has a large positive effect, and increasing the reaction time has a moderate positive effect.

[Marc] How do I interpret the value of 2.98 for the effect of the factor EPDM?

[Peter] Remember that we use coded factor levels in the analysis: a -1 for the low level of a quantitative factor such as EPDM and a $+1$ for its high level. The value of 2.98 means that switching from a low EPDM level to an intermediate level, or from an intermediate level to a high level, leads to an increase of 2.98 in surface tension. So, increasing the proportion of EPDM from low to high results in an increase in surface tension of two times 2.98, 5.96. Since the lowest value for the proportion of EPDM in your data was 1% and the highest value was 10%, that means you need to increase the proportion of EPDM by 9% to achieve that increase of surface tension of 5.96.

[Marc] That is clear. Now, when I look at the difference in intercept between the three models on either of your two slides, I can see that using the etching gas results in a surface tension that's about 13 units higher than using activation gas 1, and about 15 units higher than using activation gas 2. Right?

Peter nods, and Marc continues.

Table 9.6   Factor-effect estimates for the polypropylene experiment.

Effect	Estimate	Standard error	t Ratio	p Value
Intercept	41.0621	0.3199	128.37	< .0001
EPDM (0.01,0.1)	2.9771	0.3918	7.60	< .0001
Ethylene (0,0.15)	−3.2658	0.4730	−6.90	< .0001
EVA (yes)	1.5003	0.3299	4.55	0.0014
Flow rate (1000, 2000)	−1.3682	0.3243	−4.22	0.0022
Power (500, 2000)	10.0548	0.3245	30.98	< .0001
Time (2,15)	3.2417	0.3252	9.97	< .0001
Gas type (etching gas)	9.2681	0.4517	20.52	< .0001
Gas type (activation gas 1)	−3.6746	0.4522	−8.13	< .0001

[Marc, who has opened his laptop while talking and produced Table 9.6] Comparing the models on your two slides, I can see there is a difference in intercept of three units of surface tension. The equations have an intercept that is three units higher when EVA is present. That shows you really have to be careful when you interpret the standard output of a regression analysis. Here is the output I got.

Peter and Brad approach to have a look at Marc's screen.

[Marc] It is not hard to link most of the factor effects in this table to the prediction equations you have on your slides, but if you don't know about the codings used, you might get predictions or interpretations that are pretty far off. Nowhere in the output, there is mention of the type of coding used.

[Brad] You are right, and that is why the software has a graphical tool to help you understand the factor effects. If you scroll down a bit, you'll see it.

[Marc] Let's do that in a minute. First, please explain how I can figure out the impact of activation gas 2. There is no trace of that gas in this table.

[Peter, pointing to Table 9.6 on Marc's laptop] You can see here that the coefficient estimates for the two dummy variables corresponding to etching gas and activation gas 1 are 9.2681 and −3.6746. These two coefficients allow us to say that the difference in surface tension between the etching gas and the first activation gas is 9.2681 − (−3.6746) or 12.9427, all other things being equal. To know the impact of using the second activation gas, you need to know that there is a third *implied* coefficient. The sum of the three coefficient estimates for gas type, the ones in the output and the implied one, are constrained to sum to zero. So, the third, implied, coefficient is minus the sum of the other two. As a result, the implied coefficient for activation gas 2 is −{9.2681 + (−3.6746)}, which is −5.5936. That coefficient is the smallest of the three. So, the third activation gas leads to the lowest surface tension. About two units lower than the first activation gas, and 15 units lower than the etching gas.

[Brad] What you say, Peter, is completely correct, but it assumes that the coding used for the levels of the categorical factor is effects-type coding.

[Marc] Never heard about effects-type coding before. Can you explain how that works?

Table 9.7 Significance tests for the covariates and the experimental factors in the polypropylene experiment.

Factor	Degree(s) of freedom	Sum of squares	F Ratio	p Value
EPDM (0.01, 0.1)	1	105.91	57.73	< .0001
Ethylene (0, 0.15)	1	87.47	47.68	0.0001
EVA	1	37.93	20.68	0.0014
Flow rate (1000, 2000)	1	32.65	17.80	0.0022
Power (500, 2000)	1	1761.27	960.01	< .0001
Time (2, 15)	1	182.34	99.39	< .0001
Gas type	2	783.77	213.60	< .0001

[Peter] Sure. Let's use your three-level gas type factor as an example. To capture its effect, we need two dummy variables in the model. Whenever we have a categorical factor, we need as many dummy variables as the number of factor levels minus one. There are three gas types, so we need three minus one dummy variables. That explains why you can see only two estimates for the gas type in the software's output. The first dummy variable you need takes the value 1 for all runs with etching gas, 0 for all runs with activation gas 1, and −1 for all runs with activation gas 2. The second dummy variable takes the value 0 for all runs with etching gas, 1 for all runs with activation gas 1, and −1 for all runs with activation gas 2. If you use this coding, then you constrain the three effects of your categorical factor to sum to 0, which is one of the typical solutions for collinearity. And if you have three effects that sum to 0, it is not difficult to calculate the third effect once you know the first two.

[Marc] That's true. Another interesting thing my table shows is that all the factors have a significant effect, as all the p values in the table are smaller than 0.05.

[Peter] Here, a little caveat is in order because two of the p values are related to a three-level categorical factor. These p values are not so useful, since they depend on the coding you use for the dummy variables. If you want a meaningful p value for the three-level categorical factor, you need to look at the next table in the output.

Peter leans forward and presses the "Page Down" key on Marc's laptop, and Table 9.7 appears.

[Peter] Here we go. In this table, we have significance tests for every factor and these tests do not depend on the coding. The last line of that table contains a meaningful p value for the gas type. You can see that the p value is really small, so that the gas type is strongly significant. To obtain this p value, the software tested simultaneously whether the two coefficients of the dummy variables for gas type are significantly different from zero.

[Marc] That's new to me, and I wonder why the software doesn't suppress the individual p values then. Anyway, let us now have a look at the graphical tool that Brad mentioned.

[Peter] OK. But let me first summarize my two slides of equations. What we learn from them is that a higher proportion of EPDM, a lower proportion of ethylene, the presence of EVA, a lower flow rate, a higher power, and a longer reaction time all lead to an increase in surface tension, and therefore to a better adhesion. We have also seen than the etching gas gives better results than the first activation gas, which is itself better than the second activation gas.

[Brad] That means we can maximize the output by using the etching gas at a flow rate of 1000 sccm and a power of 2000 W for 15 minutes.

[Marc] My boss will not be very happy with that solution. Etching gas is incredibly expensive. We use it as little as possible. Also, 15 minutes is way too long. Whenever possible, we would like to use just 2 minutes because this allows us to process 7.5 times more products per unit of time. With a longer reaction time, we will never be able to keep up with the demand.

[Peter, browsing through the slides he had prepared for the meeting] Having been here a couple of times before, I expected you would say such a thing. Here, I have a slide for a scenario where you want to use activation gas 1, the better of the two cheap activation gases, and a 2-minute reaction time.

[Marc] Amazing, Peter. You have the right mind-set to start a new career in the gas plasma industry.

[Peter] Thanks, but I prefer the variety of projects our consulting brings. Looking at my slide, you can see two predicted surface tensions. Each of them is based on the assumption that the gas plasma treatment is done using the first activation gas for just 2 minutes. I have fixed the flow rate to 1000 sccm because low values for that factor give the highest surface tension.

[Marc] That is a good choice too, because a low flow rate is cheaper than a high flow rate.

[Peter] For the power and for the EPDM and ethylene proportions, I used intermediate values: 1250 W for the power, 5% for the proportion of EPDM and 7.5% for the proportion of ethylene.

[Marc] Sounds reasonable.

[Peter] The predictions I got were 36.68 mN/m for polypropylene samples containing EVA, and 33.68 mN/m for samples not containing EVA. The 95% prediction intervals were [34.87 mN/m, 38.5 mN/m] and [31.98 mN/m, 35.39 mN/m], respectively. So, the prediction intervals are about 3.5 mN/m wide.

[Marc] You know, Peter, that we like the surface tension to be 32 mN/m or higher. What I would now like to do is find out what values of the EPDM and ethylene proportions give acceptable surface tensions when I use activation gas 1 and your input settings for flow rate, time, and power.

[Peter] To answer your question, it is good to have a look at contour plots of the surface tension, as a function of the EPDM and ethylene proportions. We need two contour plots here, one for samples containing EVA and one for samples not containing EVA. Wait a second.

A few minutes later, Peter has produced the contour plots in Figure 9.4.

[Peter] Here we go. We can now use this to get a rough idea about the acceptable values for the proportions of EPDM and ethylene for samples with and without EVA. Let's start with samples that contain EVA.

[Brad] If I understood everything correctly, then what you want is surface tensions of 32 mN/m or higher. One would be tempted to look at the contour line with

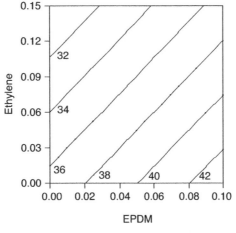

(a) Samples containing EVA

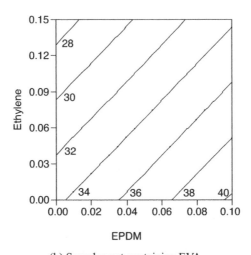

(b) Samples not containing EVA

Figure 9.4   Contour plots of the predicted surface tension as a function of the proportions of EPDM and ethylene using activation gas 1, a flow rate of 1000 sccm, a power of 1250 W, and a 2-minute reaction time.

label 32, and say that all the polypropylene samples southeast of that line will give satisfactory results. However, doing so would ignore the prediction error. What I would do is the following. The prediction intervals that you computed were about 3.5 mN/m wide, so the lower prediction limit is about 1.75 mN/m lower than the point prediction. Therefore, when we determine tolerances for the EPDM and ethylene proportions, we have to add 1.75 mN/m to the value of 32 mN/m that we consider the minimum acceptable surface tension. As a result, we have to look at the contour line corresponding to a surface tension of 33.75 mN/m. I see there is no such line in the plot Peter has made just now, but the contour line for a surface tension of 34 mN/m will do for now.

[Peter] We can do the exact math later on. Go ahead.

[Brad, walking up to a whiteboard, grabbing one of the erasable pens from its ledge and making a few calculations] The contour line crosses the points $(0, 0.06)$ and $(0.06, 0.15)$. So, the equation for the contour line is

$$\text{Ethylene proportion} = 0.06 + 1.5 \text{ EPDM proportion}.$$

For a polypropylene sample to be acceptable, its ethylene and EPDM proportions should satisfy the inequality

$$\text{Ethylene proportion} \leq \min(0.15, 0.06 + 1.5 \text{ EPDM proportion}).$$

[Marc, taking another pen from the ledge] Provided the polypropylene sample contains EVA. For samples not containing EVA, the relevant contour line goes through the origin, $(0, 0)$, and $(0.10, 0.15)$. Roughly speaking, that is. The equation of that contour line is

$$\text{Ethylene proportion} = 1.5 \text{ EPDM proportion}.$$

So, the samples should satisfy the inequality

$$\text{Ethylene proportion} \leq 1.5 \text{ EPDM proportion}.$$

[Peter] You got it.

[Brad] This procedure is based on the lower limit of 95% prediction intervals. Therefore, you will get about 97.5% conforming products and 2.5% nonconforming ones. Is that acceptable?

[Marc, scratching his head] My first reaction is to say yes. In any case, that would be much better than our current performance. Let me sleep on that. I've had enough for today. My brain is about to experience information overload. I would appreciate it if you could summarize everything we have discussed today in your report. Can you also include the detailed calculations of the tolerances for the polypropylene sample?

[Brad] Will do!

[Marc] Excellent. Now, can you show me that graphical tool you mentioned?

[Brad, displaying the plot in Figure 9.5] Sure. Here we go. The nice thing about the plot is that you need not know anything about the coding of factors to interpret the factor effects. The plot has one curve for each experimental factor. The factors' levels are on the horizontal axis, whereas the predicted response is on the vertical axis.

[Peter] The plot clearly shows the linear effects of the first six factors, as well as the effect of the categorical factor, gas type. For every factor, there is a vertical dotted line that indicates the selected factor levels. The predicted response for these levels, 36.7, is shown on the vertical axis, along with a 95% prediction interval.

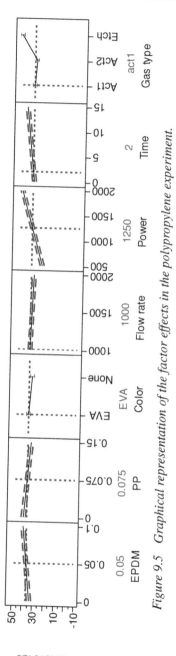

Figure 9.5 Graphical representation of the factor effects in the polypropylene experiment.

[Marc] The levels that Brad selected are the ones that match your recommendations—right Peter?

[Brad] That is true, but you can also change the factor levels interactively and observe how this affects the response.

[Peter] If there were interaction effects, then you would also see how the effect of one factor changes with the level of another factor. In our model, however, we have only main effects, so that we can't show you this nice feature now.

[Marc] In any case, this plot is a very helpful tool. Thank you very much for your help.

9.3   Peek into the black box

9.3.1   Covariates or concomitant variables

In the polypropylene case study, Peter mentions that the design problem at hand involves covariates. A synonym for a covariate is a concomitant variable. The key feature of a covariate is that its value can be measured, but that it is not under the control of the experimenter. It is a characteristic of the experimental unit in an experiment. Roughly speaking, an experimental unit is the item or the person that is the subject of each run of the experiment. In the polypropylene experiment, the experimental units were the 18 polypropylene samples, and the characteristics measured were the EPDM and ethylene proportions, and the presence of EVA. So, there were two quantitative covariates and a categorical one. In medical experiments, where the experimental units are patients, possible covariates are blood pressure, gender or age. Note that, in most textbooks, the term covariate is used only for quantitative variables. However, the rationale for using quantitative covariates is equally valid for categorical ones. Therefore, for our purposes, a covariate can be either quantitative or categorical.

In this section, we explain in detail how we deal with these kinds of variables in experimental design. First, we discuss the problem in general. Next, we focus on a specific covariate, time, and show how we can compute trend-free or trend-resistant designed experiments.

In our exposition, we assume that the values of the covariates are known before the actual running of the experiment. This way, we can take these values into account when selecting the combinations of levels for the experimental factors.

9.3.2   Models and design criteria in the presence of covariates

9.3.2.1   Models and scenarios

The models that we use for data from experiments involving covariates are identical to those that we encountered in Chapters 2–5. Therefore, the model estimation approaches and the inference procedures discussed in these chapters can be used in the presence of covariates as well.

The simplest model type that we can use is a main-effects model. If we denote the m experimental factors (that are under the control of the experimenter) by

$x_1, x_2, \ldots, x_m$ and the c covariates (which are not under the experimenter's control) by $z_1, z_2, \ldots, z_c$, then the main-effects model for experiments involving covariates is

$$Y = \beta_0 + \beta_1 x_1 + \ldots + \beta_m x_m + \gamma_1 z_1 + \ldots + \gamma_c z_c + \varepsilon,$$
$$= \beta_0 + \sum_{i=1}^{m} \beta_i x_i + \sum_{i=1}^{c} \gamma_i z_i. \tag{9.1}$$

This is the kind of model that Marc, Peter, and Brad used for the data from the polypropylene experiment. In some situations, that model will be inadequate due to the interaction of some of the experimental factors. In that case, a more appropriate model may be

$$Y = \beta_0 + \sum_{i=1}^{m} \beta_i x_i + \sum_{i=1}^{m-1} \sum_{j=i+1}^{m} \beta_{ij} x_i x_j + \sum_{i=1}^{c} \gamma_i z_i + \varepsilon. \tag{9.2}$$

Another interesting possibility is that there are interactions involving the experimental factors and the covariates. The effects of the experimental factors on the response then depend on the levels of one or more covariates. We then have to expand the model with terms that capture the interaction effects involving experimental factors, on the one hand, and covariates, on the other hand:

$$Y = \beta_0 + \sum_{i=1}^{m} \beta_i x_i + \sum_{i=1}^{m-1} \sum_{j=i+1}^{m} \beta_{ij} x_i x_j + \sum_{i=1}^{c} \gamma_i z_i$$
$$+ \sum_{i=1}^{m} \sum_{j=1}^{c} \beta_{ij}^{EC} x_i z_j + \varepsilon. \tag{9.3}$$

In this model, we use the superscript EC to emphasize the distinction between inter-action effects involving experimental factors and covariates, and interaction effects involving only experimental factors. Possibly, there are also interaction effects between the covariates. This leads to the following model:

$$Y = \beta_0 + \sum_{i=1}^{m} \beta_i x_i + \sum_{i=1}^{m-1} \sum_{j=i+1}^{m} \beta_{ij} x_i x_j + \sum_{i=1}^{c} \gamma_i z_i$$
$$+ \sum_{i=1}^{c-1} \sum_{j=i+1}^{c} \gamma_{ij} z_i z_j + \sum_{i=1}^{m} \sum_{j=1}^{c} \beta_{ij}^{EC} x_i z_j + \varepsilon. \tag{9.4}$$

Leaving interaction effects involving covariates out of the model may result in biased estimates of the factor effects. Therefore, whenever the number of experimental runs is sufficiently large, we recommend verifying whether these interaction effects are significantly different from zero, even when they are not of primary interest.

In matrix notation, each of the models (9.1)–(9.4) can be written as

$$Y = X\beta + Z\gamma + \varepsilon. \tag{9.5}$$

For the main-effects model (9.1), we have that

$$\mathbf{X} = \begin{bmatrix} 1 & x_{11} & \cdots & x_{m1} \\ 1 & x_{12} & \cdots & x_{m2} \\ \vdots & \vdots & \ddots & \vdots \\ 1 & x_{1n} & \cdots & x_{mn} \end{bmatrix}, \tag{9.6}$$

$$\mathbf{Z} = \begin{bmatrix} 1 & z_{11} & \cdots & z_{c1} \\ 1 & z_{12} & \cdots & z_{c2} \\ \vdots & \vdots & \ddots & \vdots \\ 1 & z_{1n} & \cdots & z_{cn} \end{bmatrix}, \tag{9.7}$$

$$\boldsymbol{\beta} = \begin{bmatrix} \beta_0 & \beta_1 & \cdots & \beta_m \end{bmatrix}',$$

and

$$\boldsymbol{\gamma} = \begin{bmatrix} \gamma_1 & \gamma_2 & \cdots & \gamma_c \end{bmatrix}'.$$

For the main-effects-plus-interactions model (9.4), the matrices $\mathbf{X}$ and $\mathbf{Z}$ become

$$\mathbf{X} = \begin{bmatrix} 1 & x_{11} & \cdots & x_{m1} & x_{11}x_{21} & \cdots & x_{m-1,1}x_{m1} & x_{11}z_{11} & \cdots & x_{m1}z_{c1} \\ 1 & x_{12} & \cdots & x_{m2} & x_{11}x_{21} & \cdots & x_{m-1,2}x_{m2} & x_{12}z_{12} & \cdots & x_{m2}z_{c2} \\ \vdots & \vdots & \ddots & \vdots & \vdots & \ddots & \vdots & \vdots & \ddots & \vdots \\ 1 & x_{1n} & \cdots & x_{mn} & x_{1n}x_{2n} & \cdots & x_{m-1,n}x_{mn} & x_{1n}z_{1n} & \cdots & x_{mn}z_{cn} \end{bmatrix}, \tag{9.8}$$

$$\mathbf{Z} = \begin{bmatrix} z_{11} & \cdots & z_{c1} & z_{11}z_{21} & \cdots & z_{c-1,1}z_{c1} \\ z_{12} & \cdots & z_{c2} & z_{12}z_{22} & \cdots & z_{c-1,2}z_{c2} \\ \vdots & \ddots & \vdots & \vdots & \ddots & \vdots \\ z_{1n} & \cdots & z_{cn} & z_{1n}z_{2n} & \cdots & z_{c-1,n}z_{cn} \end{bmatrix}, \tag{9.9}$$

while

$$\boldsymbol{\beta} = \begin{bmatrix} \beta_0 & \beta_1 & \cdots & \beta_m & \beta_{12} & \cdots & \beta_{m-1,m} & \beta_{12}^{EC} & \cdots & \beta_{mc}^{EC} \end{bmatrix}',$$

and

$$\boldsymbol{\gamma} = \begin{bmatrix} \gamma_1 & \gamma_2 & \cdots & \gamma_c & \gamma_{12} & \cdots & \gamma_{c-1,c} \end{bmatrix}'.$$

We can now distinguish two different but related experimental design problems:

1. In some applications, the number of available experimental units is larger than the budgeted number of observations. In the polypropylene experiment, for example, 40 different samples were available, but only 18 runs were done. The experimenter can then choose a subset of the available units, along with their covariate values, to experiment with. In such cases, the experimenter faces a twofold design problem: not only does he/she have to choose combinations of factor levels, but he/she also has to determine which experimental units to use. In technical terms, the experimenter has to optimize both **X** and **Z**.

2. In other applications, the number of observations matches the number of available experimental units. In that case, the design problem is simplified, since it is only the combinations of levels of the experimental factors that have to be chosen. So, the experimenter has to optimize **X** only. We provide an example of this kind of scenario in Section 9.3.3, where we show how to create experimental designs in the presence of time-trend effects.

9.3.2.2 Estimation

The models (9.1)–(9.4) contain two sets of effects. The first set contains the effects of the experimental factors, $x_1, x_2, \ldots, x_m$. All these effects are included in the vector β. The second set contains the effects of the covariates, $z_1, z_2, \ldots, z_c$. These effects are collected in the vector γ. In some applications (for instance, in the polypropylene experiment), estimating β and γ is equally important to the experimenter. In other applications, the experimenter's main interest is in estimating β and the covariates are included in the model to ensure that the factor-effect estimates are unbiased by possible active covariate effects, to reduce the error variance and to make more powerful inferences about the factor effects in β. In that case, the covariates are considered nuisance parameters, and the experimenter is not interested in a precise estimation of their effects contained within γ.

In any case, the ordinary least squares estimator for the parameter vectors β and γ in the models (9.1)–(9.4) is

$$\begin{bmatrix} \hat{\beta} \\ \hat{\gamma} \end{bmatrix} = \left([\mathbf{X} \quad \mathbf{Z}]'[\mathbf{X} \quad \mathbf{Z}] \right)^{-1} [\mathbf{X} \quad \mathbf{Z}]' Y,$$
$$= \begin{bmatrix} \mathbf{X'X} & \mathbf{X'Z} \\ \mathbf{Z'X} & \mathbf{Z'Z} \end{bmatrix}^{-1} \begin{bmatrix} \mathbf{X'Y} \\ \mathbf{Z'Y} \end{bmatrix}, \tag{9.10}$$

with variance–covariance matrix

$$\operatorname{var} \begin{bmatrix} \hat{\beta} \\ \hat{\gamma} \end{bmatrix} = \sigma_\varepsilon^2 \begin{bmatrix} \mathbf{X'X} & \mathbf{X'Z} \\ \mathbf{Z'X} & \mathbf{Z'Z} \end{bmatrix}^{-1}. \tag{9.11}$$

The ordinary least squares estimator in Equation (9.10) and its variance–covariance matrix in Equation (9.11) have exactly the same structure as the ordinary least squares estimator in Equation (8.5) and its variance–covariance matrix in Equation (8.6) for experiments with fixed block effects.

9.3.2.3 D-optimal designs

Minimizing the determinant of the variance–covariance matrix in Equation (9.11) results in the D-optimal design for estimating both $\boldsymbol{\beta}$ and $\boldsymbol{\gamma}$. Alternatively, you can maximize the D-criterion value

$$D = \begin{vmatrix} \mathbf{X'X} & \mathbf{X'Z} \\ \mathbf{Z'X} & \mathbf{Z'Z} \end{vmatrix}, \tag{9.12}$$

which is equal to

$$D = |\mathbf{Z'Z}| \, |\mathbf{X'X} - \mathbf{X'Z}(\mathbf{Z'Z})^{-1}\mathbf{Z'X}|. \tag{9.13}$$

What this expression shows is that the total information about the model parameters, as measured by the D-criterion value, can be decomposed into two parts: the information on the covariates' effects (measured by $|\mathbf{Z'Z}|$) and the information on the factor effects, given that you need to estimate the covariates' effects (measured by $|\mathbf{X'X} - \mathbf{X'Z}(\mathbf{Z'Z})^{-1}\mathbf{Z'X}|$).

Depending on the application, the information on the covariates' effects may or may not be fixed. If, as in the polypropylene experiment, the experimenter can choose the experimental units and the covariates to work with, then the information content on the covariate effects, $|\mathbf{Z'Z}|$, depends on the choice made for the experimental units. If, however, the experimenter has no choice about which units to use, then $|\mathbf{Z'Z}|$ is fixed. In the latter case, finding the D-optimal design requires maximizing

$$|\mathbf{X'X} - \mathbf{X'Z}(\mathbf{Z'Z})^{-1}\mathbf{Z'X}|, \tag{9.14}$$

which is less demanding computationally given that $\mathbf{Z}$, and hence also $(\mathbf{Z'Z})^{-1}$, are fixed. This computational simplification is not possible when choosing the covariates is part of the design problem.

9.3.2.4 D_s- or D_β-optimal designs

When the covariates are viewed as nuisance parameters, there is little to gain from estimating their effects as precisely as possible. In that case, it is better to choose a design optimality criterion that emphasizes precise estimation of the effects of the experimental factors $x_1, x_2, \ldots, x_m$. Instead of minimizing the determinant of the variance–covariance matrix of the estimator for $\boldsymbol{\beta}$ and $\boldsymbol{\gamma}$ in Equation (9.11), it is desirable to minimize the determinant of the variance–covariance matrix of the estimator for $\boldsymbol{\beta}$. That variance-covariance matrix equals

$$\mathrm{var}(\hat{\boldsymbol{\beta}}) = \sigma_\varepsilon^2 \{\mathbf{X'X} - \mathbf{X'Z}(\mathbf{Z'Z})^{-1}\mathbf{Z'X}\}^{-1}. \tag{9.15}$$

Minimizing the determinant of that matrix yields a D_s- or D_β-optimal design. A D_s- or D_β-optimal design can also be found by maximizing the determinant of the inverse of that matrix,

$$|X'X - X'Z(Z'Z)^{-1}Z'X|. \tag{9.16}$$

This determinant is the same as the one we have to maximize to obtain a D-optimality design with fixed covariates. Therefore, when the covariates are fixed, there is no difference between the D-optimal design and the D_s- or D_β-optimal design.

The index s in the term D_s-optimality is short for subset. The name D_s-optimality, therefore, stresses the fact that what we want is a D-optimal design for a subset of the parameters in the model. Since that subset is contained within the vector β, we can also use the term D_β-optimality.

9.3.2.5   Variance inflation

The presence of covariates in the model, in general, results in an inflation of the variances of the factor-effect estimates. This variance inflation is similar to the one we discussed in Chapter 8 on page 184. The variance inflation drops to zero whenever $X'Z$ is a zero matrix, i.e., when the covariates and the experimental factors are completely orthogonal to each other. That orthogonality is hard to achieve if the covariates are continuous variables that take many different levels.

It is worth noting, however, that D-optimal designs as well as D_β-optimal designs have a built-in incentive to attempt to achieve orthogonality and bring the variance inflation down. This is because, whenever $X'Z$ is a zero matrix, the determinant $|X'X - X'Z(Z'Z)^{-1}Z'X|$ is maximal for a given X. Roughly speaking, a D-optimal or a D_β-optimal design, therefore, results in minimum variance inflation of the factor-effect estimates and orthogonality or near-orthogonality between the experimental factors and the covariates. In other words, there will be little or no correlation between the estimates of the covariates' effects and the estimates of the factor effects if a D- or a D_β-optimal design is used.

Note that a blocking factor can be viewed as a categorical covariate. Because of the limited number of levels that this "categorical covariate" can take, it is often possible to find a design such that $X'Z$ is a zero matrix under effects-type coding for the blocking factor. We say that such a design is orthogonally blocked.

9.3.3   Designs robust to time trends

Many industrial experiments and experiments in physical and engineering sciences are conducted one run after another in time. When a single piece of equipment or a pilot plant are used, a time trend may affect the results. For instance, the build-up of deposits in a test engine or injection molding equipment may have a systematic impact on the response. Aging of a catalyst or wear in a grinding or cutting process may also result in a trend in the responses.

In such cases, we should include the time as a covariate in our model. If the time trend is linear, then an appropriate model includes a linear effect for the covariate time:

$$Y = \beta_0 + \sum_{i=1}^{m} \beta_i x_i + \sum_{i=1}^{m-1} \sum_{j=i+1}^{m} \beta_{ij} x_i x_j + \gamma t + \varepsilon, \qquad (9.17)$$

where t is the time at which the run is performed. If a quadratic time trend is expected, then a quadratic time-trend effect has be to added:

$$Y = \beta_0 + \sum_{i=1}^{m} \beta_i x_i + \sum_{i=1}^{m-1} \sum_{j=i+1}^{m} \beta_{ij} x_i x_j + \gamma_1 t + \gamma_2 t^2 + \varepsilon. \qquad (9.18)$$

These models are special cases of the models in Equations (9.1)–(9.4). Therefore, the methodology introduced in Section 9.3.2 can be transferred completely to problems involving time trends.

Of all possible covariates, time trends have by far received the largest share of attention in the literature. This has led to a specific vocabulary for experiments involving time-trend effects. Because the order of the runs is very important for the quality of a designed experiment, the term optimal run order is often used as a synonym for an optimal design in the presence of a time-trend effect. If the rows of the matrix $\mathbf{X}$ are arranged so that the settings of the experimental factors and the columns corresponding to their interaction or higher-order effects are completely orthogonal to the time-trend effects, then the design or run order is trend-robust, trend-free, or trend-resistant.

The first three columns of Table 9.8 show a design with three factors that is trend-robust in case a main-effects model is assumed and the time trend is linear. The design is a 2^3 factorial design, as it has all eight combinations of three two-level

Table 9.8   Trend-robust (and D- and D_β-optimal) design for a main-effects model in three factors and a linear time trend.

x_1	x_2	x_3	Time (uncoded)	Time t (coded)	$x_1 t$	$x_2 t$	$x_3 t$
1	−1	−1	1	−1.000	−1.000	1.000	1.000
−1	1	−1	2	−0.714	0.714	−0.714	0.714
−1	1	1	3	−0.429	0.429	−0.429	−0.429
1	−1	1	4	−0.143	−0.143	0.143	−0.143
1	1	1	5	0.143	0.143	0.143	0.143
−1	−1	1	6	0.429	−0.429	−0.429	0.429
−1	−1	−1	7	0.714	−0.714	−0.714	−0.714
1	1	−1	8	1.000	1.000	1.000	−1.000
				Sum	0.000	0.000	0.000

Table 9.9   Relative variances of main-effect estimates in the presence or absence of a linear time trend.

Effect	Presence	Absence
Intercept	0.125	0.125
x_1	0.125	0.125
x_2	0.125	0.125
x_3	0.125	0.125
Time t	0.292	—

factors x_1, x_2, and x_3. That the design is trend-robust is relatively easy to check. To do so, we need to calculate the cross-products $x_1 t$, $x_2 t$, and $x_3 t$, and verify that they sum to zero. The required calculations are shown in the last five columns of the table. Because the eight time points in this example are fixed, the design in Table 9.8 is D-optimal as well as D_β-optimal.

Because of the trend-robust nature of the design, the variances of the main-effect estimates of x_1, x_2, and x_3 are not inflated compared to a situation in which there is no time trend. This is shown in Table 9.9, where the relative variances for the main-effect estimates from Table 9.8's trend-robust 2^3 design in the presence of a linear time trend are compared to the relative variances obtained from a 2^3 design in the absence of a time trend. For each of the factor-effect estimates, the variance is 1/8 or 0.125, both in the presence and absence of the time trend.

As soon as the model involves interaction effects or a higher-order time trend, it becomes substantially harder to find trend-robust designs. Table 9.10 shows a D- and D_β-optimal design for a main-effects-plus-two-factor-interaction model involving three factors. Along with the levels of the three factors, we have also displayed the columns

Table 9.10   D- and D_β-optimal design for a main-effects-plus-two-factor-interaction model in three factors and a linear time trend.

x_1	x_2	x_3	$x_1 x_2$	$x_1 x_3$	$x_2 x_3$	Time t (coded)	$x_1 t$	$x_2 t$	$x_1 x_2 t$	$x_2 x_3 t$
1	1	−1	1	−1	−1	−1.000	−1.000	−1.000	−1.000	1.000
−1	−1	−1	1	1	1	−0.714	0.714	0.714	−0.714	−0.714
−1	1	1	−1	−1	1	−0.429	0.429	−0.429	0.429	−0.429
1	−1	1	−1	1	−1	−0.143	−0.143	0.143	0.143	0.143
−1	−1	1	1	−1	−1	0.143	−0.143	−0.143	0.143	−0.143
1	−1	−1	−1	−1	1	0.429	0.429	−0.429	−0.429	0.429
1	1	1	1	1	1	0.714	0.714	0.714	0.714	0.714
−1	1	−1	−1	1	−1	1.000	−1.000	1.000	−1.000	−1.000
						Sum	0.000	0.571	−1.714	0.000

Table 9.11 Relative variances of main-effect and interaction-effect estimates in the presence or absence of a linear time trend.

Effect	Presence	Absence
Intercept	0.125	0.125
x_1	0.125	0.125
x_2	0.127	0.125
x_3	0.127	0.125
$x_1 x_2$	0.143	0.125
$x_1 x_3$	0.143	0.125
$x_2 x_3$	0.125	0.125
Time t	0.383	—

of $\mathbf{X}$ corresponding to the interaction effects. We need these columns for checking the trend-robustness of the design. It turns out that the design is not completely trend-robust. The last four columns of Table 9.10 show that the design is trend-robust for estimating the main effect of x_1 and the interaction effect involving x_2 and x_3, but not for estimating the main effect of x_2 and the interaction effect involving x_1 and x_2.

We can also see that by looking at Table 9.11, which compares the relative variances of the factor-effect estimates in the presence and in the absence of time trends. There is no variance inflation for the main effect of x_1 and the interaction effect involving x_2 and x_3, a minor inflation for the main effect of x_2 and some more for the interaction effect involving x_1 and x_2. The minor variance inflation for the main effect of x_2 is the result of the sum of all $x_2 t$ values not being zero, while the larger variance inflation for the interaction involving x_1 and x_2 is due to the sum of the $x_1 x_2 t$ values being even more different from zero.

In general, we obtain variance inflation factors for the factor-effect estimates in this context by taking the ratios of the diagonal elements of $(\mathbf{X'X} - \mathbf{X'Z}(\mathbf{Z'Z})^{-1}\mathbf{Z'X})^{-1}$ to those of $(\mathbf{X'X})^{-1}$. The inverses of the variance inflation factors are efficiency factors, similar to those we used in Chapter 7 for a blocked experiment.

The extent to which a design is trend-robust for all the factor effects in the model can be summarized using the measure

$$\text{Trend-robustness} = \left(\frac{|\mathbf{X'X} - \mathbf{X'Z}(\mathbf{Z'Z})^{-1}\mathbf{Z'X}|}{|\mathbf{X'X}|} \right)^{\frac{1}{p}}, \qquad (9.19)$$

where p is the number of parameters contained within the vector $\boldsymbol{\beta}$. This overall measure of trend-robustness compares the information content of a design in the presence of a time trend (quantified by the numerator) and its information content in the absence of a time trend (quantified by the denominator). If a design is trend-robust for all the model terms, then the two information contents are equal and the trend-robustness of the design is 100%. The trend-robustness of the design in Table 9.10 is 96.19%.

9.3.4   Design construction algorithms

There are two different approaches to construct optimal experimental designs when covariate information is available, one for each of the two scenarios described on page 209. In general, the design matrix for a problem with covariates is

$$
\mathbf{D} = \begin{bmatrix}
x_{11} & \cdots & x_{m1} & z_{11} & \cdots & z_{c1} \\
x_{12} & \cdots & x_{m2} & z_{12} & \cdots & z_{c2} \\
\vdots & \ddots & \vdots & \vdots & \ddots & \vdots \\
x_{1n} & \cdots & x_{mn} & z_{1n} & \cdots & z_{cn}
\end{bmatrix}.
$$

The design construction algorithm for both scenario goes through the design matrix, row by row, to improve the design's optimality criterion value.

For the scenario where the number of available experimental units is equal to the budgeted number of runs, this row-wise procedure is simplest, because the part of the design matrix corresponding to the covariates is then fixed. In this case, the coordinate-exchange algorithm operates only on the columns associated with ordinary experimental factors, in exactly the same way we described in Section 2.3.9.

For the scenario where the number of available experimental units is larger than the budgeted number of runs, we have to both optimize the levels of the experimental factors, and select the best possible values or combinations of values for the covariates from the list of available ones. We optimize the levels of the experimental factors coordinate by coordinate, as in the original coordinate-exchange algorithm. However, for optimizing the values for the coordinates, we replace every vector $(z_{1i}, \ldots, z_{ci})$ in the design matrix by all possible vectors of covariate values that are available (ensuring that each experimental unit is used only once). If one of these replacements results in an improvement of the optimality criterion, the best of these improvements is saved.

9.3.5   To randomize or not to randomize

The usual advice when running an experimental design is to randomize the order of the runs and to assign the factor-level combinations randomly to the experimental units. In the presence of covariates or time trends, it is desirable to use a systematic assignment to experimental units or a systematic run order because this ensures, for every single experiment, that the factor effects are unbiased by the covariates and estimated with maximum precision. That is different from a randomization approach which also offers protection against biased estimates, but only when averaged over a large number of experiments. For one particular experiment, the randomization approach may yield a run order which is not at all trend-robust or an assignment to experimental units that results in substantial variance inflation for the factor-effect estimates.

In this context, we would like to stress that there are other ways in which we can randomize than by using a random order for the runs. One possible way of randomizing is by assigning the experimental factors randomly to the columns of a design. On top of that, we can also assign the high and low levels of the factors randomly to the labels for these levels. So, using a systematic run order or a systematic assignment to experimental units does not necessarily imply that there is no randomization at all.

9.3.6 Final thoughts

In the presence of time trends, it may be difficult to decide whether the trend is linear or quadratic. Since designs that work well for quadratic time trends also work well for linear time trends, we recommend including both linear and quadratic time-trend effects in the a priori model in such cases.

A key feature of trend-robust and nearly trend-robust designs is that they involve large numbers of changes in factor levels when going from one run to the next. When the levels of the factors are hard or time-consuming to set or to change, then running an experiment with a trend-robust design is considered costly. This will leave practitioners who prefer not resetting the factor levels after every run with the feeling that trend-robust designs are practically infeasible. In this context, we would like to stress that, for the experimental observations to be independent, it is required to reset the factor levels independently for every run, even when the level of a factor is the same in two successive runs. Failing to do so results in correlated observations and makes a proper data analysis substantially more involved, and, in some cases, impossible. When it is hard to reset factor levels independently for every run, we recommend using split-plot designs or strip-plot designs. These designs are discussed in detail in Chapters 10 and 11. The joint presence of time-trend effects and hard-to-change factors necessitates the construction of trend-robust split-plot or strip-plot designs.

9.4   Background reading

The general problem of covariate information in experimental design received much less attention in the literature than the search for good experimental designs and run orders of experimental designs in the presence of time-trend effects. Harville (1974) and Nachtsheim (1989) discussed the design of experiments in the presence of general covariates. Joiner and Campbell (1976) gave four interesting examples of experiments involving a time trend, while Hill (1960) summarized the early literature on trend-robust designs. Daniel and Wilcoxon (1966) proposed trend-robust run orders for both full factorial and fractional factorial experiments. Follow-up work, in which a trend-robust run order was sought for a given set of factor-level combinations, was done by Cheng and Jacroux (1988), Cheng (1990), John (1990), Cheng and Steinberg (1991), and Mee and Romanova (2010). Atkinson and Donev (1996) proposed an algorithm for simultaneous selection of factor-level combinations and choice of run order.

The construction of optimal run orders in the presence of a time trend as well as a blocking factor was discussed by Tack and Vandebroek (2002) and Goos et al. (2005). Carrano et al. (2006) discuss a split-plot experiment involving time-trend effects. The fact that trend-robust run orders involve many factor-level changes from one run to the next (and may therefore be inconvenient and costly) inspired Tack and Vandebroek (2001, 2003, 2004) to search for designs that are trend-robust as well as cost-efficient.

A line of research that is related to the search for trend-robust designs is the study of optimal experimental designs in the presence of serial correlation. Experimental designs are often run one run after another, with short periods of time between the successive observations. Therefore, it is likely that the responses exhibit serial correlation. In that case, the order in which the runs of the experimental design are carried out also has an impact on the precision of the factor-effect estimates. The problem of finding run orders of standard experimental designs in the presence of serial correlation has received attention from Constantine (1989), who sought efficient run orders of factorial designs, Cheng and Steinberg (1991), who propose the reverse fold-over algorithm to construct highly efficient run orders for two-level factorial designs, Martin et al. (1998b), who presented some results on two-level factorial designs, and Martin et al. (1998a), who discussed the construction of efficient run orders for multilevel factorial designs. Zhou (2001) presented a robust criterion for finding run orders of factorial and fractional factorial designs that are robust against the presence of serial correlation. All these authors concentrated on main-effects and main-effects-plus-interactions models. Optimal run orders of designs for estimating second-order response surface models are discussed in Garroi et al. (2009).

9.5 Summary

It often happens that, prior to running an experiment, the investigator has information about certain characteristics of the experimental units. The statistical concept of an experimental unit is somewhat difficult, but, often, it is just the material being processed in each run of the experiment. When you think that the experimental units' characteristics will have a substantial effect on the response, then it is natural to take them into account when you design your experiment. We call the factors that measure the characteristics of the experimental units covariate factors. Such factors may be either continuous or categorical. They are different from other continuous or categorical experimental factors in that you cannot directly control covariate factors. In this way, covariate factors are similar to blocking factors. Designing the experiment around the levels of the covariate factors reduces the standard error of the factor-effect estimates compared with an experiment that ignores the covariate information and uses complete randomization of the experimental units.

In the case study, the runs of the polypropylene experiment used material with different but known constituents, the covariate factors, that was already in stock. This was more convenient than ordering new material with strict specifications on the constituents.

In the "Peek into the Black Box," we demonstrate that the same approach works with time as a covariate factor. It is worth considering time as a covariate factor when you expect the response to experience drift over time.

10

A split-plot design

10.1 Key concepts

1. The levels of certain key factors in an experiment may be hard to change from one experimental run to the next. In such cases, it is desirable to group the runs so that the levels of such factors only vary between groups of runs. This results in a special case of a blocked experiment, called a split-plot experiment.

2. Because the levels of hard-to-change factors do not change within a group of runs, split-plot experiments are not completely randomized designs. Rather, there are two randomizations: one randomization for the order of the groups of runs and a secondary randomization for the order of the runs within each group.

3. The model for a split-plot experiment is a mixed model (involving both fixed and random effects). Such models are properly analyzed using generalized least squares (GLS) rather than using ordinary least squares.

4. The various variance optimality criteria (D, A, and I) have generalizations to deal with the split-plot structure.

Our case study considers experiments run in a wind tunnel where it is possible to vary the levels of certain factors without shutting down the tunnel. Other factors have levels that can only be changed by shutting down the tunnel. Since shutting down and restarting a wind tunnel is a major undertaking, it is economically infeasible to allow only one experimental run per wind-tunnel run just to accommodate a completely randomized design.

In this chapter, we discuss how to construct designs that explicitly consider the grouping of runs where the levels of one or more factors remain constant within each group. We also show how to analyze such designs, called split-plot designs, and compare the resulting inference with the results one would obtain by analyzing the data as if they came from a completely randomized design.

Optimal Design of Experiments: A Case Study Approach, First Edition. Peter Goos and Bradley Jones.
© 2011 John Wiley & Sons, Ltd. Published 2011 by John Wiley & Sons, Ltd.

10.2   Case: the wind tunnel experiment

10.2.1   Problem and design

Brad and Peter are driving on I64 from Richmond to Norfolk, on their way to the Langley Full Scale Tunnel (LFST) at NASA's Langley Research Center in Hampton, Virginia.

[Peter] I've never been anywhere near a wind tunnel.

[Brad] Me neither. The one we are touring today is not just a wind tunnel. It is *the* LFST. It has a rich history. Its first version was built in the late 1920s, so it is nearly 80 years old.

[Peter] Woah! For a wind tunnel, 80 years is a respectable age.

[Brad] Definitely. Actually, it is so respectable an age that there are rumors that it may be decommissioned.

Brad takes exit 261B and mentions that the tunnel was used for all World War II aircraft, various submarines, and the Mercury entry capsule.

[Peter] So, what are we going to do there?

[Brad] Race car modification. The Langley facility has a long tradition of testing the NASCAR race cars. If I remember correctly, the car they are about to test is the NASCAR Winston Cup Chevrolet Monte Carlo stock car.

[Peter] That's a mouth full. If, as you say, they have a long tradition experimenting with these cars, what do they hope to learn from us.

[Brad] Marc Cavendish, the guy who is in charge of the NASCAR tests attended your presentation on experiments with hard-to-change factors at the Fall Technical Conference in Roanoke last October. On the phone, he said that talk had been an eye-opener for him. He had discovered that, by using a split-plot design, he can do a 50-run study in only 80% of the time it takes to do a 28-run completely randomized design.

[Peter, more excited now that he hears the term split-plot design] Aha. I see. That is why they want us.

Brad parks the car. Fifteen minutes later, he and Peter have passed the security check and are welcomed by Dr. Cavendish. While walking to the wind tunnel, Dr. Cavendish sketches his background. He earned a PhD in Aerospace Engineering from Old Dominion University and has been serving as Chief Engineer at the LFST since 1999.

[Dr. Cavendish] I feel lucky to work here where I have supervised dozens of wind tunnel experiments. I have taken several courses in design of experiments and the design textbooks on my bookshelves are not just for show. A few of them have many "yellow stickies" marking all the really good examples.

[Peter] The experiment you are about to conduct must have a few truly challenging features then. Otherwise, a DOE expert such as yourself would not need any help.

[Dr. Cavendish] Right. But let me sketch the history of DOE at LFST before we dive into the technical details. At the time, my introduction of statistical design and analysis of experiments in the wind tunnel testing here was a major step forward.

The traditional approach to testing was to vary one factor at a time while all other factors were held constant. This approach failed to detect often-important input factor interaction effects. Obviously, we missed a lot of opportunities for improvement.

[Brad] Does management appreciate your new approach?

[Dr. Cavendish] Yes. My boss has been very generous and, now, he also allows me quite a bit of latitude in setting up each test.

When they enter the wind tunnel, Dr. Cavendish leads Peter and Brad to a sports car.

[Dr. Cavendish] Here is our newest baby. That's the car we are going to study.

[Brad] On the phone, you said that the experiment involved hard-to-change and easy-to-change factors.

[Dr. Cavendish] That is true. Based on the seven years of experience we have with this type of car, we have chosen four factors to vary during the experiment: front ride height, rear ride height, yaw angle, and closing or opening of the car's lower cooling inlet.

[Peter] What is the yaw angle?

[Dr. Cavendish] It's the angle between a vehicle's heading and some reference heading. In the wind tunnel, we can easily adjust that angle because it is electronically automated.

[Brad] So, yaw angle is a factor whose levels are easy to change. I guess opening and closing a car's lower cooling inlet is also easy to do.

[Dr. Cavendish] True. We do that by covering or uncovering all or part of the cooling inlet with some tape. Changing the front and the rear ride heights is another matter. To do so, we need to stop the tunnel and make changes to the ride heights at each wheel. Additionally, the wheel loads must be checked to ensure that they do not exceed specified tolerances, and that they are balanced.

[Peter] That definitely sounds as if it is more involved than applying or pulling off some tape.

[Dr. Cavendish] Right. So, we have two factors that are hard to change and two that are easy to change.

Dr. Cavendish opens the briefcase he's been carrying along, produces two sheets of paper and hands them over to Peter and Brad. Both sheets contain Table 10.1.

[Dr. Cavendish] Here, you have the list of experimental factors, along with the levels we have in mind for each of them. The center levels for the front and rear ride heights are 3.5 and 35 inch, respectively, because NASCAR only allows small variations around these values.

Table 10.1   Factors and levels for the wind tunnel experiment.

Factor (label)	Type	Low level	Center level	High level
Front ride height (FRH)	Hard-to-change	3.0 in	3.5 in	4.0 in
Rear ride height (RRH)	Hard-to-change	34 in	35 in	36 in
Yaw angle (yaw)	Easy-to-change	$-3.0°$	$-1.0°$	$+1.0°$
Grille tape coverage (tape)	Easy-to-change	0%	50%	100%

[Peter] What kind of design do you have in mind?

[Dr. Cavendish] The approach we used in recent years for a four-factor experiment was to implement a completely randomized design, often a central composite design with 28 runs. That design was quite cumbersome. To ensure that our factor-effect estimates are not contaminated by lurking variables, we did the 28 runs in a random order. Between two runs, we even reset the levels of each factor to make sure that all the runs were independent.

[Brad, whistling] Wow! I've seldom met experimenters who were willing to reset the factor levels between each pair of runs. In any case, that is the kind of effort that is required to justify the use of ordinary least squares for estimating the factor effects.

[Dr. Cavendish] There is no doubt that resetting the factor levels for every run is useful, but the engineers are not very enthusiastic about that part of the designed experiments I have them carry out. It takes an awful lot of time, especially for factors whose levels are as hard to change as the front and rear ride height of this Chevy here.

[Brad] And now, after you attended Peter's talk at the Fall Technical Conference, you would like to leave the well-paved path of completely randomized designs.

[Dr. Cavendish] Right. What I remember from the talk is that it is possible to not reset the hard-to-change factors for every run and still get useful results.

[Peter] That is true provided you take into account the hard-to-change nature of the front and the rear ride heights when you design your experiment and when you analyze the data.

[Dr. Cavendish] And how do I do that?

[Peter] The design you need is a split-plot design. A split-plot design essentially involves a master design for the hard-to-change factors, and for each run of the master design you have a design for the easy-to-change factors. For your study, we need a master design for the two ride heights, and, for each combination of ride heights in the master design, we need a design for the easy-to-change factors, the yaw angle and the grille tape coverage.

[Dr. Cavendish] How do I run this design?

[Peter] To start, you randomly select a front and a rear ride height for the car from the master design and adjust the car according to these two heights. As soon as this has been done, you test the combinations of the yaw angle and the grille tape coverage in a random order. Next, you randomly select another combination of front and rear ride height from the master design and you again test several combinations of levels of the easy-to-change factors, the yaw angle and the grille tape coverage, in a random order.

[Brad] The idea is that you repeat these steps until you have gone through all the runs of the master design and you have reset the front and the rear ride heights the desired number of times.

[Dr. Cavendish] So, that approach involves two randomizations. First, you randomize the order of the settings of the hard-to-change factors, and, for each of these settings, you randomize the settings of the easy-to-change factors.

[Peter, nodding] Using this procedure, your experiment consists of groups of dependent runs.

[Dr. Cavendish] I can see that. The runs that you perform as a group, without resetting the ride heights, are definitely not independent. The runs within such a group would be more like each other than runs from different groups.

[Brad] So, the dependence between the runs within each group results in correlated responses.

[Dr. Cavendish] And that means one of the key assumptions behind ordinary least squares regression is violated. I learned about that at the Fall Technical Conference. You mentioned generalized least squares or GLS as an alternative to ordinary least squares or OLS estimation. Correct?

[Peter] That's right.

[Dr. Cavendish] Am I right to assume that analyzing the data with GLS is a piece of cake for you two?

Peter and Brad nod, and Dr. Cavendish continues.

[Dr. Cavendish] Great. We can worry about the data analysis later. What I need sooner rather than later is a design. I have been comparing various experimental plans in terms of cost, and I figured out that we can do a pretty large experiment if we reset the ride heights a limited number of times. For instance, if we set the front and the rear ride heights ten times and test five combinations of yaw angle and grille tape coverage at each of the ten ride-height combinations, we have a 50-run experiment that is 20% less expensive than the completely randomized central composite design with 28 runs.

[Peter] So, we need a split-plot design with ten settings of the hard-to-change factors and five combinations of levels of the easy-to-change factors for each of them. Do you have an experimental design in mind already?

[Dr. Cavendish] A couple. I tried to produce a few sensible design options, but I would like to have your opinion on what I did. One of the ideas I had was to use a 2^2 factorial design plus center point for the front and rear ride heights. That would be the master design. Then, I would cross that design with two replicates of it for the yaw angle and the grille tape coverage.

Dr. Cavendish opens his briefcase again and hands Peter and Brad each a new sheet of paper, containing the design in Figure 10.1. The large square in the figure defines the possible levels for the hard-to-change factors, the front and rear ride heights. The small squares define the possible levels for the easy-to-change factors, yaw angle and grille tape coverage. In total, the graphical representation of the design involves ten small squares in five positions. This means that five different combinations of levels of the hard-to-change factors are used. These five combinations form a 2^2 factorial design with a center point, the master design Dr. Cavendish suggests. Each of the five combinations is used twice. The bullets in the small squares define what combinations of levels of the easy-to-change factors are used. The easy-to-change factor-level combinations at each combination of the hard-to-change factor levels also form a 2^2 factorial design with a center point.

[Brad, looking at the design] Why include so many center points?

[Dr. Cavendish] For two reasons. The engineers feel uncomfortable with the idea of just looking at the end points of each factor's range. They like to have some tests in the middle in case there is something nonlinear going on. Also, the 2^2 factorial

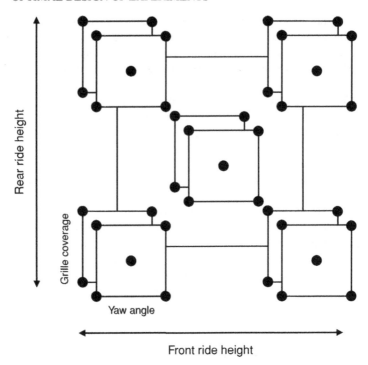

Figure 10.1   First design option considered by Dr. Cavendish.

design for the yaw angle and the grill coverage only has four runs, and, in our split-plot design, we need five runs at each setting of the hard-to-change factors.

[Peter] I buy your first argument to include center runs, but, at the same time, I think that there is a problem with using the factorial design plus center point. While it is true that you can test for curvature, you cannot estimate the four quadratic effects with your design. Ten settings for the hard-to-change factors and 50 runs in total is plenty for estimating a full quadratic model, including the quadratic effects of the four factors in your study.

[Dr. Cavendish] I had anticipated you'd say that. That's why I brought this other design using a 3^2 factorial design for the front and rear ride heights.

Dr. Cavendish produces a third sheet of paper from his briefcase and hands it over to Peter and Brad. It contains Figure 10.2.

[Brad] I like this one better!

[Dr. Cavendish] Sure. With that design, I can estimate all the quadratic effects. The 3^2 factorial design for the front and rear ride heights, the master design, allows me to estimate the quadratic effects of the two hard-to-change factors, while the two different designs for the yaw angle and the grille tape coverage allow for estimating the quadratic effects of the easy-to-change factors.

[Brad] What I like about the design compared to the previous one is that you can now estimate all the effects you are interested in. That is a major leap forward.

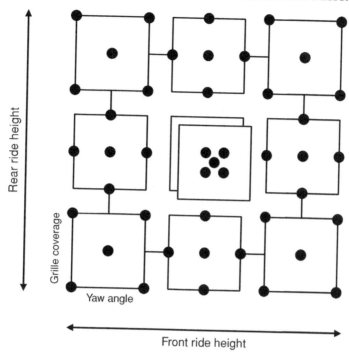

Figure 10.2   Second design option considered by Dr. Cavendish.

However, I don't know where you got the idea for the five runs at the center of the 3^2 design.

[Dr. Cavendish] Actually, there are ten runs at the center of the 3^2 design, split over two independent settings of the front and the rear ride heights. I found the idea in a few journal articles. It seemed important to have replicates at the center of the design region to be able to use GLS.

[Brad] Only if you don't use software specifically programmed to fit data from any kind of split-plot design. Our methods do not require replicated observations to compute estimates of the variance components.

[Peter] Right. The ten center points you have in your design are essentially meant to estimate the two variance components in the split-plot model. It is crucial that you can estimate these variance components, but they are not usually of primary concern. We think most investigators are more interested in precise estimates of the factor effects.

[Dr. Cavendish, nodding] Go on.

[Peter] Your ten center points mean that you are spending only 80% of your experimental resources on the estimation of things you are interested in, the factor effects.

[Brad] What we can do for you is generate a design that pays more attention to the estimation of the factor effects.

[Dr. Cavendish] That sounds appealing for sure.

[Peter] Let me try to find a good design for you.

Peter boots up his laptop and opens his favorite DOE software package.

[Peter] I am planning to generate an I-optimal split-plot design. You will see it is an easy thing to do with modern software.

Dr. Cavendish closely follows each of Peter's interactions with the software. Peter first defines four quantitative factors, and indicates that two of these have levels that are hard to change. Next, he inputs the number of independent resettings of the two hard-to-change factors, that is, the number of runs for the master design. Finally, he enters the total number of runs. Soon, his laptop is computing an I-optimal design for a full quadratic model in the four factors. While they are waiting, Dr. Cavendish asks a few questions about the software Peter is using and the capabilities of alternative software packages for generating optimal split-plot designs. Peter explains that the I-optimal design is the design that has the smallest average prediction variance over the experimental region defined by the factors' lower and upper bounds. They are interrupted by Brad who draws their attention to the fact that the software has stopped running and displayed the I-optimal split-plot design in Table 10.2 on the screen.

[Dr. Cavendish] That was fast.

[Peter] Let us have a look at the design.

[Brad] I can already see one similarity between the I-optimal design and the second design you proposed, Marc. The I-optimal design also uses the 3^2 factorial design plus one additional center point for the hard-to-change factors. So, the two master designs are identical.

[Peter] Can you make a graphical representation of the design?

[Brad] Give me a second.

A minute or so later, Brad has produced Figure 10.3.

[Brad] Here you go. Note that the large white circle in the center of the picture represents three replicates at the center point, and that the circled black bullet represents two replicates.

[Peter] The I-optimal design has only five center runs instead of ten. Another difference that I can immediately see is that there are only 11, instead of 18, center runs for the easy-to-change factors.

[Dr. Cavendish] I see, but I am not quite sure I like your design. The designs I suggested look much better. They are symmetric. Look at my second design. It has symmetric subdesigns at every setting of the front and rear ride heights. This is not at all true for the I-optimal design.

[Peter] The symmetry of your design is appealing, but my mother taught me that it is not wise to choose a partner just for her looks.

[Dr. Cavendish, nodding] Right. But then, why should I dance with your I-optimal design?

[Brad] I think the I-optimal design will allow for more precise estimation of the factor effects and for more precise predictions. Peter, can you compute the D-efficiency of Marc's second design relative to the I-optimal one.

While Peter inputs the factor settings for the design in Figure 10.2 in his software, Brad explains that the relative D-efficiency expresses how well all the factor effects are estimated jointly with one design compared to another.

Table 10.2   I-optimal split-plot design with ten settings of the hard-to-change factors front and rear ride height and 50 runs in total for the wind tunnel experiment.

Hard-to-change factor setting	Front ride height	Rear ride height	Yaw angle	Grille coverage
1	−1	−1	−1	−1
1	−1	−1	1	1
1	−1	−1	0	0
1	−1	−1	1	−1
1	−1	−1	−1	1
2	0	−1	1	1
2	0	−1	0	0
2	0	−1	1	−1
2	0	−1	0	1
2	0	−1	−1	0
3	1	−1	1	−1
3	1	−1	1	1
3	1	−1	−1	−1
3	1	−1	−1	1
3	1	−1	0	0
4	−1	0	−1	0
4	−1	0	0	1
4	−1	0	1	0
4	−1	0	0	−1
4	−1	0	−1	1
5	0	0	0	−1
5	0	0	1	0
5	0	0	0	0
5	0	0	−1	1
5	0	0	0	0
6	0	0	1	0
6	0	0	0	0
6	0	0	0	0
6	0	0	0	0
6	0	0	−1	−1
7	1	0	0	−1
7	1	0	1	1
7	1	0	0	0
7	1	0	0	1
7	1	0	−1	0

(*continued overleaf*)

Table 10.2   (*continued*)

Hard-to-change factor setting	Front ride height	Rear ride height	Yaw angle	Grille coverage
8	−1	1	−1	−1
8	−1	1	1	−1
8	−1	1	0	0
8	−1	1	1	1
8	−1	1	−1	1
9	0	1	−1	0
9	0	1	0	−1
9	0	1	0	0
9	0	1	0	1
9	0	1	1	1
10	1	1	−1	1
10	1	1	1	−1
10	1	1	−1	−1
10	1	1	1	0
10	1	1	0	1

[Peter] Here we are. The relative D-efficiency of Marc's design is 89.2%. In other words, the joint estimation of the 15 parameters in the full quadratic model, i.e., the intercept, four main effects, six two-factor interaction effects, and four quadratic effects, is 12.1% better for the I-optimal design.

[Brad] You could also say that we have 12.1% more information for the same budget. Can you also show us the variances of the individual factor-effect estimates?

[Peter, clicking his mouse a few times and producing Table 10.3] Sure. Here we have a side-by-side comparison of the variances for Marc's design and for the I-optimal design.

[Brad] The I-optimal design does a better job at estimating all the factor effects, except for the main effects of the front and the rear right heights and the interaction effect involving the two ride heights. For these three effects, the two designs give exactly the same variance.

[Dr. Cavendish, looking carefully at the table] My design is better for estimating the intercept.

[Brad] That is true, but nailing down the intercept is usually less important than getting precise estimates of the factor effects.

[Dr. Cavendish] I admit that, for our race car experiment, what you say is true because what we are interested in is minimizing the drag and maximizing the efficiency. If the goal is the maximization or minimization of a function, the constant is irrelevant.

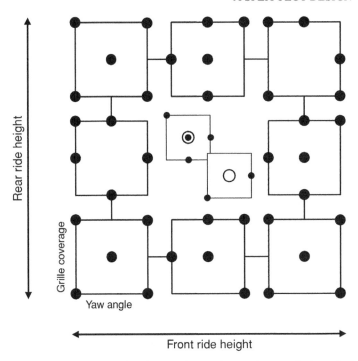

Figure 10.3   I-optimal design for the wind tunnel experiment. ⊙ *is a design point replicated twice.* ○ *is a design point replicated thrice.*

[Peter] I'd say that a precise estimation of the intercept is only required when you want to find factor settings that achieve a certain nominal target response. Anyway, the average variance of the estimates is smaller for the I-optimal design, whether we take into account the intercept or not. Therefore, the I-optimal design is better in terms of the A-optimality criterion as well.

[Dr. Cavendish, ignoring Peter's technical comment on the A-optimality] Interestingly, the main effects of the yaw angle and the grille tape coverage are estimated more precisely than those of the front and rear ride heights. I presume that this is due to the fact that the yaw angle and the grille tape coverage were reset independently for every run, whereas the ride heights were set only ten times?

[Brad] Correct. The larger variance for the estimates of the hard-to-change factors' effects is the price you pay for not resetting these factors every time.

[Peter] The table shows this in three ways. Have a look at the variances of the factor-effect estimates we obtain from the I-optimal design, for instance. The variances for the main effects of the ride heights are 0.2, as opposed to 0.032 for the yaw angle and the grille tape coverage. Also, the variance for the estimate of the interaction effect involving the front and the rear ride heights, 0.3, is larger than the variances of 0.046 for the estimates of the other interaction effects. Finally,

Table 10.3 Comparison of the variances of the factor-effect estimates obtained from Marc's design in Figure 10.2 and the I-optimal design in Table 10.2 and Figure 10.3, assuming $\sigma_\gamma^2 = \sigma_\varepsilon^2 = 1$.

Effect	Marc	I-optimal
Intercept	0.429	0.454
Front ride height (FRH)	0.200	0.200
Rear ride height (RRH)	0.200	0.200
Yaw angle	0.042	0.032
Grille coverage	0.042	0.032
FRH × RRH	0.300	0.300
FRH × Yaw angle	0.050	0.046
FRH × Grille coverage	0.050	0.046
RRH × Yaw angle	0.050	0.046
RRH × Grille coverage	0.050	0.046
Yaw angle × Grille coverage	0.063	0.042
FRH × FRH	0.554	0.523
RRH × RRH	0.554	0.523
Yaw angle × Yaw angle	0.125	0.102
Grille coverage × Grille coverage	0.125	0.102
Average (including intercept)	0.189	0.180
Average (excluding intercept)	0.172	0.160

the variances of the estimates for the quadratic effects of the ride heights are larger than those for the yaw angle's and grille tape coverage's quadratic effects.

[Dr. Cavendish] I can see the same pattern for my design.

[Brad, nodding] Sure. You will observe the same pattern for virtually every design involving hard-to-change factors that you do not reset as often as the other factors.

[Dr. Cavendish] I can also see that the interaction effects involving one hard-to-change factor and one easy-to-change factor are estimated as well as the interaction effect involving the two easy-to-change factors. All of these effects are estimated with a variance of less than 0.05, despite the fact that some involve factors that were reset a few times only.

[Peter] Good point! That is one of the interesting features of a split-plot design. Interaction effects involving a hard-to-change and an easy-to-change factor can often be estimated as precisely as the easy-to-change factors' main effects and interaction effects.

Dr. Cavendish remains silent for a moment while absorbing all the information concerning the comparison of the different designs.

[Dr. Cavendish] OK. I think you have convinced me that your I-optimal design is better at estimating the factor effects than my design. Is it also better when it comes to making predictions?

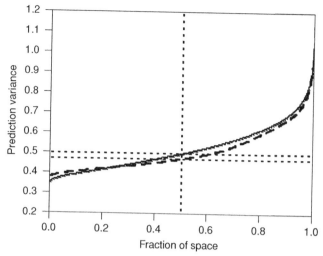

Figure 10.4   Side-by-side Fraction of Design Space plots for Marc's design in Figure 10.2 (solid line) and for the I-optimal design in Table 10.2 and Figure 10.3 (dashed line).

[Peter, clicking a few buttons on his laptop] By definition, the I-optimal design should give the smallest average prediction variance, since the I-optimal design minimizes the average prediction variance over the experimental region. Let me see. Your design gives an average variance of prediction of 0.526, whereas the I-optimal design leads to an average variance 0.512.

[Dr. Cavendish] I am not impressed by the difference.

[Peter, who continues working with his laptop until he produces Figure 10.4] Me neither. This shows that the design you constructed is really good. Here we have a plot that compares the performance of our two designs in some more detail. This figure compares the Fraction of Design Space plot for your design to that for the I-optimal design. The performance of your design, Marc, is given by the solid line, whereas the dashed line shows the performance of the I-optimal design.

[Brad] You can see that the solid line is lower on the left side of the plot, between 0 and 0.20. That means that your design gives a smaller prediction variance than the I-optimal in 20% of the experimental region. In the rest of the experimental region, the I-optimal design gives the smaller prediction variances.

[Peter] My educated guess is that your design predicts very well in the center of the experimental region because your design has a lot of runs at the center point. In other words, your design gives a lot of information about the center of the experimental region.

[Dr. Cavendish] Do the horizontal reference lines correspond to the median prediction variances for the two designs?

[Brad] They do. The median prediction variance for the I-optimal design is about 0.47, while it is 0.5 for your design.

[Dr. Cavendish] That is an interesting plot. And it makes my design look competitive.

[Peter] Sure. You did your homework well. Our I-optimal design is only a tiny bit better than yours.

[Brad] Congratulations!

[Dr. Cavendish] Don't patronize me. I have been going through the literature and puzzling for many, many hours to produce my design. Then, the two of you walk in and generate a better design in a few seconds.

[Peter, grinning] Sorry, but it hasn't always been that easy. When I did my PhD about the optimal design of experiments involving hard-to-change factors, all you could do was go through the literature and spend some time putting together a sensible design. Now that software companies have implemented my PhD work and wrapped it with a nice graphical user interface, the life of industrial experimenters has become much easier.

[Dr. Cavendish, looking at his watch] All right. I have to get going. I am inclined to use your I-optimal design for our experiment with the race car here. Why don't you send me the design by e-mail later today? I expect to get back to you with the data in about ten days.

10.2.2   Data analysis

A fortnight later, Brad and Peter receive an e-mail from Dr. Cavendish with a description of the experiment that he ran in NASA's Langley Research Center, along with a spreadsheet containing the data, some explanations about the measured responses and a preliminary analysis for one of the responses. Here is the e-mail:

Dear Brad and Peter,

We ended up using the I-optimal design that you generated upon your visit of the Langley research facility for our experiment with the Winston Cup race car (did you ever notice that race car is a palindrome?). I have attached a spreadsheet with your design and the four responses we obtained. We measured the coefficient of drag (C_D) and two coefficients of lift, one for the downforce over the front axle (C_{LF}) and one for the downforce over the rear axle (C_{LR}). Our fourth response, the lift over drag ratio or the efficiency, was derived from the other three, as follows:

$$\text{Efficiency} = \frac{-(C_{LF} + C_{LR})}{C_D}.$$

This ratio is a measure of the efficiency. All of the four responses are important to us, so we would like you to fit your split-plot model to each of them separately. As you will see in Table 10.4 of the attached spreadsheet, the coefficient of drag ranged from 0.367 to 0.435 in our experiment. The coefficients of lift ranged from -0.175 to -0.074 (front) and from -0.303 to -0.202 (rear). The efficiency (or lift over drag ratio) lay between 0.705 and 1.159.

Table 10.4 Data from the wind tunnel experiment, obtained using the I-optimal split-plot design in Table 10.2 and Figure 10.3.

Hard-to-change setting	Front ride height (inch)	Rear ride height (inch)	Yaw angle	Grille coverage (%)	Drag (C_D)	Lift front (C_{LF})	Lift rear (C_{LR})	Efficiency
1	3	34	$-3.0°$	0	0.402	-0.105	-0.246	0.873
1	3	34	$+1.0°$	100	0.367	-0.141	-0.214	0.969
1	3	34	$-1.0°$	50	0.384	-0.127	-0.240	0.959
1	3	34	$+1.0°$	0	0.378	-0.088	-0.223	0.821
1	3	34	$-3.0°$	100	0.391	-0.156	-0.242	1.019
2	3.5	34	$+1.0°$	100	0.375	-0.132	-0.213	0.921
2	3.5	34	$-1.0°$	50	0.392	-0.119	-0.243	0.923
2	3.5	34	$+1.0°$	0	0.388	-0.092	-0.227	0.821
2	3.5	34	$-1.0°$	100	0.386	-0.142	-0.229	0.958
2	3.5	34	$-3.0°$	50	0.404	-0.119	-0.251	0.914
3	4	34	$+1.0°$	0	0.399	-0.074	-0.208	0.705
3	4	34	$+1.0°$	100	0.386	-0.122	-0.210	0.861
3	4	34	$-3.0°$	0	0.419	-0.086	-0.231	0.756
3	4	34	$-3.0°$	100	0.408	-0.130	-0.222	0.861
3	4	34	$-1.0°$	50	0.401	-0.099	-0.214	0.780
4	3	35	$-3.0°$	50	0.406	-0.136	-0.270	0.999
4	3	35	$-1.0°$	100	0.389	-0.154	-0.258	1.060
4	3	35	$+1.0°$	50	0.383	-0.120	-0.228	0.910
4	3	35	$-1.0°$	0	0.398	-0.102	-0.238	0.854
4	3	35	$-3.0°$	100	0.402	-0.160	-0.275	1.082
5	3.5	35	$-1.0°$	0	0.406	-0.104	-0.238	0.841
5	3.5	35	$+1.0°$	50	0.390	-0.122	-0.224	0.885
5	3.5	35	$-1.0°$	50	0.402	-0.126	-0.238	0.905
5	3.5	35	$-3.0°$	100	0.410	-0.160	-0.260	1.025
5	3.5	35	$-1.0°$	50	0.401	-0.130	-0.246	0.936
6	3.5	35	$+1.0°$	50	0.392	-0.117	-0.213	0.844
6	3.5	35	$-1.0°$	50	0.402	-0.123	-0.235	0.891
6	3.5	35	$-1.0°$	50	0.402	-0.128	-0.235	0.902
6	3.5	35	$-1.0°$	50	0.403	-0.125	-0.225	0.871
6	3.5	35	$-3.0°$	0	0.420	-0.112	-0.238	0.833
7	4	35	$-1.0°$	0	0.415	-0.098	-0.228	0.786
7	4	35	$+1.0°$	100	0.394	-0.145	-0.202	0.879
7	4	35	$-1.0°$	50	0.410	-0.124	-0.229	0.860
7	4	35	$-1.0°$	100	0.405	-0.145	-0.221	0.905
7	4	35	$-3.0°$	50	0.421	-0.125	-0.242	0.870

(*continued overleaf*)

Table 10.4   (*continued*)

Hard-to-change setting	Front ride height (inch)	Rear ride height (inch)	Yaw angle	Grille coverage (%)	Drag (C_D)	Lift front (C_{LF})	Lift rear (C_{LR})	Efficiency
8	3	36	−3.0°	0	0.419	−0.124	−0.286	0.980
8	3	36	+1.0°	0	0.394	−0.102	−0.255	0.907
8	3	36	−1.0°	50	0.402	−0.134	−0.286	1.045
8	3	36	+1.0°	100	0.386	−0.155	−0.266	1.094
8	3	36	−3.0°	100	0.412	−0.175	−0.303	1.159
9	3.5	36	−3.0°	50	0.423	−0.148	−0.275	1.000
9	3.5	36	−1.0°	0	0.414	−0.110	−0.263	0.901
9	3.5	36	−1.0°	50	0.409	−0.139	−0.266	0.990
9	3.5	36	−1.0°	100	0.405	−0.164	−0.265	1.059
9	3.5	36	+1.0°	100	0.393	−0.161	−0.242	1.025
10	4	36	−3.0°	100	0.428	−0.168	−0.256	0.991
10	4	36	+1.0°	0	0.413	−0.107	−0.235	0.828
10	4	36	−3.0°	0	0.435	−0.118	−0.254	0.853
10	4	36	+1.0°	50	0.408	−0.140	−0.237	0.923
10	4	36	−1.0°	100	0.415	−0.161	−0.253	0.997

Note that we did not run the experiment in the order given in the table. We did what you suggested: each time, we randomly selected one of the ten combinations of front and rear ride height first, and then used the five combinations for the yaw angle and the grille tape coverage corresponding to it. If you want, I can dig up the exact sequence in which we used the hard-to-change and the easy-to-change factor-level combinations from the log, once I get back to my office (I am traveling this week).

What I also did is perform a preliminary analysis with my own software and ordinary least squares to get a rough idea about the results. I am sure that what I did is not the correct thing to do because, in using ordinary least squares, I ignored the correlation structure in the responses due to the split-plot design, but I couldn't stop myself from doing it. I guess I was just too curious to see interpretable results. Anyway, you can find the simplified model I got for the efficiency after dropping insignificant terms in Table 10.5. I started by estimating the full model involving all main effects, two-factor interaction effects and quadratic effects, and dropped the insignificant terms from the model one by one, starting with the term that had the largest p value. The results from my analysis suggest that I should set the front ride height to 3 inches, the rear ride height to 36 inches, the yaw angle to −3° and the grille tape coverage to 100%. This

Table 10.5   Factor-effect estimates, standard errors, degrees of freedom, and p values obtained by Dr. Cavendish from the wind tunnel data for the efficiency response in Table 10.4 after simplifying the model by dropping terms with insignificant effects, using ordinary least squares.

Effect	Estimate	Standard error	DF	t Ratio	p Value
β_0	0.9014	0.0046	42	194.83	$<.0001$
β_1	−0.0607	0.0038	42	−16.05	$<.0001$
β_2	0.0529	0.0038	42	14.04	$<.0001$
β_3	−0.0237	0.0038	42	−6.28	$<.0001$
β_4	0.0756	0.0037	42	20.32	$<.0001$
β_{22}	0.0241	0.0060	42	4.03	0.0002
β_{13}	0.0097	0.0045	42	2.15	0.0374
β_{14}	−0.0103	0.0044	42	−2.34	0.0244

The indices 1, 2, 3 and 4 in the first column of the table refer to the front ride height, the rear ride height, the yaw angle and the grille tape coverage, respectively.

would give me an efficiency of 1.1586 ± 0.0201. I am anxious to see how different your results will be from mine.

Best wishes,
Marc

Later that day, Peter analyzes the data from the wind tunnel experiment and gets back to Dr. Cavendish by e-mail:

Dear Marc,

Here is a quick reaction, with the most important results. We will send you a more comprehensive account of our analysis in one of the next few days.

Let me first comment on your preliminary analysis by means of ordinary least squares. One important feature of the estimates that you obtained is that they are unbiased, assuming your model includes the right terms. So, your factor-effect estimates are certainly not worthless. You can see that by looking at the simplified model that we obtained using generalized least squares, taking into account the correlations in the responses. Table 10.6, which is attached to this e-mail, contains the details for our model. For instance, the estimate for the main effect of the front ride height (β_1 in the tables) is the same in our two analyses. All the other estimates differ, but not by much. The main effect of the grille tape coverage (β_4 in the tables) is, for instance, 0.0756 in your analysis and 0.0743 in our analysis. That is the good news.

The standard errors and the degrees of freedom used for the significance tests in your analysis are, however, wrong. In your analysis, for

Table 10.6   Factor-effect estimates, standard errors, degrees of freedom and p values obtained by Peter and Brad from the wind tunnel data for the efficiency response in Table 10.4 after simplifying the model by dropping terms with insignificant effects, using generalized least squares.

Effect	Estimate	Standard error	DF	t Ratio	p Value
β_0	0.9160	0.0068	6.99	135.38	$<.0001$
β_1	-0.0607	0.0087	6.99	-6.94	0.0002
β_2	0.0524	0.0087	6.99	5.99	0.0005
β_3	-0.0246	0.0028	35.07	-8.82	$<.0001$
β_4	0.0743	0.0028	35.18	26.85	$<.0001$
β_{13}	0.0102	0.0033	35.03	3.08	0.0040
β_{14}	-0.0107	0.0033	35.07	-3.29	0.0023
β_{24}	0.0078	0.0033	35.08	2.39	0.0226

The indices 1, 2, 3 and 4 in the first column of the table refer to the front ride height, the rear ride height, the yaw angle and the grille tape coverage, respectively.

instance, all the estimates for the main effects $\beta_1 - \beta_4$ have nearly the same standard error (0.0037 or 0.0038), even though the true standard errors of the estimates of β_1 and β_2 (the main effects of the front and the rear ride heights) are larger than those of β_3 and β_4 (the main effects of the yaw angle and the grille tape coverage). Similarly, your analysis employed about the same standard errors for all the interaction effects, even though the true standard error of the estimate for β_{12} (the interaction effect involving the two ride heights) is larger than the true standard error of all the other interaction-effect estimates. Finally, your analysis also used incorrect standard errors for the estimates of the quadratic effects: the true standard errors of the estimates of β_{11} and β_{22} (the quadratic effects of the front and the rear ride heights) are larger than those of β_{33} and β_{44} (the quadratic effects of the yaw angle and the grille tape coverage). I have included a side-by-side comparison of the results of your analysis (using ordinary least squares) and ours (using generalized least squares, which gives correct standard errors) in Table 10.7 to make this more clearly visible. Perhaps the most spectacular differences between the two analyses in the table are that the quadratic effect of the rear ride height (β_{22}) is significantly different from zero in your analysis but not in ours, and that the interaction effect involving the rear ride height and the grille tape coverage (β_{24}) is significant in our analysis but not in yours. It is worth noting that our simplified models in Tables 10.5 and 10.6 show the same differences.

Essentially, when you use an ordinary least squares analysis for split-plot data, two things happen. First, you are more likely to erroneously detect (main, quadratic, and interaction) effects of the hard-to-change factors. Second, you are also more likely to miss the easy-to-change factors' (main, quadratic and interaction) effects. The reason for the incorrect inference concerning the hard-to-change factors is that the ordinary least

Table 10.7 Comparison of the ordinary and generalized least squares results for the full quadratic model for the efficiency response.

Effect	Ordinary least squares					Generalized least squares				
	Estimate	Standard error	DF	t Ratio	p Value	Estimate	Standard error	DF	t Ratio	p Value
β_0	0.9109	0.0060	45	151.64	< .0001	0.9114	0.0117	4.21	77.67	< .0001
β_1	−0.0609	0.0037	45	−16.54	< .0001	−0.0609	0.0079	3.98	−7.70	0.0016
β_2	0.0523	0.0037	45	14.19	< .0001	0.0522	0.0079	3.98	6.60	0.0028
β_3	−0.0241	0.0037	45	−6.53	< .0001	−0.0247	0.0027	31.03	−9.04	< .0001
β_4	0.0758	0.0036	45	20.95	< .0001	0.0745	0.0027	31.19	27.50	< .0001
β_{12}	0.0042	0.0045	45	0.93	0.3592	0.0042	0.0097	3.97	0.44	0.6833
β_{13}	0.0104	0.0044	45	2.37	0.0236	0.0106	0.0033	31.03	3.24	0.0028
β_{14}	−0.0107	0.0043	45	−2.47	0.0184	−0.0111	0.0032	31.08	−3.46	0.0016
β_{23}	−0.0022	0.0044	45	−0.49	0.6280	−0.0015	0.0033	31.04	−0.47	0.6418
β_{24}	0.0066	0.0043	45	1.54	0.1334	0.0078	0.0032	31.12	2.44	0.0208
β_{34}	−0.0016	0.0043	45	−0.37	0.7135	0.0000	0.0033	31.39	−0.01	0.9940
β_{11}	−0.0079	0.0062	45	−1.27	0.2125	−0.0075	0.0127	4.07	−0.58	0.5896
β_{22}	0.0286	0.0062	45	4.64	< .0001	0.0291	0.0127	4.07	2.28	0.0834
β_{33}	−0.0081	0.0063	45	−1.30	0.2033	−0.0075	0.0047	31.21	−1.59	0.1212
β_{44}	−0.0044	0.0065	45	−0.68	0.5016	−0.0064	0.0048	31.11	−1.34	0.1914

The indices 1, 2, 3 and 4 in the first column of the table refer to the front ride height, the rear ride height, the yaw angle and the grille tape coverage, respectively.

squares analysis underestimates the variance of the estimates of their effects, which leads to too large t ratios and too many rejections of the null hypothesis. The reason for the incorrect inference concerning the easy-to-change factors is that the ordinary least squares analysis overestimates the variance of the estimates of their effects, which leads to too small t ratios and too few rejections of the null hypothesis.

I could go on for quite a while on the technical reasons why the variances for the hard-to-change factors' effects are larger than those for the easy-to-change factors' effects. However, let me just give you an intuitive explanation: you set the front ride height and the rear ride height independently only ten times, whereas you independently set the yaw angle and the grille tape coverage 50 times. Therefore, you can never get the same precision in your estimates for effects involving hard-to-change factors only (in your case the ride heights) as for effects involving the easy-to-change factors (yaw angle and grille tape coverage).

One last thing I would like to say on the two different analyses is that your ordinary least squares analysis uses 42 degrees of freedom for all the significance tests in your simplified model (see the column labeled DF in Table 10.5), whereas our generalized least squares analysis uses about seven degrees of freedom for the effects of the hard-to-change factors and about 35 degrees of freedom for the effects of the easy-to-change factors (see the column labeled DF in Table 10.6). So, here too, the ordinary least squares analysis treats all the effects alike, whereas the generalized least squares' degrees of freedom match the way in which the experiment was run better: there are fewer independent resettings for the hard-to-change factors than for the easy-to-change factors, and, hence, there are fewer independent units of information on the hard-to-change factors than on the easy-to-change factors. That is why there are fewer degrees of freedom for the main effects, the quadratic effects and the interaction effect of the two ride heights in the generalized least squares analyses I presented.

Let me know if you want to learn more about the exact details of the split-plot analysis involving generalized least squares! Now, I will give an overview of the models that we obtained for the four responses in your experiment. I have taken the liberty to add a fifth response as well, the total lift $C_{LF} + C_{LR}$.

As you asked, we have generated suitable models for each of the responses. The significant effects are all listed in Table 10.8. As you will see, the four main effects are significant in all models except one. The main effect of grille tape coverage is not significantly different from zero in the model for the rear lift. For all of the responses, there are a few statistically significant interaction effects. Most of these interaction effects are, however, practically insignificant. There are also a few statistically significant quadratic effects, but these are also of little practical relevance. You can see that by looking at the absolute magnitudes of the estimates of the interaction effects and the quadratic effects in Table 10.8.

Table 10.8 Simplified models for the five responses in the wind tunnel experiment, along with factor-effect estimates, their standard errors and the *p*-values.

Effect	Efficiency Estimate	Standard error	p Value	Drag Estimate	Standard error	p Value	Front lift Estimate	Standard error	p Value	Rear lift Estimate	Standard error	p Value	Total lift Estimate	Standard error	p Value
Intercept	0.9160	0.0068	< .0001	0.4015	0.0003	< .0001	−0.1263	0.0007	< .0001	−0.2378	0.0032	< .0001	−0.3682	0.0027	< .0001
FRH	−0.0607	0.0087	0.0002	0.0086	0.0003	< .0001	0.0043	0.0009	0.0032	0.0122	0.0025	0.0030	0.0164	0.0035	0.0022
RRH	0.0524	0.0087	0.0005	0.0088	0.0003	< .0001	−0.0123	0.0009	< .0001	−0.0169	0.0025	0.0005	−0.0291	0.0035	0.0001
YA	−0.0246	0.0028	< .0001	−0.0117	0.0001	< .0001	0.0063	0.0005	< .0001	0.0145	0.0010	< .0001	0.0208	0.0011	< .0001
GC	0.0743	0.0028	< .0001	−0.0049	0.0001	< .0001	−0.0247	0.0005	< .0001	−0.0007	0.0010	0.5029	−0.0253	0.0011	< .0001
FRH × RRH							−0.0055	0.0011	0.0025						
FRH × YA	0.0102	0.0033	0.0040	0.0006	0.0001	< .0001	−0.0024	0.0006	0.0005	−0.0029	0.0012	0.0217	−0.0053	0.0013	0.0004
FRH × GC	−0.0107	0.0033	0.0023							0.0027	0.0012	0.0263	0.0035	0.0013	0.0105
RRH × YA				−0.0005	0.0001	0.0001									
RRH × GC	0.0078	0.0033	0.0226	0.0009	0.0001	< .0001				−0.0033	0.0012	0.0076	−0.0045	0.0013	0.0015
YA × GC				−0.0004	0.0001	0.0013									
RRH × RRH										−0.0104	0.0040	0.0412			
YA × YA				0.0009	0.0002	< .0001									
GC × GC										0.0035	0.0017	0.0482			

The problem that you now face is that it is impossible to find settings for the four factors that optimize the different responses simultaneously. I have played around with the models for the five responses a little bit. When I assume that you want to minimize the drag and the total lift and to maximize the efficiency, then setting the front and the rear ride height to their lowest and highest possible values, respectively (i.e., 3.0 inches for the front ride height and 36 inches for the rear ride height), setting the yaw angle to +1.0 and using 100% of grille tape coverage seems a reasonable thing to do. If you'd like me to, I can give a demonstration of the various options to do multi-response process optimization. We have some useful interactive tools to help you with that.

Peter

10.3   Peek into the black box

10.3.1   Split-plot terminology

The model we use for analyzing data from a split-plot experiment is essentially the same as the model for an experiment run in b blocks with k observations each. What is different between a split-plot experiment and a blocked experiment is the terminology used and the reason for using each type of experiment.

In a split-plot experiment, the blocks are usually called whole plots, whereas the runs are referred to as sub-plots or split-plots. This terminology reflects the agricultural origin of the split-plot design. This is explained in Attachment 10.1. Like each block in a blocked experiment, every whole plot in a split-plot experiment is a set of experimental runs for which the responses are correlated. The reason for using a split-plot design is, however, different than that for using a blocked experiment. In many experiments, there are one or more experimental factors whose settings are somehow hard to change. Experimenters then do not like a completely randomized design, in which the order of the runs is determined at random and in which it is required to reset the levels of the experimental factors independently for every run, because they would have to reset the level of the hard-to-change-factor(s) for every run. This results in a time-consuming and/or expensive experiment. When hard-to-change factors are present, experimenters much prefer to group the runs that have the same level for each hard-to-change factor and test the different settings for the easy-to-change factors one after another. The grouping of the runs in a split-plot experiment is thus caused by the hard-to-change nature of some of the experimental factors. In a blocked experiment, the grouping is done because not all the runs are done under homogeneous circumstances, and the groups are chosen so that within the groups the circumstances are as homogeneous as possible. Finally, because the hard-to-change factors cause the grouping of runs into whole plots, they are called the whole-plot factors. The easy-to-change factors are named sub-plot factors. By construction, the levels of the hard-to-change factors remain constant within every whole plot, whereas the levels of the easy-to-change factors change from run to run.

Attachment 10.1   Split-plot designs.

The terminology used in the context of split-plot designs derives from its initial agricultural applications, where experiments are carried out on different plots of land. For example, in experiments for investigating the effect of different fertilizers and varieties on the yield of crops, the fertilizers are often sprayed from planes and large plots of land must then be treated with the same fertilizer. Crop varieties can be planted on smaller plots of land that are obtained by dividing up the large plots that were sprayed together. The larger plots of land are called *whole plots*, whereas the smaller plots are named *split-plots* or *sub-plots*. Because the levels of the factor fertilizer are applied to whole plots, that factor is called the *whole-plot factor* of the experiment. The second factor, crop variety, is applied to the sub-plots, and is referred to as the *sub-plot factor* of the experiment. A schematic representation of a split-plot design involving fertilizer as the whole-plot factor and variety as the sub-plot factor is given in Figure 10.5. In the figure, the levels of the whole-plot factor are represented by F_1, F_2, and F_3, and those of the sub-plot factor by V_1, V_2, and V_3.

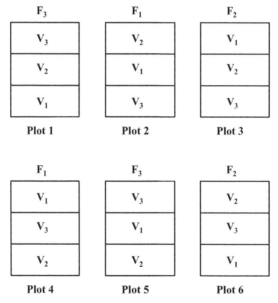

Figure 10.5   Classical agricultural split-plot design where F_1, F_2, and F_3 and V_1, V_2, and V_3 represent the levels of the factors fertilizer and variety, respectively. The design has six whole plots (labeled Plot 1–6), each of which is divided into three sub-plots.

In classical split-plot designs, all combinations of levels of the sub-plot factors occur within each combination of levels of the whole-plot factors. This is not required in split-plot response surface designs, such as the wind tunnel experiment, because models for response surface designs involve fewer parameters than classical analysis of variance models for split-plot data, and because restricted maximum likelihood is a more flexible method for estimating variance components than the traditional analysis of variance.

10.3.2 Model

The model we use for analyzing data from a split-plot experiment with b whole plots of k runs is

$$Y_{ij} = \mathbf{f}'(\mathbf{x}_{ij})\boldsymbol{\beta} + \gamma_i + \varepsilon_{ij}, \tag{10.1}$$

where Y_{ij} is the response measured at the jth run or sub-plot in the ith whole plot, $\mathbf{x}_{ij}$ is a vector that contains the levels of all the experimental factors at the jth run in the ith whole plot, $\mathbf{f}'(\mathbf{x}_{ij})$ is that vector's model expansion, and $\boldsymbol{\beta}$ contains the intercept and all the factor effects that are in the model. The term γ_i represents the ith whole-plot effect and ε_{ij} is the residual error associated with the jth run in whole plot i. Each whole plot corresponds to an independent setting of the hard-to-change factors, and each run or sub-plot corresponds to one independent setting of the easy-to-change factors.

We like to stress the fact that there are two types of factors in a split-plot experiment by using two different symbols. For the N_w hard-to-change factors, we use the symbols $w_1, \ldots, w_{N_w}$ or $\mathbf{w}$. For the N_s easy-to-change factors, we use the symbols $s_1, \ldots, s_{N_s}$ or $\mathbf{s}$. The split-plot model then is

$$Y_{ij} = \mathbf{f}'(\mathbf{w}_i, \mathbf{s}_{ij})\boldsymbol{\beta} + \gamma_i + \varepsilon_{ij}, \tag{10.2}$$

where $\mathbf{w}_i$ gives the settings of the hard-to-change factors in the ith whole plot and $\mathbf{s}_{ij}$ gives the settings of the easy-to-change factors at the jth run within the ith whole plot. This way of writing the model is instructive as it emphasizes the fact that the levels of the hard-to-change factors do not change within a whole plot.

We call the hard-to-change factors also *whole-plot factors* because their levels are applied to whole plots, and we also use the term *sub-plot factors* for all the easy-to-change factors because their levels are applied to sub-plots. In a model for split-plot data, we distinguish three types of effects: whole-plot effects, sub-plot effects, and whole-plot-by-sub-plot interaction effects. The *whole-plot effects* are the main effects and quadratic effects of the hard-to-change factors and the interaction effects involving only hard-to-change factors. The *sub-plot effects* are the main effects and quadratic effects of the easy-to-change factors and the interaction effects involving only easy-to-change factors. Finally, the *whole-plot-by-sub-plot interaction effects* are interaction effects involving one hard-to-change factor and one easy-to-change factor.

As in the model for data from blocked experiments in Equation (7.5), we assume that the whole-plot effects γ_i and the residual errors ε_{ij} are random effects that are all independent and normally distributed with zero mean, that the whole-plot effects γ_i have variance σ_γ^2, and that the residual errors ε_{ij} have variance σ_ε^2. The dependence between the k responses from the ith whole plot, $Y_{i1}, \ldots, Y_{ik}$, is given by the symmetric $k \times k$ matrix

$$
\Lambda = \begin{bmatrix}
\sigma_\gamma^2 + \sigma_\varepsilon^2 & \sigma_\gamma^2 & \cdots & \sigma_\gamma^2 \\
\sigma_\gamma^2 & \sigma_\gamma^2 + \sigma_\varepsilon^2 & \cdots & \sigma_\gamma^2 \\
\vdots & \vdots & \ddots & \vdots \\
\sigma_\gamma^2 & \sigma_\gamma^2 & \cdots & \sigma_\gamma^2 + \sigma_\varepsilon^2
\end{bmatrix}. \tag{10.3}
$$

The dependence structure of all the responses from a split-plot experiment is given by

$$
V = \begin{bmatrix}
\Lambda & 0_{k \times k} & \cdots & 0_{k \times k} \\
0_{k \times k} & \Lambda & \cdots & 0_{k \times k} \\
\vdots & \vdots & \ddots & \vdots \\
0_{k \times k} & 0_{k \times k} & \cdots & \Lambda
\end{bmatrix}.
$$

The interpretation of the elements of V is similar to that for the elements of Λ. The diagonal elements of V are the variances of the responses Y_{ij}, and the off-diagonal elements are covariances between pairs of responses. The nonzero off-diagonal elements of V all correspond to pairs of responses from within a given whole plot. All zero elements of V correspond to a pair of runs from two different whole plots. The derivation of these results is completely the same as for experiments with random blocks (see Section 7.3.1).

Because the responses in the model in Equation (10.2) are correlated, it is better to use generalized least squares or GLS estimates, which account for the correlation and are generally more precise than ordinary least squares estimates. The GLS estimates are calculated from

$$
\hat{\beta} = (X'V^{-1}X)^{-1}X'V^{-1}Y. \tag{10.4}
$$

As explained in Section 7.3.1, the variances σ_γ^2 and σ_ε^2 (and therefore also the matrix V) are unknown, so that we have to estimate them. We can do this by using restricted maximum likelihood (REML) estimation (see Section 7.3.3 and Attachment 7.1 on page 150). We can then substitute the estimates of σ_γ^2 and σ_ε^2 in the GLS estimator in Equation (10.4), which yields the feasible GLS estimator

$$
\hat{\beta} = (X'\hat{V}^{-1}X)^{-1}X'\hat{V}^{-1}Y. \tag{10.5}
$$

We obtain the standard errors and the degrees of freedom for the significance tests using the Kenward–Roger approach described in Section 7.3.4. In the main software packages, the split-plot analysis is a standard option.

10.3.3   Inference from a split-plot design

A convenient feature of the GLS approach is that it automatically detects which of the effects are estimated more precisely than others. The approach also associates appropriate degrees of freedom with every factor-effect estimate, so that a correct analysis is guaranteed.

This is very important for split-plot designs because, compared to the effects of the easy-to-change factors, all the effects of the hard-to-change factors are estimated imprecisely, and, as a result, the significance tests for these effects are not as powerful. Obviously, this is due to the fact that the levels of these factors are not reset independently very often. For the effects of the easy-to-change factors, which are reset for every run in every whole plot, the estimates are usually more precise and, as a result, the significance tests are usually more powerful. It is therefore generally more likely to detect significant effects of the easy-to-change factors than it is for the effects of the hard-to-change factors when a split-plot design is used.

Three comments must be made at this point. The first one is about the interaction effects involving a hard-to-change factor and an easy-to-change factor, whereas the second is about interaction effects involving two easy-to-change factors. The final comment is concerned with the quadratic effects.

1. An interesting feature of split-plot designs is that they allow precise estimation of and powerful inference for the interaction effects involving a hard-to-change factor and an easy-to-change factor (i.e., the whole-plot-by-sub-plot interaction effects), even though the levels of the hard-to-change factors are reset independently a limited number of times only. The reason why is shown in Table 10.9. In the table, a split-plot design with one hard-to-change factor w and three easy-to-change factors s_1, s_2, and s_3 is given. The design involves eight independent resettings of the hard-to-change factor's levels, or eight whole plots. In the table, we also displayed the two-factor interaction effect contrast for ws_1, ws_2, and ws_3. It is easy to verify that, within each whole plot, the levels of ws_1, ws_2, and ws_3 change from run to run. The ws_1, ws_2, and ws_3 columns are therefore similar to the main-effect columns for the easy-to-change factors s_1, s_2, and s_3. As a result, the split-plot design contains as much information about the three interaction effects involving the hard-to-change factor w as about the main effects of the easy-to-change factors. Hence, the estimates of the interaction effects involving w will be as precise as the estimates of the easy-to-change factors' main effects, and the inference will be equally powerful as well. This is shown in Table 10.10, where we displayed the variances of all the main-effect and interaction-effect estimates for the design in Table 10.9, assuming that $\sigma_\gamma^2 = \sigma_\varepsilon^2 = 1$. The fact that precise information is available for all the two-factor interaction effects involving any hard-to-change factor and an easy-to-change factor is typical for split-plot designs. The degrees of freedom for the significance tests of these interaction effects will also be similar, and often identical, to those of the main effects of the easy-to-change factors.

Table 10.9   Split-plot design with eight whole plots of two experimental runs for one hard-to-change factor w and three easy-to-change factors s_1, s_2, and s_3, along with the corresponding two-factor interaction effect contrast columns of the model matrix $\mathbf{X}$.

Whole plot	Main effects columns				Whole-plot-by-sub-plot interaction columns			Sub-plot interaction columns		
	w	s_1	s_2	s_3	ws_1	ws_2	ws_3	s_1s_2	s_1s_3	s_2s_3
1	−1	−1	−1	−1	+1	+1	+1	+1	+1	+1
1	−1	+1	+1	+1	−1	−1	−1	+1	+1	+1
2	+1	−1	+1	−1	−1	+1	−1	−1	+1	−1
2	+1	+1	−1	+1	+1	−1	+1	−1	+1	−1
3	−1	−1	+1	+1	+1	−1	−1	−1	−1	+1
3	−1	+1	−1	−1	−1	+1	+1	−1	−1	+1
4	+1	−1	−1	−1	−1	−1	−1	+1	+1	+1
4	+1	+1	+1	+1	+1	+1	+1	+1	+1	+1
5	−1	−1	+1	−1	+1	−1	+1	−1	+1	−1
5	−1	+1	−1	+1	−1	+1	−1	−1	+1	−1
6	+1	−1	+1	+1	−1	+1	+1	−1	−1	+1
6	+1	+1	−1	−1	+1	−1	−1	−1	−1	+1
7	−1	−1	−1	+1	+1	+1	−1	+1	−1	−1
7	−1	+1	+1	−1	−1	−1	+1	+1	−1	−1
8	+1	−1	−1	+1	−1	−1	+1	+1	−1	−1
8	+1	+1	+1	−1	+1	+1	−1	+1	−1	−1

2. Usually, split-plot designs also allow precise estimation and powerful inference for the interactions involving two easy-to-change factors. This is a logical consequence of the fact that we reset the levels of the easy-to-change factors independently for every run. In one particular situation, however, the interaction effects involving two easy-to-change factors are estimated just as imprecisely as the main effect(s) of the hard-to-change factor(s). This is also illustrated in Table 10.9, which has contrast columns for the three two-factor interaction effects involving the easy-to-change factors s_1, s_2, and s_3. From the s_1s_2, s_1s_3, and s_2s_3 contrast columns, you can see that, within one whole plot, the levels of these interaction contrasts do not change, just like the level of the hard-to-change factor w. The two-factor interactions involving the easy-to-change

Table 10.10   Variances of the effect estimates for the split-plot design in Table 10.9 assuming that $\sigma_\gamma^2 = \sigma_\varepsilon^2 = 1$.

Effect	Variance
Intercept	0.1875
w	0.1875
s_1	0.0625
s_2	0.0625
s_3	0.0625
ws_1	0.0625
ws_2	0.0625
ws_3	0.0625
$s_1 s_2$	0.1875
$s_1 s_3$	0.1875
$s_2 s_3$	0.1875

factors, therefore, behave like the hard-to-change factor's main effect in the analysis. This is shown in Table 10.10, where we displayed the variances of all the effect estimates for the design in Table 10.9 assuming that $\sigma_\gamma^2 = \sigma_\varepsilon^2 = 1$. The table shows that the variances for the estimates of the interaction effects involving the easy-to-change factors are identical to that of the estimate of the main effect of the hard-to-change factor w. As a result, the power for the significance tests concerning these interaction effects will be as low as that for the hard-to-change factor's main effect. Also, the degrees of freedom for the corresponding significance tests will be the same as those for the main effect of the hard-to-change factor.

3. The estimation precision and the power of the significance tests for quadratic effects of easy-to-change factors is to a large extent similar to that of the main effects and the quadratic effects of the hard-to-change factors and of the interaction effects involving only hard-to-change factors. This can be seen from Table 10.11, where we show the first, third, and sixth whole plot of the I-optimal split-plot in Table 10.2, along with two extra columns for the squares of the settings of the easy-to-change factors yaw angle and grille tape coverage. Unlike the columns for the main effects of the easy-to-change factors, the two columns for the quadratic effects only have two different values. Moreover, in each of the three whole plots displayed, one of the two levels appears four times, while the other appears only once. As a result, the columns for the quadratic effects are very nearly constant within these whole plots, so that, to some extent, these effects resemble hard-to-change factors' effects, whose columns in the model matrix $\mathbf{X}$ are completely constant within each whole plot.

The first and third of these comments are illustrated in the right panel of Table 10.7, where the results from a correct GLS analysis of the efficiency response for the

Table 10.11   Excerpt from the I-optimal split-plot design in Table 10.2 with two extra columns for the quadratic effects of the easy-to-change factors yaw angle and grille tape coverage.

Hard-to-change factor setting	Front ride height	Rear ride height	Yaw angle	Grille coverage	$(\text{Yaw angle})^2$	$(\text{Grille coverage})^2$
1	−1	−1	−1	−1	+1	+1
1	−1	−1	1	1	+1	+1
1	−1	−1	0	0	0	0
1	−1	−1	1	−1	+1	+1
1	−1	−1	−1	1	+1	+1
3	1	−1	1	−1	+1	+1
3	1	−1	1	1	+1	+1
3	1	−1	−1	−1	+1	+1
3	1	−1	−1	1	+1	+1
3	1	−1	0	0	0	0
6	0	0	1	0	0	0
6	0	0	0	0	0	0
6	0	0	0	0	0	0
6	0	0	0	0	0	0
6	0	0	−1	−1	+1	+1

wind tunnel experiment are given. In the table, it can be verified that the standard errors for estimates of the whole-plot-by-sub-plot interaction effects β_{13}, β_{14}, β_{23}, and β_{24} (which are all 0.0032 or 0.0033) are about equally large as the standard errors of the estimates of the main effects of the easy-to-change factors yaw angle and grille tape coverage, β_3 and β_4 (which are both equal to 0.0027). The estimates of the quadratic effects of the easy-to-change factors, β_{33} and β_{44}, have standard errors of 0.0047 and 0.0048, respectively. These standard errors are higher than those of any other estimate of an effect involving at least one easy-to-change factor. So, of all the effects involving easy-to-change factors in the wind tunnel experiment, the quadratic effects are those that resemble the hard-to-change factors' effects most in terms of precision of the estimation.

10.3.4   Disguises of a split-plot design

It is not always obvious that an experiment has a split-plot structure. In the following scenarios, some of the experimental factors are somehow hard to change or hard to reset independently, and experimenters often use a split-plot design.

1. When different combinations of levels of some experimental factors are tested while the levels of some other experimental factors are not reset, then the

resulting design is a split-plot design. This happens, for example, when experimenters try out different combinations during one run of an oven. The oven temperature is not reset for all these observations. It is therefore a whole-plot factor of the experiment. The other factors are sub-plot factors.

2. In two-stage experiments, it sometimes happens that some of the experimental factors are applied in the first stage, whereas others are applied in the second stage. In those cases, the first stage of the experiment involves the production of various batches of experimental material using different combinations of levels of the first-stage experimental factors. These batches are then split in sub-batches after the first stage. The sub-batches undergo different treatments (dictated by the levels of the second-stage factors) at each run in the second stage of the experiment. The factors applied in the first stage are the whole-plot factors, whereas those applied in the second stage are the sub-plot factors.

3. Experiments for simultaneously investigating the impact of ingredients of a mixture and of process factors, i.e., mixture-process variable experiments, are frequently conducted as split-plot experiments (Cornell (1988)). This can happen in two ways. First, several batches can be prepared that undergo different process conditions. In that case, the ingredients of the mixture are the whole-plot factors and the factors determining the tested process conditions act as sub-plot factors. Second, processing conditions can be held fixed while consecutively trying out different mixtures. In that case, the ingredients of the mixture are the sub-plot factors of the design, while the process factors are the whole-plot factors.

4. When experimental designs need to be run sequentially, experimenters are often reluctant to change the levels of one or more factors because this is impractical, time-consuming, or expensive. All the observations that are obtained by not resetting the levels of the hard-to-change factors then play the role of a whole plot. The hard-to-change factors are the whole-plot factors of the design, whereas the remaining factors are sub-plot factors.

5. A special case of an experiment involving hard-to-change factors is a prototype experiment, in which the levels of some of the factors define a prototype, whereas the levels of other factors determine operating conditions under which the prototype is tested. As changing the levels of a prototype factor involves assembling a new prototype, these levels are held constant for several observations under different operating conditions. In that type of experiment, the factors related to the prototype are the whole-plot factors, whereas the ones connected to the operating conditions serve as sub-plot factors.

6. Robust product experiments, i.e., experiments aimed at designing products that work well under different environmental conditions, often involve crossed arrays. In that case, they are frequently run as split-plot designs. This is because the environmental factors (also named noise factors) are usually hard to control and changing their levels is cumbersome. These factors then act as whole-plot

factors, while the other factors, the control factors, act as sub-plot factors. An interesting feature of split-plot designs for robust product experimentation is, therefore, that control-by-noise interaction effects can be estimated more precisely than by using a completely randomized experiment. This is because the control-by-noise interaction effects are whole-plot-by-sub-plot interaction effects. As we explained in the first comment in Section 10.3.3, we can estimate this type of interaction effect very precisely when using a split-plot design.

10.3.5   Required number of whole plots and runs

To find an optimal design for a split-plot experiment, we have to make various decisions. One important decision is the determination of the number of independent resettings (or, in the split-plot jargon, the number of whole plots) that will be used for the hard-to-change factors. To a large extent, that number drives the total cost of the experiment and the time required to conduct it. Therefore, many researchers want to keep the number of independent resettings of the hard-to-change factors to a strict minimum. In general, the fewer times the levels of the hard-to-change factors are reset, the poorer the estimation of the hard-to-change factors' effects and the more difficult it becomes to make valid statements about these effects. If the number of independent resettings is too small, the entire model estimation and inference is jeopardized.

It is usually doable to calculate the required number of resettings. To do so, we must determine the number of pure whole-plot effects. In the wind tunnel experiment, for instance, there are two hard-to-change factors and the goal of the experiment is to estimate a full quadratic model. Nine of the effects in that model involve at least one hard-to-change factor: the two main effects β_1 and β_2, the interaction effect β_{12}, the two quadratic effects β_{11} and β_{22}, and the four whole-plot-by-sub-plot interaction effects $\beta_{13}, \beta_{14}, \beta_{23}$, and β_{24}. We have seen in Section 10.3.3 that, generally, the whole-plot-by-sub-plot interaction effects behave like easy-to-change factor effects, and, therefore, they can be ignored when determining the required number of resettings of the hard-to-change factors. The whole-plot-by-sub-plot interaction effects are, therefore, not pure whole-plot effects. The remaining five effects are whole-plot effects, and estimating them requires five resettings of the hard-to-change factors. On top of that, we need an independent resetting to estimate the whole-plot error variance σ_γ^2 and an independent resetting to estimate the constant, or intercept, β_0. In total, we thus need at least seven whole plots in the wind tunnel experiment. Dr. Cavendish ended up using ten independent resettings instead of seven. This has a substantial beneficial impact on the quality of his factor-effect estimates and on the inferences he can make concerning the hard-to-change factors' effects.

When the plan is to run a two-level split-plot experiment with two observations per whole plot, as in Table 10.9, the sub-plot-by-sub-plot interaction effects should be considered pure whole-plot effects, for the reasons given in Section 10.3.3. In that case, the required number of whole plots is higher than what would be expected at first sight. Failing to recognize this may result in the whole-plot error variance σ_γ^2 not being estimable. In that case, a proper (GLS) analysis of the split-plot data is impossible.

The total number of runs should at least be equal to the total number of terms in the model, plus two. The two additional runs are needed to allow for the estimation of σ_γ^2 and σ_ε^2. For all the easy-to-change factors' effects and all their interaction effects with the hard-to-change factors to be estimable, the total number of runs minus the number of resettings of the hard-to-change factors should be larger than their number, plus one (because a proper data analysis requires estimation of σ_ε^2 as well). Most often, these minimum numbers of whole plots and runs will allow estimation of the a priori model.

For certain combinations of numbers of runs within the whole plots and total numbers of runs, however, this procedure for determining the minimum numbers of whole plots and runs underestimates the truly required numbers. This is because, especially for small total numbers of runs and small or odd numbers of runs within each whole plot, the inevitable nonorthogonality of optimal split-plot designs for models including interaction effects and quadratic effects may make either one of the variances σ_γ^2 and σ_ε^2 or one of the factor effects nonestimable. If a design with minimal numbers of runs and whole plots is wanted, we recommend simulating data and performing a split-plot analysis on the simulated data to verify that σ_γ^2, σ_ε^2 and all the factor effects are indeed estimable, before conducting the actual experiment. For powerful statistical inference, we recommend using more whole plots and more runs than the minimum required number.

It should be clear that the choice of the number of independent resettings of the hard-to-change factors as well as the total number of runs is crucial for successful data analysis. The total number of runs is usually spread equally across the different resettings of the hard-to-change factors. This is not strictly required, but it is certainly most elegant. In some cases, the experimenter cannot choose the number of runs at each setting of the hard-to-change factors. For instance, when the hard-to-change factor of a split-plot experiment is oven temperature and four runs can be simultaneously conducted in the oven, then the number of runs at every resetting of the oven temperature is automatically four.

10.3.6 Optimal design of split-plot experiments

Once the number of resettings of the hard-to-change factors has been fixed, as well as the total number of runs and the number of runs at each setting of the hard-to-change factors, we have to determine factor levels for each run. Obviously, when doing so, we must take into account the hard-to-change nature of some of the experimental factors. The challenge is to find the factor-level combinations that result in the most precise GLS estimates possible and/or guarantee the most precise predictions. A D-optimal split-plot design is the set of factor-level combinations that minimizes the determinant of $(\mathbf{X}'\mathbf{V}^{-1}\mathbf{X})^{-1}$, or maximizes the determinant of the information matrix $\mathbf{X}'\mathbf{V}^{-1}\mathbf{X}$. We call the determinant $|\mathbf{X}'\mathbf{V}^{-1}\mathbf{X}|$ the D-optimality criterion value. A D-optimal design guarantees precise estimates for the effects of the experimental factors. An I-optimal split-plot design minimizes the average prediction variance

$$\frac{\int_\chi \mathbf{f}'(\mathbf{x})(\mathbf{X}'\mathbf{V}^{-1}\mathbf{X})^{-1}\mathbf{f}(\mathbf{x})d\mathbf{x}}{\int_\chi d\mathbf{x}} \tag{10.6}$$

over the experimental region χ. The difference between this average prediction variance and the one in Equation (4.8) is that we now take into account the correlation structure of the data from a split-plot experiment, through the matrix $\mathbf{V}$. For the wind tunnel experiment, Peter and Brad preferred the I-optimal design over the D-optimal design because Dr. Cavendish wanted to use the models built as tools for prediction.

A technical problem with finding a D- or I-optimal split-plot design is that the matrix $\mathbf{V}$, and therefore also the D- and I-optimality criteria, depend on the unknown variances σ_γ^2 and σ_ε^2. Fortunately, the optimal split-plot designs do not depend on the absolute magnitude of these two variances, but only on their relative magnitude. Therefore, software to generate optimal split-plot designs requires input only on the relative magnitude of σ_γ^2 and σ_ε^2. As σ_γ^2 is often larger than σ_ε^2, we suggest specifying that the variance ratio $\sigma_\gamma^2/\sigma_\varepsilon^2$ is at least one when generating D- or I-optimal split-plot design. For the purpose of generating an excellent design, an educated guess of the variance ratio is good enough because a design that is optimal for one variance ratio is also optimal for a broad range of variance ratios smaller and larger than the specified one. Moreover, whenever different variance ratios lead to different designs, the quality of these designs is almost identical. Goos (2002) recommends using a variance ratio of one for finding optimal split-plot designs in the absence of detailed a priori information about it.

10.3.7 A design construction algorithm for optimal split-plot designs

In addition to the specification of the number of runs, the a priori model, the number of factors, an indication of whether a factor is continuous, categorical or a mixture ingredient, any additional constraints on factor-level combinations and the number of starting designs to consider, the coordinate-exchange algorithm for finding D- or I-optimal split-plot designs requires the designation of the factors that are hard-to-change, the number, b, and size, k, of the whole plots, and the expected ratio $\sigma_\gamma^2/\sigma_\varepsilon^2$ of the two variance components.

As with the algorithms we described in the previous chapters, the body of the split-plot coordinate-exchange algorithm has two parts. The first part involves the creation of a starting design. The second is the iterative improvement of this design until no further improvement is possible. Improvements are measured by increases in the D-optimality criterion $|\mathbf{X}'\mathbf{V}^{-1}\mathbf{X}|$ or decreases in the I-optimality criterion defined in Equation (10.6).

The starting design is formed column by column. For easy-to-change or sub-plot factor columns, the levels for each run, i.e., for each row of the design matrix $\mathbf{D}$, are chosen randomly. For hard-to-change or whole-plot factor columns, one random level is chosen for any given whole plot, and that level is assigned to all the runs corresponding to that whole plot, i.e., to all the rows of the design matrix $\mathbf{D}$ corresponding to that whole plot. This procedure gives the starting design the desired split-plot structure,

$$
D =
\left[
\begin{array}{ccccccc}
w_{11} & w_{21} & \cdots & w_{N_w1} & s_{111} & s_{211} & \cdots & s_{N_s11} \\
w_{11} & w_{21} & \cdots & w_{N_w1} & s_{112} & s_{212} & \cdots & s_{N_s12} \\
\vdots & \vdots & \ddots & \vdots & \vdots & \vdots & \ddots & \vdots \\
w_{11} & w_{21} & \cdots & w_{N_w1} & s_{11k} & s_{21k} & \cdots & s_{N_s1k} \\
w_{12} & w_{22} & \cdots & w_{N_w2} & s_{121} & s_{221} & \cdots & s_{N_s21} \\
w_{12} & w_{22} & \cdots & w_{N_w2} & s_{122} & s_{222} & \cdots & s_{N_s22} \\
\vdots & \vdots & \ddots & \vdots & \vdots & \vdots & \ddots & \vdots \\
w_{12} & w_{22} & \cdots & w_{N_w2} & s_{12k} & s_{22k} & \cdots & s_{N_s2k} \\
& & & \vdots & & & & \\
w_{1b} & w_{2b} & \cdots & w_{N_wb} & s_{1b1} & s_{2b1} & \cdots & s_{N_sb1} \\
w_{1b} & w_{2b} & \cdots & w_{N_wb} & s_{1b2} & s_{2b2} & \cdots & s_{N_sb2} \\
\vdots & \vdots & \ddots & \vdots & \vdots & \vdots & \ddots & \vdots \\
w_{1b} & w_{2b} & \cdots & w_{N_wb} & s_{1bk} & s_{2bk} & \cdots & s_{N_sbk}
\end{array}
\right],
$$

where w_{ij} represents the level of the ith hard-to-change factor in whole plot j and s_{ijl} is the level of the ith easy-to-change factor at the lth run of the jth whole plot. In the absence of constraints on the factor levels, the algorithm generates random levels on the interval $[-1, +1]$ for all factors that are continuous. For all factors that are categorical, the algorithm chooses one of its possible levels at random. In the presence of constraints on the factor levels, the algorithm uses the approach sketched in Section 5.3.3. For mixture ingredients, the procedure for generating random values is the same as in Section 6.3.5.

Improvements are made to the starting design by considering changes in the design matrix D on an element-by-element basis. The procedure for changing any given element depends on whether that element is the level of a hard-to-change factor or the level of an easy-to-change factor.

For the level of a continuous easy-to-change factor, the optimality criterion value is evaluated over a discrete number of values spanning the range of that factor. For the level of a categorical easy-to-change factor, the optimality criterion value is evaluated over all possible levels of that factor. If the best value of the optimality criterion improves the current best value, then the current best value is replaced and the current level of the factor in the design is replaced by the level corresponding to the best value of the optimality criterion.

For the level of a hard-to-change factor, the procedure is more involved. This is because changing the level of a hard-to-change factor for one run requires also changing its level for all other runs in the same whole plot. For a continuous hard-to-change factor, the optimality criterion value is evaluated over a discrete number of levels spanning the range of the factor. For the level of a categorical hard-to-change factor, the optimality criterion value is evaluated over all possible levels of that factor. Again, if the best value of the optimality criterion value improves the current best value, then the current best value is replaced and the current level of the factor in the design is replaced by the level corresponding to the best value of the optimality criterion. This is done for all k runs in the whole plot.

This element-by-element procedure continues until a complete cycle through the entire design matrix is completed. Then, another complete cycle through the design is performed. This continues until no changes are made in a whole pass.

10.3.8   Difficulties when analyzing data from split-plot experiments

Extensive simulation studies have shown that the probability of detecting nonzero effects of the hard-to-change factors is very low for split-plot designs that have small numbers of resettings of the hard-to-change factors. We call the number of resettings small when it is one, two, or three units larger than the number of whole-plot effects plus one (because σ_γ^2 must be estimated as well). Some researchers suggest that drawing trustworthy conclusions about the hard-to-change factors' effects (except for the interaction effects involving a hard-to-change factor and an easy-to-change factor) is impossible from the data from such split-plot designs, and that it might be better to assume that each of the whole-plot effects is significant.

Small numbers of resettings of the hard-to-change factors also lead to estimation problem for the whole-plot error variance σ_γ^2. As a matter of fact, σ_γ^2 is sometimes estimated as negative or zero (depending on the default options in the software), in which case the standard errors of the whole-plot effects are underestimated and too many degrees of freedom are used for their significance tests. In such cases, we recommend investigating how sensitive the factor-effect estimates, their standard errors and their p values are to the value of σ_γ^2. One way to do so is to plug different strictly positive values for σ_γ^2 into the matrix $\mathbf{V}$ in the GLS estimator in Equation (10.4) and its variance–covariance matrix $(\mathbf{X'V^{-1}X})^{-1}$. Another approach you can take is to use the Bayesian approach of Gilmour and Goos (2009).

10.4   Background reading

Jones and Nachtsheim (2009) provide a review of different methods for the design and analysis of split-plot experimental designs. More details about the optimal design of split-plot experiments can be found in Chapters 6, 7, and 8 of Goos (2002). Goos (2006) compares various construction methods for split-plot experimental designs. The GLS analysis of split-plot data is discussed in detail in Goos et al. (2006) and Langhans et al. (2005), and problems with that approach when there are few resettings of the hard-to-change factors are circumvented by using a Bayesian approach in Gilmour and Goos (2009). Algorithms for the construction of tailor-made split-plot designs are discussed in Trinca and Gilmour (2001), Goos and Vandebroek (2003), and Jones and Goos (2007). Prototype experiments run using split-plot designs are discussed in Bisgaard and Steinberg (1997). Box and Jones (1992) discuss the usefulness of split-plot designs for robust product experiments.

The combinatorial construction of split-plot designs is outlined in Bingham and Sitter (1999a, 1999b, 2001, 2003) and Bingham et al. (2004).

Federer and King (2007) provide a thorough review of agricultural applications of split-plot designs and variations on split-plot designs.

Split-plot designs belong to the broader family of multistratum designs, the design of which is discussed in Trinca and Gilmour (2001). The family of multistratum designs also includes the two-way split-plot designs (also known as strip-plot or strip-block designs) that are discussed in detail in Chapter 11, and split-split-plot designs that are discussed in Jones and Goos (2009).

10.5 Summary

The famous statistical consultant, Cuthbert Daniel, once claimed that all industrial experiments are split-plot experiments. We would not venture so far. Still, logistic considerations very often result in the need to perform the runs of an experiment in groups, where one or more hard-to-change factors' levels stay constant within each group. The resulting experimental designs are split-plot designs.

Until 2005, no commercial software existed to support the construction of split-plot designs. In addition, the software capable of a correct analysis of data from split-plot experiments required expert understanding to use. These holes in the software are now filled, so practitioners can take principled account of logistic constraints in conducting designed experiments.

Unfortunately, there are still statistical consultants who advise their clients to completely randomize every design. This often results in the runs being sorted for convenience without the knowledge of the person who does the statistical analysis. The consequences of analyzing a split-plot experiment as if it were completely randomized are twofold. First, the main effects of the hard-to-change factors, as well as their quadratic effects and interactions effects involving only hard-to-change factors, appear to be more statistically significant than they actually are. This leads to false positives. Second, the effects involving the other, easy-to-change, factors appear to be less statistically significant. This leads to false negatives.

It often happens that a split-plot experiment is both cheaper to run and simultaneously more statistically efficient than the completely randomized alternative. Using a split-plot design sometimes also offers the possibility to increase the number of runs substantially without a dramatic increase of the experimental cost. Therefore, we advise identifying hard-to-change factors during the preparation for any experiment, and investigating the possible added value of a split-plot design.

11

A two-way split-plot design

11.1　Key concepts

1. Modern production processes often proceed in two or more steps. Each of these steps may involve factors whose levels are hard to change. If it is possible to do so, the most statistically efficient design involves reordering the experimental units between steps. Such a design is called a two-way split-plot design (also known as a strip-plot design among other names).

2. Two-way split-plot experiments are extensions of split-plot experiments to more than one random effect, which is a comparatively new development with substantial practical import.

The case study in this chapter considers a process with two stages, each of which involves hard-to-change factors. The use of a two-way split-plot design dramatically reduces the time and experimental resource requirements of this study compared to an ordinary split-plot design and especially compared to a completely randomized design.

11.2　Case: the battery cell experiment

11.2.1　Problem and design

Peter and Brad fly into Madison, Wisconson, and rent a car. An hour later, Brad parks in the visitor's parking area of Rayovac's Portage, Wisconsin, facility. Inside the building, he and Peter are welcomed by Alvaro Pino, who is the head of the quality department at the facility and leads the two visitors to a meeting room that offers a view of the North–South runway of Portage Municipal Airport.

Optimal Design of Experiments: A Case Study Approach, First Edition. Peter Goos and Bradley Jones.
© 2011 John Wiley & Sons, Ltd. Published 2011 by John Wiley & Sons, Ltd.

[Pino] Well, gentlemen, the reason I called you is that we have serious trouble keeping open circuit voltage or OCV within specifications for one type of battery cells we are manufacturing here. The OCV has a direct impact on the functionality of the cells. Cells with high OCV cannot be sent to customers since the cells will be self-discharging and have low performance.

[Peter] Any idea about what is causing the trouble?

[Pino] We've done a brainstorming session with the engineers and a few operators to discuss the problems.

[Brad] Did anything come out of that?

[Pino] Yes, we agreed on a list of potential factors for further investigation. Our list has six factors. Four of the factors are related to the assembly process. The two others are related to the subsequent storage or curing stage. I could give you the names and the levels of the factors, but my boss wants me to share as few details with you as possible. He has a mild form of paranoia.

[Brad] No worries. Often, we can help without knowing all the details. Go on.

[Pino] We would like to run a screening experiment for our six factors, and start with two levels for each factor. We believe this is good enough to start with. If we can get an idea of the factors' main effects and their two-factor interaction effects, we should be getting close to a solution. The most obvious design to use is a 2^6 factorial design. However, a completely randomized full factorial design would require 320 days to run.

[Peter] Why is that?

[Pino] The trouble is that a complete curing cycle takes 5 days. If we were to run a 64-run completely randomized design, the curing conditions would need to be changed 64 times. Sixty-four times 5 gives you 320 days in total.

[Brad] Obviously, you can't wait for 320 days to solve your problem with the OCV.

[Pino] Nope. We have an idea about how to reduce the time required to do the 64-run design in just 20 days. To see how it works, it is important to understand the flow of the production. Batteries are assembled and emerge at the end of the first stage, the assembly process, one by one. The four factors associated with the first stage can, in principle, be changed between the production of individual batteries. However, it is more economical to produce a lot, often 2000 batteries, for each factorial level combination. At the end of the assembly stage, each lot of batteries is placed in trays before moving to the curing process. The curing process, which takes place in a temperature- and humidity-controlled chamber and lasts at least 5 days, can, in principle, also process individual batteries.

[Peter] Hmm. A temperature- and humidity-controlled chamber you say. May I infer that the two curing factors are temperature and humidity?

[Pino, smiling] Oops. Slip of the tongue. Anyway, processing individual batteries in the curing stage would be extremely wasteful. A much more sensible strategy is to process several lots simultaneously under the same set of curing conditions.

While he's talking, Pino walks up to a whiteboard on the wall opposite the window and produces the scheme in Figure 11.1. As he continues his exposition, Table 11.1 gradually appears as well.

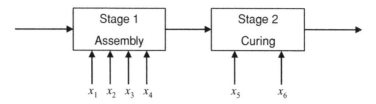

Figure 11.1   Battery cell production process.

[Pino] For every setting of the assembly factors, we produce a lot of 2000 batteries. As there are four assembly factors, this gives us 2^4 or 16 lots. Now, our plan is to split randomly each of these lots into four sublots of 500 batteries each, and randomly allocate each of these four sublots to one of the four settings of the curing factors. All 16 sublots allocated to the same setting of the curing factors can then be processed together. This would require as few as four curing cycles. Times five gives us 20 days.

[Brad] That is pretty neat.

[Pino] In terms of experimental cost, it is. But we hesitate to run that design because it is not completely randomized. In fact, it looks quite a bit like one of those

Table 11.1   Pino's strip-plot design, also known as a two-way split-plot design, for the battery cell experiment.

Assembly					Curing			
				Factors	Cycle 1	Cycle 2	Cycle 3	Cycle 4
		Factors		x_5	-1	$+1$	-1	$+1$
Lot	x_1	x_2	x_3	x_4	x_6	-1	-1	$+1$	$+1$
1	-1	-1	-1	-1		Sublot 1	Sublot 2	Sublot 4	Sublot 3
2	$+1$	-1	-1	-1		Sublot 2	Sublot 4	Sublot 1	Sublot 3
3	-1	$+1$	-1	-1		Sublot 3	Sublot 1	Sublot 4	Sublot 2
4	$+1$	$+1$	-1	-1		Sublot 4	Sublot 3	Sublot 1	Sublot 2
5	-1	-1	$+1$	-1		Sublot 1	Sublot 4	Sublot 3	Sublot 2
6	$+1$	-1	$+1$	-1		Sublot 1	Sublot 3	Sublot 2	Sublot 4
7	-1	$+1$	$+1$	-1		Sublot 2	Sublot 1	Sublot 3	Sublot 4
8	$+1$	$+1$	$+1$	-1		Sublot 4	Sublot 1	Sublot 2	Sublot 3
9	-1	-1	-1	$+1$		Sublot 1	Sublot 2	Sublot 3	Sublot 4
10	$+1$	-1	-1	$+1$		Sublot 4	Sublot 3	Sublot 1	Sublot 2
11	-1	$+1$	-1	$+1$		Sublot 1	Sublot 2	Sublot 3	Sublot 4
12	$+1$	$+1$	-1	$+1$		Sublot 1	Sublot 4	Sublot 2	Sublot 3
13	-1	-1	$+1$	$+1$		Sublot 1	Sublot 4	Sublot 3	Sublot 2
14	$+1$	-1	$+1$	$+1$		Sublot 2	Sublot 4	Sublot 1	Sublot 3
15	-1	$+1$	$+1$	$+1$		Sublot 2	Sublot 1	Sublot 4	Sublot 3
16	$+1$	$+1$	$+1$	$+1$		Sublot 4	Sublot 3	Sublot 2	Sublot 1

split-plot designs I've been reading about. But it is not quite the same thing either. I think our design is what I'd call a two-way split-plot design.

[Peter] You're completely right. This is a two-way split-plot design. That name has been used in the literature for designs like yours, along with half a dozen other names.

[Pino] Really?

[Peter] Sure. The most commonly used names for your design are strip-plot or strip-block design, and criss-cross design. But I've seen your name here and there too. I like the name two-way split-plot design because it seems more descriptive than the other names. You have a split-plot design when you look at the rows, and another split-plot design when you look at the columns.

[Pino] That's what we figured out the other day. I guess that our data will be correlated in two ways then. The responses that come from the same lot will be correlated. But on top of that, there will be correlation between responses obtained from a given curing cycle. So, our two-way split-plot design gives us two-way correlation, and a complicated analysis.

[Brad] With modern software, you needn't worry. The kind of correlation pattern you face can be easily dealt with if you use generalized least squares or GLS for estimating the factor effects.

[Pino] I am familiar with GLS from looking at a few papers about split-plot designs. We need the following formula for estimating the factor effects, right?

Pino writes the formula

$$\hat{\boldsymbol{\beta}} = (\mathbf{X}'\mathbf{V}^{-1}\mathbf{X})^{-1}\mathbf{X}'\mathbf{V}^{-1}\mathbf{Y}$$

on a flip chart, and looks at Brad and Peter for confirmation.

[Peter] That's it. The only thing is that the $\mathbf{V}$ matrix for your design looks different from that of an ordinary split-plot design. Your $\mathbf{V}$ matrix will be a bit more involved because of the two-way correlation.

[Brad] There is really nothing complicated anymore in doing the correct analysis. Basically, all that's required in modern software is that you identify the row and column corresponding to each observation.

[Peter] That's true, but I can see a problem with your proposal.

[Pino] And what's that?

[Peter] To use the GLS approach for data from a two-way split-plot design, you need to estimate three variance components. One variance component for the row-to-row variation, one for the column-to-column variation, and one for the residual error variance. The estimation of the second of these is problematic because your design involves only four columns. That is, it involves only four independent settings of the curing factors. This gives you only three degrees of freedom and you have to estimate four things related to the columns: the main effects of the two column factors, their interaction effect, and the column-to-column variation. To cut a long story short, you can estimate the main effects and the two-factor interaction effect of the column factors, but there is no way you can test whether these effects are significantly different from zero.

[Pino] What would you suggest instead?

[Peter] I guess I would consider a design with six or eight columns, or curing cycles, instead of four. That would give you some degrees of freedom for estimating the column-to-column variation, and perform significance tests.

[Brad] Allowing for significance testing comes at a substantial cost, Peter. Using six curing cycles increases the time required to run the experiment from 20 to 30 days. Using eight curing cycles doubles the run time of the experiment.

[Pino] I think that we could live with six curing cycles and 30 days of experimentation. Certainly, if it is possible to use the smallest of our curing chambers.

[Peter] What is that chamber's capacity?

[Pino] We can accommodate 4000 batteries in that one.

[Brad] That means eight sublots of 500 batteries instead of 16. If you use six curing cycles of eight sublots each, you end up with 48 runs. That is plenty for estimating all main effects and two-factor interaction effects.

[Peter] If you are happy with 16 settings for the assembly factors and six curing cycles of eight sublots, I am sure we can generate a nice design for you.

[Pino] Wait a minute. Forty-eight is not a power of two. How can you generate such a design?

[Brad] We use a computerized search for an optimal design. That search is not restricted to run sizes that are powers of two.

[Pino] Can you show me how that works?

[Brad, grabbing his laptop and firing it up] Sure. Here we go.

Brad demonstrates in his favorite software how the design problem can be specified, and starts the computerized search.

[Brad] That'll take a few minutes.

While the computer is running, Pino offers to get coffee and re-appears after a few minutes with three mugs, some cream and sugar.

[Pino] Here we go. Has your machine produced a design already?

[Brad, turning his laptop to Pino] Yes, here it is. I have already put it in a format similar to that of your table on the whiteboard.

The screen of Brad's laptop shows Table 11.2.

[Peter] The design involves only 48 sublots, three for every setting of the assembly factors. And each of the six curing cycles involves eight of the sublots. That's exactly what we wanted.

[Pino] Can we get statistically independent estimates of the factor effects with that design?

[Brad] Let's see.

After a few mouse clicks and typing in a few commands, Table 11.3 flashes to the screen.

[Brad] I generated this table assuming that the three variance components Peter mentioned earlier are all one. It contains the variance of each of the 22 factor-effect estimates for your experiment. There are a lot of interesting things that you can learn from this table. Perhaps the most important thing you can see is that there are three groups of effects. One group of effects has a variance of about 0.085. These are

Table 11.2 D-optimal 48-run two-way split-plot design with 16 rows and six columns.

	Assembly				Curing					
		Factors			Factors					
					x_5	Cycle 1	Cycle 2	Cycle 3	Cycle 4	Cycle 5	Cycle 6

						Cycle 1	Cycle 2	Cycle 3	Cycle 4	Cycle 5	Cycle 6
					x_5	−1	+1	−1	+1	−1	+1
					x_6	−1	−1	+1	+1	+1	+1
Lot	x_1	x_2	x_3	x_4							
1	−1	−1	−1	−1		Sublot 1		Sublot 3	Sublot 2		Sublot 2
2	+1	−1	−1	−1			Sublot 3			Sublot 1	Sublot 3
3	−1	+1	−1	−1			Sublot 1		Sublot 1	Sublot 2	
4	+1	+1	−1	−1		Sublot 2		Sublot 3		Sublot 1	
5	−1	−1	+1	−1			Sublot 2				Sublot 3
6	+1	−1	+1	−1		Sublot 1		Sublot 2	Sublot 3		
7	−1	+1	+1	−1		Sublot 3		Sublot 2	Sublot 1		
8	+1	+1	+1	−1			Sublot 2			Sublot 1	Sublot 3
9	−1	−1	−1	+1			Sublot 2		Sublot 3	Sublot 3	Sublot 1
10	+1	−1	−1	+1		Sublot 1		Sublot 2	Sublot 2		
11	−1	+1	−1	+1		Sublot 1		Sublot 3			
12	+1	+1	−1	+1			Sublot 1		Sublot 1	Sublot 2	Sublot 3
13	−1	−1	+1	+1		Sublot 3		Sublot 2			
14	+1	−1	+1	+1			Sublot 2			Sublot 1	Sublot 3
15	−1	+1	+1	+1			Sublot 2			Sublot 3	Sublot 1
16	+1	+1	+1	+1		Sublot 3		Sublot 1	Sublot 2		

Table 11.3 Variances of the factor-effect estimates for the D-optimal 48-run two-way split-plot design.

Effect	Variance	Type
Intercept	0.2734	–
x_1	0.0859	Row factor effect
x_2	0.0859	Row factor effect
x_3	0.0859	Row factor effect
x_4	0.0859	Row factor effect
x_5	0.2237	Column factor effect
x_6	0.2109	Column factor effect
x_1x_2	0.0833	Row factor effect
x_1x_3	0.0833	Row factor effect
x_1x_4	0.0833	Row factor effect
x_1x_5	0.0227	Row × Column interaction effect
x_1x_6	0.0234	Row × Column interaction effect
x_2x_3	0.0833	Row factor effect
x_2x_4	0.0833	Row factor effect
x_2x_5	0.0227	Row × Column interaction effect
x_2x_6	0.0234	Row × Column interaction effect
x_3x_4	0.0833	Row factor effect
x_3x_5	0.0227	Row × Column interaction effect
x_3x_6	0.0234	Row × Column interaction effect
x_4x_5	0.0227	Row × Column interaction effect
x_4x_6	0.0234	Row × Column interaction effect
x_5x_6	0.2237	Column factor effect

the main effects of the assembly factors, x_1–x_4, and the six two-factor interaction effects involving these factors. In the table, I have labeled the effects in that group "Row factor effects." A second group of effects contains the main effects of the two curing factors x_5 and x_6, and their interaction effect. These effects are estimated with a variance of about 0.22. I labeled them "Column factor effects."

[Peter] That variance is substantially larger than that for the effects of the assembly factors, simply because there are fewer columns in the design than there are rows. In other words, fewer independent resettings of the curing factors than of the assembly factors. Fewer independent resettings result in a higher variance, that is, a smaller precision.

[Brad] Right. The last group of effect estimates, which I labeled "Row × column interaction effects," has much lower variances than the two other groups. For that group, which involves interaction effects involving one of the assembly factors and one of the curing factors, the variances are all around 0.023.

[Pino] Is it generally the case that the estimates of these interaction effects have small variances?

[Peter] Yes, it is a key feature of strip-plot designs, or two-way split-plot designs as you called them. Interaction effects involving a row factor and a column factor are estimated precisely, or, in the statistical jargon, efficiently.

[Pino] Interesting. But you still didn't answer my question concerning the independence of the factor-effect estimates.

[Brad] Oh, right. Thank you for reminding me. The table with the variances indirectly gives us the answer. Within each group of effects, the variances are not all exactly the same. That is an indication that the effect estimates are not statistically independent.

[Pino] Hmm. That feels wrong.

[Brad] How come?

[Pino] Well, in my design courses, and in every textbook I've seen, the emphasis is on orthogonality and the importance of obtaining independent estimates of the factor effects. Are you suggesting that I run a design that does not give me independent estimates?

[Brad, pointing to Table 11.3] Yes, I am. Look at the variances for the estimates of the row-by-column interaction effects. They are all around 0.023, which is only a tiny little bit higher than $\frac{1}{48}$, which is the smallest possible variance for these effects.

[Pino] What is $\frac{1}{48}$ in decimals?

[Peter] Brad likes to show off that he can calculate these things in his head. $\frac{1}{48}$ is around 0.021.

[Brad, ignoring Peter's comment] Similarly, the variances of the estimates of the row factor effects are all $\frac{1}{12}$, that is, 0.0833, or slightly larger than that. With classical analysis of variance calculus, you can show that $\frac{1}{12}$ is the variance you can expect for the row factor-effect estimates from a perfectly orthogonal two-way split-plot design.

[Peter] The fact that the variances in our table are so close to $\frac{1}{48}$ or $\frac{1}{12}$ shows that the dependence of the estimates is negligible.

While Pino is processing that information, Brad has already produced a few pieces of additional information. This time, Table 11.4 appears on the screen of his laptop.

[Brad] Look at this! Only six pairs of factor-effect estimates are dependent. Out of a total of $22 \times 21/2$ pairs, which is 231. That is pretty good. The six nonzero correlations occur between the main-effect estimates of the first five factors and the

Table 11.4   Nonzero correlations between the factor-effect estimates for the 48-run two-way split-plot design in Table 11.2.

Effects	Correlation
x_1 and x_1x_6	−0.174
x_2 and x_2x_6	−0.174
x_3 and x_3x_6	−0.174
x_4 and x_4x_6	−0.174
x_5 and x_5x_6	−0.371
Intercept and x_6	−0.293

estimates of their two-factor interaction effects with the sixth factor, and between the main-effect estimate of the sixth factor and the estimate of the intercept. So, any main-effect estimate is independent from any other main-effect estimate. Also, most of the interaction effects can be estimated independently. In absolute value, the largest absolute correlation is, let me see, only 0.371. Again assuming the three variance components are all one.

[Peter] A correlation of roughly $\frac{1}{3}$ is definitely no cause for alarm.

[Pino] I see. I will run your design, and get back to you with the results. One final question though. You mentioned a perfectly orthogonal two-way split-plot design that gives variances of $\frac{1}{12}$. Why did your computer not return this orthogonal design?

[Peter] Because there is no orthogonal two-way split-plot design with 48 runs, 16 rows, and six columns that gives independent estimates for all main effects and all two-factor interaction effects. The orthogonal design Brad mentioned is essentially a theoretical construct, a hypothetical benchmark, that helps us judge how good a design is.

11.2.2 Data analysis

Five weeks later, Peter and Brad return to Rayovac's Portage facility to present the results of their statistical analysis of the data from the two-way split-plot design that was run by Alvaro Pino's special task force. The new meeting again takes place in the room overlooking the airport, but this time Pino has invited several others involved. Brad sips his coffee, and starts with his presentation. First, he gives a brief recap of the design that was used and shows the data in Table 11.5. The response, OCV, in the table is the average OCV coded by (Volts $- 1.175$) $\times$ 1000 per sublot.

[Brad] I started by fitting a main-effects-plus-two-factor-interactions model to the data. That model involves 22 unknown parameters, six main effects, 15 two-factor interaction effects, and an intercept. Not all of these effects were significant, so I did a backward selection by dropping the insignificant effects one by one, starting with the least significant one. Here is the model I ended up with.

The final model Brad obtained appears on the screen. It is given by the equation

$$OCV = 30.25 - 3.44x_1 + 3.19x_4 - 0.19x_5 - 14.38x_6$$
$$-2.27x_1x_5 + 2.31x_1x_6 + 1.69x_4x_6.$$

[Brad] When you look at this model, you will see that there is one massive effect. All the other significant effects are much smaller and comparable to each other.

[Pino] Is that main effect of x_5 truly significantly different from zero?

[Brad, displaying Table 11.6] No, it's not. As you can see from this table, the p value for that effect is 0.91. So, it is pretty negligible.

[Pino] Then, if it is negligible, why keep the corresponding term in the model?

[Brad] There are a few theoretical, technical reasons, but it is mainly a matter of taste whether or not you include the effect in the model. Most statisticians prefer models that include the main effects of every factor that appears in a significant interaction term. We say that such models exhibit strong heredity.

Table 11.5   Data from the D-optimal 48-run two-way split-plot design for the battery cell experiment.

Row	Column	x_1	x_2	x_3	x_4	x_5	x_6	OCV
1	1	−1	−1	+1	+1	−1	−1	55
1	2	−1	−1	+1	+1	+1	+1	26
1	3	−1	−1	+1	+1	−1	+1	26
2	4	+1	−1	−1	−1	+1	−1	28
2	5	+1	−1	−1	−1	+1	+1	9
2	6	+1	−1	−1	−1	−1	+1	12
3	1	−1	−1	−1	−1	−1	−1	49
3	2	−1	−1	−1	−1	+1	+1	17
3	3	−1	−1	−1	−1	−1	+1	8
4	4	+1	+1	−1	+1	+1	−1	28
4	5	+1	+1	−1	+1	+1	+1	14
4	6	+1	+1	−1	+1	−1	+1	15
5	1	−1	+1	−1	+1	−1	−1	53
5	2	−1	+1	−1	+1	+1	+1	27
5	3	−1	+1	−1	+1	−1	+1	24
6	4	+1	−1	+1	+1	+1	−1	43
6	5	+1	−1	+1	+1	+1	+1	29
6	6	+1	−1	+1	+1	−1	+1	26
7	1	+1	−1	−1	+1	−1	−1	51
7	2	+1	−1	−1	+1	+1	+1	14
7	3	+1	−1	−1	+1	−1	+1	18
8	4	+1	+1	+1	−1	+1	−1	40
8	5	+1	+1	+1	−1	+1	+1	11
8	6	+1	+1	+1	−1	−1	+1	12
9	1	+1	+1	+1	+1	−1	−1	38
9	2	+1	+1	+1	+1	+1	+1	16
9	3	+1	+1	+1	+1	−1	+1	24
10	4	−1	+1	−1	−1	+1	−1	41
10	5	−1	+1	−1	−1	+1	+1	19
10	6	−1	+1	−1	−1	−1	+1	5
11	1	+1	−1	+1	−1	−1	−1	44
11	2	+1	−1	+1	−1	+1	+1	12
11	3	+1	−1	+1	−1	−1	+1	15
12	4	−1	+1	+1	+1	+1	−1	52
12	5	−1	+1	+1	+1	+1	+1	27
12	6	−1	+1	+1	+1	−1	+1	5
13	1	−1	+1	+1	−1	−1	−1	60
13	2	−1	+1	+1	−1	+1	+1	19
13	3	−1	+1	+1	−1	−1	+1	11

Table 11.5   (*continued*)

Row	Column	x_1	x_2	x_3	x_4	x_5	x_6	OCV
14	4	-1	-1	$+1$	-1	$+1$	-1	44
14	5	-1	-1	$+1$	-1	$+1$	$+1$	10
14	6	-1	-1	$+1$	-1	-1	$+1$	7
15	1	$+1$	$+1$	-1	-1	-1	-1	39
15	2	$+1$	$+1$	-1	-1	$+1$	$+1$	4
15	3	$+1$	$+1$	-1	-1	-1	$+1$	5
16	4	-1	-1	-1	$+1$	$+1$	-1	49
16	5	-1	-1	-1	$+1$	$+1$	$+1$	24
16	6	-1	-1	-1	$+1$	-1	$+1$	17

[Peter] The main technical reason to use models with heredity is that the predictions then are independent of the coding you use for the factors in the experiment. So, you get the exact same predictions with coded factor levels and with uncoded factor levels if you stick to models with heredity.

[Brad] Anyway, whether or not you leave the main effect in the model has hardly any impact on the predictions you make with it.

[Pino] Speaking of predictions, you can now use the model to predict what OCV levels we might be able to achieve if we optimize our process.

[Brad] Sure. The minimum OCV predicted value I've been able to achieve with the estimated model is 7.42. For that purpose, you need to set x_1, x_5, and x_6 at their high levels, and x_4 at its low level.

[Pino] That sounds promising. I guess we have to conduct a couple of confirmatory runs to test the settings you suggest.

For a few seconds, the meeting room is silent, but then the silence is suddenly broken by a tall lady who stands up in the back of the room.

[Hermans] Hi there, I'm Vanessa. Vanessa Hermans. You just said that the minimum OCV value we could achieve is around 7.42. However, when you showed the

Table 11.6   Simplified model for the data from the D-optimal 48-run two-way split-plot design for the battery cell experiment.

Effect	Estimate	Standard error	DF	t Ratio	p Value
Intercept	30.25	1.89	4.21	16.02	0.0001
x_1	-3.44	1.09	12.69	-3.14	0.0080
x_4	3.19	1.09	12.69	2.91	0.0124
x_5	-0.19	1.60	2.69	-0.12	0.9148
x_6	-14.38	1.66	2.57	-8.66	0.0057
$x_1 x_5$	-2.27	0.62	26.69	-3.66	0.0011
$x_1 x_6$	2.31	0.63	24.82	3.69	0.0011
$x_4 x_6$	1.69	0.63	24.82	2.70	0.0124

Table 11.7   Best linear unbiased predictions (BLUPs) of the row and column effects for the data from the battery cell experiment in Table 11.5.

Effect	Prediction	Effect	Prediction	Effect	Prediction
Row 1	2.33	Row 9	−1.75	Column 1	2.90
Row 2	−0.52	Row 10	−1.58	Column 2	0.64
Row 3	−0.11	Row 11	1.86	Column 3	−0.11
Row 4	−3.89	Row 12	−2.39	Column 4	−2.90
Row 5	1.63	Row 13	3.61	Column 5	2.25
Row 6	5.65	Row 14	−2.51	Column 6	−2.79
Row 7	−0.58	Row 15	−3.49		
Row 8	2.74	Row 16	−0.99		

data at the beginning of this meeting, there were OCV values as low as five I believe. Why isn't your model capable of producing settings that give OCV values as low as that?

[Peter] That's a question I'd like to answer, Brad. If you don't mind of course.

[Brad] Go for it, Peter.

Peter walks up to Brad's laptop and puts Table 11.5 back on display.

[Peter] When we look at the data table, we can find three OCV values of five, and even one value as low as four. Now, two of the fives were obtained in the sixth curing cycle, indicated as Column 6 in the data table. The OCV value of four and the third five was obtained for the 15th setting of the assembly process, which is indicated as row 15 in the table. Now, it turns out that this particular curing cycle and this particular run of the assembly process both systematically led to lower OCV values. So low that they cannot be explained by the six factors in the experiment.

[Hermans] How can we see that?

[Peter] Well, a nice thing about the analysis we did is that we can estimate how big the differences were between the six curing cycles and between the 16 runs of the assembly process, after correcting for the effects of the experimental factors. For that, we need to compute the best linear unbiased predictions or BLUPs of the random effects corresponding to the columns and the rows of the design.

Peter browses through the output of the data analysis on Brad's laptop until he finds Table 11.7. He magnifies the font size of the table and projects it on the screen.

[Peter] Here we go. These are the two types of BLUPs for the data of our two-way split-plot design. The first 16 BLUPs, in the left and middle parts of the table, correspond to the 16 rows of our design. They estimate the random effects of the 16 independent settings of the assembly process. The last six BLUPS, in the right part of the table, correspond to the columns, and give an estimate of the random effect of the six runs of the curing process. You can see that some of these BLUPs are well above zero, while other are well below zero. That indicates that certain rows and columns have substantially higher or lower OCV values than others.

[Hermans] I can see that Column 6 has a BLUP of −2.79 and that Row 15 has one of −3.49. These are among the lowest of all BLUPs. What does that mean?

[Peter] You can interpret the BLUPs as residuals, but then for the rows and the columns rather than for the individual runs. Therefore, you can interpret them as differences between rows and columns that cannot be explained by the model.

[Hermans] Is what you're saying that the OCV values of four and five were the result of pure chance then?

[Peter] It is the result of both a systematic effect of the settings of the assembly factors and chance. There is no doubt that the settings for the assembly process in Row 15 result in pretty low OCV values, but random chance produced a few results that are extra low. A similar argument can be made for the fives in Column 6.

After Hermans, satisfied with Peter's answer, has gone back to her seat, it is Pino who asks the next question.

[Pino] I suppose these BLUPs are linked to the three variance components you mentioned the other day? Brad, you didn't say anything about the three variance components.

[Brad] Good point! The BLUPs are indeed linked to the variance components, albeit in an intricate way. The estimates for the variance components related to the rows and columns are very similar. I get 12.88 for the variance component that corresponds to the rows, and 12.62 for the variance component that corresponds to the columns. The estimate of the random error variance is 16.72.

[Peter] That means that there are substantial differences between different restarts of the assembly process and between different curing cycles.

Several weeks later, Alvaro Pino and Peter bump into each other during a coffee break at the Quality and Productivity Research Conference held at the University of Wisconsin.

[Peter] Hi Mr. Pino. How are you? How did the confirmatory runs go after your battery cell experiment.

[Pino] Excellent I would say. We managed to reduce the production line defective rate to 1%. That was an improvement of about 80%.

[Peter] Great. Let's drink to that!

11.3   Peek into the black box

Two-way split-plot designs have a long history in agricultural research, but they are lesser known to experimenters in industry. In agriculture, the experimental factors in these designs are often treated as categorical, and the designs involve replicates of complete factorial designs. In industry, it is more common to use fractions of complete factorial designs, often without replication. The typical layout of an agricultural two-way split-plot design is discussed in Attachment 11.1.

Attachment 11.1   Two-way split-plot designs or strip-plot designs.

Two-way split-plot designs are known under various names, the best known of which are strip-plot or strip-block design. Alternatives are two-way whole-plot design or criss-cross design. The first applications of strip-plot designs appeared in agricultural experimentation and involved a simple full factorial design. Typically, these initial applications involved replicates of the design on different fields. Each field was then subdivided in ab cells arranged in a rows and b columns. The numbers a and b are the numbers of levels of the two experimental factors, A and B. The a levels of factor A were randomly applied to the rows, and the b levels of factor B were randomly applied to the columns. A different random assignment was used for the different fields. A pictorial representation of a two-way split-plot design is given in Figure 11.2 for an experiment involving two fertilizers, four herbicides, and two fields. The two levels of the factor fertilizer are applied to the columns, whereas the four levels of the factor herbicide are applied to the rows. The rows and columns are sometimes referred to as strips, which explains the names strip-plot or strip-block design that are also used for this experimental design.

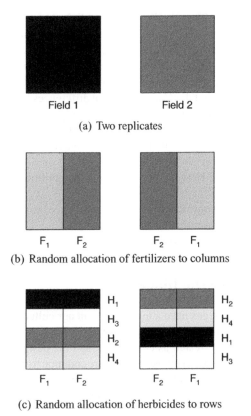

Figure 11.2 Two-way split-plot design with two replicates for studying two fertilizers (F_1 and F_2) and four herbicides (H_1, H_2, H_3 and H_4).

A key feature of the initial applications of strip-plot designs is that they are fully replicated on different fields. This ensures that the variance components associated with rows, columns and cells in the design are estimable, so that formal hypothesis tests can be conducted for the main effects of the two experimental factors and for the interaction effects. Full replication is not required if the numbers of rows and columns on a field are two or more units larger than the numbers of factor effects related to the rows and columns.

11.3.1 The two-way split-plot model

The model for analyzing data from a two-way split-plot experiment run with r rows and c columns requires an additive effect for every row and column. The model for the response Y_{ij} of the run in row i and column j is

$$Y_{ij} = \mathbf{f}'(\mathbf{x}_{ij})\boldsymbol{\beta} + \gamma_i + \delta_j + \varepsilon_{ij}, \tag{11.1}$$

where $\mathbf{x}_{ij}$ is a vector that contains the levels of all the experimental factors at the run in row i and column j, $\mathbf{f}'(\mathbf{x}_{ij})$ is its model expansion, and $\boldsymbol{\beta}$ contains the intercept and all the factor effects. As opposed to the random block effects model in Equation (7.5) and the split-plot model in Equation (10.2), the two-way split-plot model has three random effects. The first, γ_i, represents the ith row effect. The second, δ_j, is the effect of the jth column, and the third, ε_{ij}, is the residual error associated with the run in row i and column j.

We assume that the random effects γ_i, δ_j, and ε_{ij} are all independent and normally distributed with zero mean, that the row effects have variance σ_γ^2, that the column effects have variance σ_δ^2, and that the residual errors have variance σ_ε^2. The variance of a response Y_{ij} is therefore

$$\text{var}(Y_{ij}) = \text{var}(\gamma_i + \delta_j + \varepsilon_{ij}) = \text{var}(\gamma_i) + \text{var}(\delta_i) + \text{var}(\varepsilon_{ij}) = \sigma_\gamma^2 + \sigma_\delta^2 + \sigma_\varepsilon^2.$$

The covariance between each pair of responses Y_{ij} and $Y_{ij'}$ from the same row i is

$$\text{cov}(Y_{ij}, Y_{ij'}) = \text{cov}(\gamma_i + \delta_j + \varepsilon_{ij}, \gamma_i + \delta_{j'} + \varepsilon_{ij'}) = \text{cov}(\gamma_i, \gamma_i) = \text{var}(\gamma_i) = \sigma_\gamma^2.$$

Likewise, the covariance between each pair of responses Y_{ij} and $Y_{i'j}$ from the same column j is

$$\text{cov}(Y_{ij}, Y_{i'j}) = \text{cov}(\gamma_i + \delta_j + \varepsilon_{ij}, \gamma_{i'} + \delta_j + \varepsilon_{i'j}) = \text{cov}(\delta_j, \delta_j) = \text{var}(\delta_j) = \sigma_\delta^2.$$

Finally, the covariance between each pair of responses Y_{ij} and $Y_{i'j'}$ from different rows i and i' and different columns j and j' equals

$$\text{cov}(Y_{ij}, Y_{i'j'}) = \text{cov}(\gamma_i + \delta_j + \varepsilon_{ij}, \gamma_{i'} + \delta_{j'} + \varepsilon_{i'j'}) = 0.$$

These results indicate that we assume that responses obtained from runs in either the same row or the same column are correlated, and that responses from runs in different rows and columns are uncorrelated. In the two-way split-plot model, we thus have a two-way correlation pattern.

The total number of runs, n, is usually a common multiple of the number of rows, r, and the number of columns, c. For the design in Table 11.1, n equals 64, which is a common multiple of the number of rows, 16, and the number of columns, four. For the design in Table 11.2, n is 48, which is a common multiple of the number of rows, 16, and the number of columns, six.

11.3.2   Generalized least squares estimation

As the independence assumption of the ordinary least squares estimator is violated in the two-way split-plot model, it is again better to use generalized least squares or GLS estimates, which account for the correlation. As in the random block effects model and in the split-plot model, the GLS estimator is calculated from

$$\hat{\beta} = (\mathbf{X}'\mathbf{V}^{-1}\mathbf{X})^{-1}\mathbf{X}'\mathbf{V}^{-1}Y, \tag{11.2}$$

where the matrix $\mathbf{V}$ describes the covariance between the responses. To derive the exact nature of $\mathbf{V}$, we rewrite the model for analyzing data from two-way split-plot experiments in Equation (11.1) as

$$Y = \mathbf{X}\beta + \mathbf{Z}_1\gamma + \mathbf{Z}_2\delta + \varepsilon, \tag{11.3}$$

where Y is the $n \times 1$ vector of responses, $\mathbf{X}$ is the $n \times p$ model matrix,

$$\gamma = [\gamma_1 \quad \gamma_2 \quad \cdots \quad \gamma_r]',$$
$$\delta = [\delta_1 \quad \delta_2 \quad \cdots \quad \delta_c]',$$

and ε is the $n \times 1$ vector of random errors. The matrices $\mathbf{Z}_1$ and $\mathbf{Z}_2$ are $n \times r$ and $n \times c$ matrices of zeros and ones. The former has a one in its ith row and jth column if the ith run belongs to the jth row of the design. The latter has a one in its ith row and jth column if the ith run belongs to the jth column of the design. Because we assume that all the random effects are independent, the random-effects vectors γ, δ, and ε have variance–covariance matrices $\text{var}(\gamma) = \sigma_\gamma^2 \mathbf{I}_r$, $\text{var}(\delta) = \sigma_\delta^2 \mathbf{I}_c$, and $\text{var}(\varepsilon) = \sigma_\varepsilon^2 \mathbf{I}_n$, respectively, where $\mathbf{I}_r$, $\mathbf{I}_c$, and $\mathbf{I}_n$ are identity matrices of dimensions r, c, and n, respectively.

Under all these assumptions, it is possible to derive a compact expression for the variance–covariance matrix of the response vector:

$$\begin{aligned} \mathbf{V} &= \text{var}(Y), \\ &= \text{var}(\mathbf{X}\beta + \mathbf{Z}_1\gamma + \mathbf{Z}_2\delta + \varepsilon) \\ &= \text{var}(\mathbf{Z}_1\gamma + \mathbf{Z}_2\delta + \varepsilon) \end{aligned}$$

Table 11.8 Twelve-run two-way split-plot design with four rows
and four columns for two row factors and one column factor.

Run	Row	Column	x_1	x_2	x_3
1	1	1	-1	-1	-1
2	1	2	-1	-1	1
3	1	3	-1	-1	1
4	2	4	1	-1	-1
5	2	1	1	-1	-1
6	2	2	1	-1	1
7	3	3	1	1	1
8	3	4	1	1	-1
9	3	1	1	1	-1
10	4	2	-1	1	1
11	4	3	-1	1	1
12	4	4	-1	1	-1

$$= \text{var}(\mathbf{Z}_1\boldsymbol{\gamma}) + \text{var}(\mathbf{Z}_2\boldsymbol{\delta}) + \text{var}(\boldsymbol{\varepsilon}),$$
$$= \mathbf{Z}_1[\text{var}(\boldsymbol{\gamma})]\mathbf{Z}_1' + \mathbf{Z}_2[\text{var}(\boldsymbol{\delta})]\mathbf{Z}_2' + \text{var}(\boldsymbol{\varepsilon}),$$
$$= \mathbf{Z}_1(\sigma_\gamma^2\mathbf{I}_r)\mathbf{Z}_1' + \mathbf{Z}_2(\sigma_\gamma^2\mathbf{I}_c)\mathbf{Z}_2' + \sigma_\varepsilon^2\mathbf{I}_n,$$
$$= \sigma_\gamma^2\mathbf{Z}_1\mathbf{Z}_1' + \sigma_\delta^2\mathbf{Z}_2\mathbf{Z}_2' + \sigma_\varepsilon^2\mathbf{I}_n.$$

It is difficult to give general expressions for $\mathbf{Z}_1$, $\mathbf{Z}_2$, and $\mathbf{V}$. Therefore, we illustrate
their construction using the small 12-run design with four rows and four columns for
two row factors and one column factor in Table 11.8. An alternative representation
of that design, which is D-optimal for estimating a main-effects model, is shown in
Table 11.9. Because $n = 12$ and $r = c = 4$ for this design, $\mathbf{Z}_1$ and $\mathbf{Z}_2$ have 12 rows

Table 11.9 Alternative representation of the 12-run two-way split-plot design with
four rows and four columns in Table 11.8.

	Stage 1		Stage 2			
	Factors		Factor	Column 1	Column 2	Column 3	Column 4
Lot	x_1	x_2	x_3	-1	$+1$	$+1$	-1
1	-1	-1		Run 1	Run 2	Run 3	
2	$+1$	-1		Run 5	Run 6		Run 4
3	$+1$	$+1$		Run 9		Run 7	Run 8
4	-1	$+1$			Run 10	Run 11	Run 12

and four columns:

$$
\mathbf{Z}_1 = \begin{bmatrix} 1 & 0 & 0 & 0 \\ 1 & 0 & 0 & 0 \\ 1 & 0 & 0 & 0 \\ 0 & 1 & 0 & 0 \\ 0 & 1 & 0 & 0 \\ 0 & 1 & 0 & 0 \\ 0 & 0 & 1 & 0 \\ 0 & 0 & 1 & 0 \\ 0 & 0 & 1 & 0 \\ 0 & 0 & 0 & 1 \\ 0 & 0 & 0 & 1 \\ 0 & 0 & 0 & 1 \end{bmatrix} \quad \text{and} \quad \mathbf{Z}_2 = \begin{bmatrix} 1 & 0 & 0 & 0 \\ 0 & 1 & 0 & 0 \\ 0 & 0 & 1 & 0 \\ 0 & 0 & 0 & 1 \\ 1 & 0 & 0 & 0 \\ 0 & 1 & 0 & 0 \\ 0 & 0 & 1 & 0 \\ 0 & 0 & 0 & 1 \\ 1 & 0 & 0 & 0 \\ 0 & 1 & 0 & 0 \\ 0 & 0 & 1 & 0 \\ 0 & 0 & 0 & 1 \end{bmatrix}.
$$

As a result,

$$
\mathbf{V} = \begin{bmatrix}
\sigma_Y^2 & \sigma_\gamma^2 & \sigma_\gamma^2 & 0 & \sigma_\delta^2 & 0 & 0 & 0 & \sigma_\delta^2 & 0 & 0 & 0 \\
\sigma_\gamma^2 & \sigma_Y^2 & \sigma_\gamma^2 & 0 & 0 & \sigma_\delta^2 & 0 & 0 & 0 & \sigma_\delta^2 & 0 & 0 \\
\sigma_\gamma^2 & \sigma_\gamma^2 & \sigma_Y^2 & 0 & 0 & 0 & \sigma_\delta^2 & 0 & 0 & 0 & \sigma_\delta^2 & 0 \\
0 & 0 & 0 & \sigma_Y^2 & \sigma_\gamma^2 & \sigma_\gamma^2 & 0 & \sigma_\delta^2 & 0 & 0 & 0 & \sigma_\delta^2 \\
\sigma_\delta^2 & 0 & 0 & \sigma_\gamma^2 & \sigma_Y^2 & \sigma_\gamma^2 & 0 & 0 & \sigma_\delta^2 & 0 & 0 & 0 \\
0 & \sigma_\delta^2 & 0 & \sigma_\gamma^2 & \sigma_\gamma^2 & \sigma_Y^2 & 0 & 0 & 0 & \sigma_\delta^2 & 0 & 0 \\
0 & 0 & \sigma_\delta^2 & 0 & 0 & 0 & \sigma_Y^2 & \sigma_\gamma^2 & \sigma_\gamma^2 & 0 & \sigma_\delta^2 & 0 \\
0 & 0 & 0 & \sigma_\delta^2 & 0 & 0 & \sigma_\gamma^2 & \sigma_Y^2 & \sigma_\gamma^2 & 0 & 0 & \sigma_\delta^2 \\
\sigma_\delta^2 & 0 & 0 & 0 & \sigma_\delta^2 & 0 & \sigma_\gamma^2 & \sigma_\gamma^2 & \sigma_Y^2 & 0 & 0 & 0 \\
0 & \sigma_\delta^2 & 0 & 0 & 0 & \sigma_\delta^2 & 0 & 0 & 0 & \sigma_Y^2 & \sigma_\gamma^2 & \sigma_\gamma^2 \\
0 & 0 & \sigma_\delta^2 & 0 & 0 & 0 & \sigma_\delta^2 & 0 & 0 & \sigma_\gamma^2 & \sigma_Y^2 & \sigma_\gamma^2 \\
0 & 0 & 0 & \sigma_\delta^2 & 0 & 0 & 0 & \sigma_\delta^2 & 0 & \sigma_\gamma^2 & \sigma_\gamma^2 & \sigma_Y^2
\end{bmatrix},
$$

where the diagonal elements are $\sigma_Y^2 = \sigma_\gamma^2 + \sigma_\delta^2 + \sigma_\varepsilon^2$. The ith diagonal element of that matrix is the variance of the ith response, σ_Y^2. The element in the ith row and jth column gives the covariance between the ith and the jth run of the design. The fifth element in the first row of $\mathbf{V}$, for instance, is the covariance between the responses of the first and the fifth run. These runs belong to the same column according to Table 11.8, which is why their covariance is σ_δ^2. Similarly, the third element in the first column of $\mathbf{V}$ gives the covariance between the responses of the third run and the first one. The covariance is σ_γ^2 because these runs belong to the first row of the design.

The covariance between the first and last run's responses is zero because these runs are neither in the same column nor in the same row of the design in Table 11.8.

The use of the GLS estimator requires the estimation of the three variance components in $\mathbf{V}$, namely, σ_γ^2, σ_δ^2, and σ_ε^2. The default estimation method for variance components in statistical packages is restricted maximum likelihood (REML; see Section 7.3.3 and Attachment 7.1 on page 150). For determining the degrees of freedom for the usual significance tests, we recommend the Kenward–Roger method (see Section 7.3.4) that has been implemented in several packages.

11.3.3   Optimal design of two-way split-plot experiments

To find an optimal two-way split-plot design with a given number of rows and columns, we have to determine the levels of the row factors and the levels of the column factors so that the GLS estimates are as precise as possible. The way in which we do this is by minimizing the determinant of $(\mathbf{X}'\mathbf{V}^{-1}\mathbf{X})^{-1}$, or by maximizing the determinant of the information matrix $\mathbf{X}'\mathbf{V}^{-1}\mathbf{X}$. We call the determinant $|\mathbf{X}'\mathbf{V}^{-1}\mathbf{X}|$ the D-optimality criterion value. The design that maximizes the D-optimality criterion value is the D-optimal two-way split-plot design. This results both in small variances of the parameter estimates and in small covariances between the parameter estimates.

As with the random block effects model in Chapter 7 and the (one-way) split-plot model in Chapter 10, the optimal two-way split-plot design depends on the relative magnitudes of the variance components in $\mathbf{V}$. The difference is that there are now three variance components instead of two, and that, as a result, two variance ratios have to be specified instead of one. Since both σ_γ^2 and σ_δ^2 are often larger than the run-to-run variance, σ_ε^2, we suggest specifying that the variance ratios $\sigma_\gamma^2/\sigma_\varepsilon^2$ and $\sigma_\delta^2/\sigma_\varepsilon^2$ are at least one when generating a D-optimal two-way split-plot design. An educated guess of the variance ratios is good enough because D-optimal designs are not sensitive to the specified values. The optimal designs shown in Tables 11.2 and 11.8 were generated assuming that $\sigma_\gamma^2/\sigma_\varepsilon^2 = \sigma_\delta^2/\sigma_\varepsilon^2 = 1$, but they are optimal for any values of these ratios between 0.1 and 10. So, the designs are optimal for any variance ratios one is likely to encounter in practice.

11.3.4   A design construction algorithm for D-optimal two-way split-plot designs

In addition to the specification of the number of runs, the a priori model, the number of factors, an indication of whether a factor is continuous, categorical or a mixture ingredient, any additional constraints on factor-level combinations, and the number of starting designs to consider, the coordinate-exchange algorithm for finding D-optimal two-way split-plot designs requires the designation of the factors applied to the rows of the design and the factors applied to the columns of the design, the numbers of rows and columns, and the expected ratios $\sigma_\gamma^2/\sigma_\varepsilon^2$ and $\sigma_\delta^2/\sigma_\varepsilon^2$ describing the relative magnitudes of the three variance components.

The body of the two-way split-plot design construction algorithm has two parts. The first part involves again the creation of a starting design. The second is the iterative improvement of this design until no further improvement is possible. Improvements are measured by increases in the D-optimality criterion, $|\mathbf{X}'\mathbf{V}^{-1}\mathbf{X}|$.

The starting design is formed column by column. For any factor applied to the rows of the two-way split-plot design, one random level is chosen for any given row of the design. That level is assigned to all the runs corresponding to that row. For any factor applied to the columns of the two-way split-plot design, one random level is chosen for any given column. That level is assigned to all the runs corresponding to that column. This procedure gives the starting design the desired two-way split-plot structure. In the absence of constraints on the factor levels, the algorithm generates random levels on the interval $[-1, +1]$ for all factors that are continuous. For all factors that are categorical, the algorithm chooses one of its possible levels at random. In the presence of constraints on the factor levels, the algorithms uses the approach sketched in Section 5.3.3. For mixture ingredients, the procedure for generating random values is the same as in Section 6.3.5.

Improvements are made to the starting design by considering changes in the design matrix on an element-by-element basis. The procedure for changing any given element depends on whether that element is the level of a factor applied to the rows of the two-way split-plot design or the level of a factor applied to the columns. This is because changing the level of a factor applied to the rows for a given run requires that that factor's level must also change for all other runs corresponding to the same row of the two-way split-plot design, and changing the level of a factor applied to the columns for a given run requires that that factor's level must also change for all other runs corresponding to the same column.

For the level of a continuous factor applied to the rows, the D-optimality criterion value is evaluated over a discrete number of values spanning the range of the factor. For the level of a categorical factor applied to the rows, the D-optimality criterion value is evaluated over all possible levels of that factor. If the best value of the D-optimality criterion improves the current best value, then the current best value is replaced and the current level of the factor in the design is replaced by the level corresponding to the best value of the D-optimality criterion. For the level of factors applied to the columns, the procedure is analogous.

This element-by-element procedure continues until a complete cycle through the entire design matrix is completed. Then, another complete cycle through the design is performed. This continues until no changes are made in a whole pass.

11.3.5   Extensions and related designs

In this chapter, we have discussed two-way split-plot designs for two production stages. Obviously, the methodology we presented can be extended to designs for experiments involving more than two production stages. The designs could be named multiway split-plot designs, but the name split-lot designs has become the standard term for them.

Another related design is a split-plot design that involves blocks. Sometimes, the output of a split-plot experiment has to be processed further before the actual responses can be measured. When the processing is done in blocks, then the required design and model are very similar to the two-way split-plot design and model discussed here. For instance, the variance-covariance matrix of the response vector, $\mathbf{V}$, has the same structure, and the same row-column representation of the designs can be used. The difference with two-way split-plot designs is that no factors are associated with the columns.

Split-split-plot designs are also similar to two-way split-plot designs because they can be used for two-stage production processes too. One key difference between the designs is that, in split-split-plot designs, the sublots from different runs of the first stage of the process are not grouped and processed together in the second stage. Instead, every sublot is processed individually in the second stage. An additional difference is that split-split-plot experiments also involve a third set of factors that are applied to individual outputs of the two-stage production process.

11.4   Background reading

Literature on the use of two-way split-plot or strip-plot designs in industry is scarce when compared to that on design for blocked or (one-way) split-plot experiments. Undoubtedly, the most influential reference concerning the use of two-way split-plot designs is Box and Jones (1992), who point out that these designs are suitable for many robust product experiments, where some of the factors are control or design factors and others are noise or environmental factors. The main goal of robust product experiments is to identify control-by-noise interaction effects. The main advantage of two-way split-plot designs is that they allow for an efficient estimation of these interactions if all control factors appear in the design as row factors and all noise factors are treated as column factors, or vice versa. Thus, two-way split-plot designs are inexpensive designs for estimating control-by-noise interactions efficiently.

Most other published work on two-way split-plot designs deals with the combinatorial construction of good fractions of factorial designs (see Miller (2001), Butler (2004), and Vivacqua and Bisgaard (2004, 2009)). Similar work covering two or more production stages is in Mee and Bates (1998) and Paniagua-Quiñones and Box (2008, 2009). Arnouts et al. (2010) discuss the optimal design of two-way split-plot designs.

Federer and King (2007) provide a thorough review of agricultural applications of two-way split-plot designs, as well as other variations on split-plot designs.

The combinatorial construction of split-plot experiments that require blocking is discussed in McLeod and Brewster (2004, 2006). Jones and Goos (2009) study the optimal design of split-split-plot experiments. Ferryanto and Tollefson (2010) and Castillo (2010) describe case studies of split-split-plot designs.

Trinca and Gilmour (2001) present a general algorithmic approach for setting up split-plot, split-split-plot, and nested multistratum designs. Combining this algorithm with the row–column design algorithm in Gilmour and Trinca (2003) offers the possibility to construct other multistratum designs, such as two-way split-plot designs, too.

11.5 Summary

Multistage processes are common in industry. In the past, statistical consultants often advised their clients only to experiment with one stage at a time to avoid the complexity that comes with multistage experimentation. In the same way that factorial experiments are more efficient than one-factor-at-a-time experiments, studies encompassing more than one processing step are more efficient than studies that deal with only one processing step at a time.

However, it is true that experimenting with multiple process steps involves both logistical and statistical complexity. The logistical complexity involves being able to track the experimental material through the processing steps to make sure that all the factors get set correctly.

In this chapter, we focused on two-way split-plot experiments, also known as strip-plot experiments, for two-step processes. A key feature of these experiments is that the outputs of the first processing step are split and then regrouped before being processed group per group in the second step. This is different from split-plot experiments, which can be used for two-step processes too, but where the outputs from the first step are not regrouped.

Bibliography

Anderson-Cook, C.M., Borror, C.M., and Montgomery, D.C. (2009). Response surface design evaluation and comparison. *Journal of Statistical Planning and Inference*, **139**, 629–674.

Arnouts, H., Goos, P., and Jones, B. (2010). Design and analysis of industrial strip-plot experiments. *Quality and Reliability Engineering International*, **26**, 127–136.

Atkinson, A.C., and Donev, A.N. (1989). The construction of exact D-optimum experimental designs with application to blocking response surface designs. *Biometrika*, **76**, 515–526.

Atkinson, A.C., and Donev, A.N. (1996). Experimental designs optimally balanced for trend. *Technometrics*, **38**, 333–341.

Atkinson, A.C., Donev, A.N., and Tobias, R.D. (2007). *Optimum Experimental Designs, with SAS*. New York: Oxford University Press.

Bie, X., Lu, Z., Lu, F., and Zeng, X. (2005). Screening the main factors affecting extraction of the antimicrobial substance from bacillus sp. fmbj using the Plackett–Burman method. *World Journal of Microbiology and Biotechnology*, **21**, 925–928.

Bingham, D.R., and Sitter, R.R. (1999a). Minimum-aberration two-level fractional factorial split-plot designs. *Technometrics*, **41**, 62–70.

Bingham, D.R., and Sitter, R.R. (1999b). Some theoretical results for fractional factorial split-plot designs. *Annals of Statistics*, **27**, 1240–1255.

Bingham, D.R., and Sitter, R.R. (2001). Design issues in fractional factorial split-plot designs. *Journal of Quality Technology*, **33**, 2–15.

Bingham, D.R., and Sitter, R.R. (2003). Fractional factorial split-plot designs for robust parameter experiments. *Technometrics*, **45**, 80–89.

Bingham, D.R., Schoen, E.D., and Sitter, R.R. (2004). Designing fractional factorial split-plot experiments with few whole-plot factors. *Journal of the Royal Statistical Society, Ser. C (Applied Statistics)*, **53**, 325–339. Corrigendum, **54**, 955–958.

Bisgaard, S., and Steinberg, D.M. (1997). The design and analysis of $2^{k-p} \times s$ prototype experiments. *Technometrics*, **39**, 52–62.

Box, G.E.P., and Draper, N. (1987). *Empirical Model-Building and Response Surfaces*. New York: John Wiley & Sons, Inc.

Box, G.E.P., and Jones, S.P. (1992). Split-plot designs for robust product experimentation. *Journal of Applied Statistics*, **19**, 3–26.

Box, G.E.P., Hunter, W., and Hunter, J. (2005). *Statistics for Experimenters: Design, Innovation, and Discovery*. New York: John Wiley & Sons, Inc.

Optimal Design of Experiments: A Case Study Approach, First Edition. Peter Goos and Bradley Jones.
© 2011 John Wiley & Sons, Ltd. Published 2011 by John Wiley & Sons, Ltd.

Brenneman, W.A., and Myers, W.R. (2003). Robust parameter design with categorical noise variables. *Journal of Quality Technology*, **35**, 335–341.

Butler, N.A. (2004). Construction of two-level split-lot fractional factorial designs for multi-stage processes. *Technometrics*, **46**, 445–451.

Carrano, A.L., Thorn, B.K., and Lopez, G. (2006). An integer programming approach to the construction of trend-free experimental plans on split-plot designs. *Journal of Manufacturing Systems*, **25**, 39–44.

Castillo, F. (2010). Split-split-plot experimental design in a high-throughput reactor. *Quality Engineering*, **22**, 328–335.

Cheng, C.S. (1990). Construction of run orders of factorial designs. In: *Statistical Design and Analysis of Industrial Experiments*, (edited by S Ghosh), pp. 423–439. New York: Marcel Dekker.

Cheng, C.S., and Jacroux, M. (1988). The construction of trend-free run orders of two-level factorial designs. *Journal of the American Statistical Association*, **83**, 1152–1158.

Cheng, C.S., and Steinberg, D.M. (1991). Trend robust two-level factorial designs. *Biometrika*, **78**, 325–336.

Constantine, G.M. (1989). Robust designs for serially correlated observations. *Biometrika*, **76**, 241–251.

Cook, R.D., and Nachtsheim, C.J. (1989). Computer-aided blocking of factorial and response-surface designs. *Technometrics*, **31**, 339–346.

Cornell, J.A. (1988). Analyzing data from mixture experiments containing process variables: A split-plot approach. *Journal of Quality Technology*, **20**, 2–23.

Cornell, J.A. (2002). *Experiments with Mixtures: Designs, Models, and the Analysis of Mixture Data*. New York: John Wiley & Sons, Inc.

Daniel, C., and Wilcoxon, F. (1966). Factorial 2^{p-q} plans robust against linear and quadratic trends. *Technometrics*, **8**, 259–278.

Dean, A., and Lewis, S. (2006). *Screening Methods for Experimentation in Industry, Drug Discovery, and Genetics*. New York: Springer.

Derringer, D., and Suich, R. (1980). Simultaneous optimization of several response variables. *Journal of Quality Technology*, **12**, 214–219.

Federer, W.T., and King, F. (2007). *Variations on Split Plot and Split Block Experiment Designs*. New York: John Wiley & Sons, Inc.

Fedorov, V.V. (1972). *Theory of Optimal Experiments*. New York: Academic Press.

Ferryanto, L., and Tollefson, N. (2010). A split-split-plot design of experiments for foil lidding of contact lens packages. *Quality Engineering*, **22**, 317–327.

Fisher, R.A. (1926). The arrangement of field experiments. *Journal of the Ministry of Agriculture*, **33**, 503–513.

Garroi, J.J., Goos, P., and Sörensen, K. (2009). A variable-neighbourhood search algorithm for finding optimal run orders in the presence of serial correlation. *Journal of Statistical Planning and Inference*, **139**, 30–44.

Gilmour, S.G. (2006). Factor screening via supersaturated designs. In: *Screening Methods for Experimentation in Industry, Drug Discovery, and Genetics*, (edited by A Dean and S Lewis), pp. 169–190. New York: Springer.

Gilmour, S.G., and Goos, P. (2009). Analysis of data from nonorthogonal multi-stratum designs. *Journal of the Royal Statistical Society Series C (Applied Statistics)*, **58**, 467–484.

Gilmour, S.G., and Trinca, L.A. (2000). Some practical advice on polynomial regression analysis from blocked response surface designs. *Communications in Statistics: Theory and Methods*, **29**, 2157–2180.

Gilmour, S.G., and Trinca, L.A. (2003). Row-column response surface designs. *Journal of Quality Technology*, **35**, 184–193.

Giovannitti-Jensen, A., and Myers, R.H. (1989). Graphical assessment of the prediction capability of response surface designs. *Technometrics*, **31**, 159–171.

Goos, P. (2002). *The Optimal Design of Blocked and Split-Plot Experiments*. New York: Springer.

Goos, P. (2006). Optimal versus orthogonal and equivalent-estimation design of blocked and split-plot experiments. *Statistica Neerlandica*, **60**, 361–378.

Goos, P., and Donev, A.N. (2006a). Blocking response surface designs. *Computational Statistics and Data Analysis*, **51**, 1075–1088.

Goos, P., and Donev, A.N. (2006b). The D-optimal design of blocked experiments with mixture components. *Journal of Quality Technology*, **38**, 319–332.

Goos, P., and Donev, A.N. (2007). Tailor-made split-plot designs with mixture and process variables. *Journal of Quality Technology*, **39**, 326–339.

Goos, P., and Vandebroek, M. (2001). D-optimal response surface designs in the presence of random block effects. *Computational Statistics and Data Analysis*, **37**, 433–453.

Goos, P., and Vandebroek, M. (2003). D-optimal split-plot designs with given numbers and sizes of whole plots. *Technometrics*, **45**, 235–245.

Goos, P., and Vandebroek, M. (2004). Estimating the intercept in an orthogonally blocked experiment. *Communications in Statistics: Theory and Methods*, **33**, 873–890.

Goos, P., Langhans, I., and Vandebroek, M. (2006). Practical inference from industrial split-plot designs. *Journal of Quality Technology*, **38**, 162–179.

Goos, P., Tack, L., and Vandebroek, M. (2005). The optimal design of blocked experiments in industry. In: *Applied Optimal Designs*, (edited by MPF Berger and WK Wong), pp. 247–279. New York: John Wiley & Sons, Inc.

Harrington, E.C. (1965). The desirability function. *Industrial Quality Control*, **21**, 494–498.

Harville, D.A. (1974). Nearly optimal allocation of experimental units using observed covariate values. *Technometrics*, **16**, 589–599.

Harville, D.A., and Jeske, D.R. (1992). Mean squared error of estimation or prediction under a general linear model. *Journal of the American Statistical Association*, **87**, 724–731.

Hill, H.M. (1960). Experimental designs to adjust for time trends. *Technometrics*, **2**, 67–82.

John, P.W.M. (1990). Time trends and factorial experiments. *Technometrics*, **32**, 275–282.

Johnson, M.E., and Nachtsheim, C.J. (1983). Some guidelines for constructing exact D-optimal designs on convex design spaces. *Technometrics*, **25**, 271–277.

Joiner, B.L., and Campbell, C. (1976). Designing experiments when run order is important. *Technometrics*, **18**, 249–259.

Jones, B., and DuMouchel, W. (1996). Follow-up designs to resolve confounding in multifactor experiments. Discussion. *Technometrics*, **38**, 323–326.

Jones, B., and Goos, P. (2007). A candidate-set-free algorithm for generating D-optimal split-plot designs. *Journal of the Royal Statistical Society, Ser. C (Applied Statistics)*, **56**, 347–364.

Jones, B., and Goos, P. (2009). D-optimal design of split-split-plot experiments. *Biometrika*, **96**, 67–82.

Jones, B., and Nachtsheim, C.J. (2009). Split-plot designs: What, why, and how. *Journal of Quality Technology*, **41**, 340–361.

Jones, B., and Nachtsheim, C.J. (2011a). A class of three-level designs for definitive screening in the presence of second-order effects. *Journal of Quality Technology*, **43**, 1–15.

Jones, B., and Nachtsheim, C.J. (2011b). Efficient designs with minimal aliasing. *Technometrics*, **53**, 62–71.

Kackar, R.N., and Harville, D.A. (1981). Unbiasedness of two-stage estimation and prediction procedures for mixed linear models. *Communications in Statistics: Theory and Methods*, **10**, 1249–1261.

Kackar, R.N., and Harville, D.A. (1984). Approximations for standard errors of estimators of fixed and random effects in mixed linear models. *Journal of the American Statistical Association*, **79**, 853–862.

Kenward, M.G., and Roger, J.H. (1997). Small sample inference for fixed effects from restricted maximum likelihood. *Biometrics*, **53**, 983–997.

Khuri, A.I. (1992). Response surface models with random block effects. *Technometrics*, **34**, 26–37.

Khuri, A.I. (1996). Response surface models with mixed effects. *Journal of Quality Technology*, **28**, 177–186.

Kowalski, S.M., Cornell, J.A., and Vining, G.G. (2000). A new model and class of designs for mixture experiments with process variables. *Communications in Statistics: Theory and Methods*, **29**, 2255–2280.

Kowalski, S.M., Cornell, J.A., and Vining, G.G. (2002). Split-plot designs and estimation methods for mixture experiments with process variables. *Technometrics*, **44**, 72–79.

Langhans, I., Goos, P., and Vandebroek, M. (2005). Identifying effects under a split-plot design structure. *Journal of Chemometrics*, **19**, 5–15.

Li, X., Sudarsanam, N., and Frey, D.D. (2006). Regularities in data from factorial experiments. *Complexity*, **11**, 32–45.

Loukas, Y.L. (1997). $2^{(k-p)}$ fractional factorial design via fold over: Application to optimization of novel multicomponent vesicular systems. *Analyst*, **122**, 1023–1027.

Marley, C.J., and Woods, D. (2010). A comparison of design and model selection methods for supersaturated experiments. *Computational Statistics and Data Analysis*, **54**, 3158–3167.

Martin, R.J., Eccleston, J.A., and Jones, G. (1998a). Some results on multi-level factorial designs with dependent observations. *Journal of Statistical Planning and Inference*, **73**, 91–111.

Martin, R.J., Jones, G., and Eccleston, J.A. (1998b). Some results on two-level factorial designs with dependent observations. *Journal of Statistical Planning and Inference*, **66**, 363–384.

McLeod, R,G., and Brewster, J,F. (2004). The design of blocked fractional factorial split-plot experiments. *Technometrics*, **46**, 135–146.

McLeod, R.G., and Brewster, J.F. (2006). Blocked fractional factorial split-plot experiments for robust parameter design. *Journal of Quality Technology*, **38**, 267–279.

Mee, R.W. (2009). *A Comprehensive Guide to Factorial Two-Level Experimentation*. New York: Springer.

Mee, R.W., and Bates, R.L. (1998). Split-lot designs: Experiments for multistage batch processes. *Technometrics*, **40**, 127–140.

Mee, R.W., and Romanova, A.V. (2010). Constructing and analyzing two-level trend-robust designs. *Quality Engineering*, **22**, 306–316. Corrigendum, **23**, 112.

Meyer, R.D., Steinberg, D.M., and Box, G. (1996). Follow-up designs to resolve confounding in multifactor experiments. *Technometrics*, **38**, 303–313.

Meyer, R.K., and Nachtsheim, C.J. (1995). The coordinate-exchange algorithm for constructing exact optimal experimental designs. *Technometrics*, **37**, 60–69.

Miller, A. (1997). Strip-plot configurations of fractional factorials. *Technometrics*, **39**, 153–161.

Miller, A., and Sitter, R.R. (2001). Using the folded-over 12-run Plackett–Burman design to consider interactions. *Technometrics*, **43**, 44–55.

Mitchell, T.J. (1974). An algorithm for the construction of D-optimal experimental designs. *Technometrics*, **16**, 203–210.

Montgomery, D.C. (2009). *Design and Analysis of Experiments*. New York: John Wiley & Sons, Inc.

Myers, R.H., Montgomery, D.C., and Anderson-Cook, C. (2009). *Response Surface Methodology: Process and Product Optimization using Designed Experiments*. New York: John Wiley & Sons, Inc.

Myers, R.H., Vining, G.G., Giovannitti-Jensen, A., and Myers, S.L. (1992). Variance dispersion properties of second-order response surface designs. *Journal of Quality Technology*, **24**, 1–11.

Nachtsheim, C.J. (1989). On the design of experiments in the presence of fixed covariates. *Journal of Statistical Planning and Inference*, **22**, 203–212.

Paniagua-Quiñones, C., and Box, G.E.P. (2008). Use of strip-strip-block design for multi-stage processes to reduce cost of experimentation. *Quality Engineering*, **20**, 46–52.

Paniagua-Quiñones, C., and Box, G.E.P. (2009). A post-fractionated strip-strip-block design for multi-stage processes. *Quality Engineering*, **21**, 156–167.

Patterson, H.D., and Thompson, R. (1971). Recovery of inter-block information when block sizes are unequal. *Biometrika*, **5**, 545–554.

Piepel, G.F. (1983). Defining consistent constraint regions in mixture experiments. *Technometrics* **25**, 97–101.

Piepel, G.F. (1999). Modeling methods for mixture-of-mixtures experiments applied to a tablet formulation problem. *Pharmaceutical Development and Technology*, **4**, 593–606.

Piepel, G.F., Cooley, S.K., and Jones, B. (2005). Construction of a 21-component layered mixture experiment design using a new mixture coordinate-exchange algorithm. *Quality Engineering* **17**, 579–594.

Prescott, P. (2004). Modelling in mixture experiments including interactions with process variables. *Quality Technology and Quality Management* **1**, 87–103.

Rodríguez, M., Jones, B., Borror, C.M., and Montgomery, D.C. (2010). Generating and assessing exact G-optimal designs. *Journal of Quality Technology*, **42**, 1–18.

Ruggoo, A., and Vandebroek, M. (2004). Bayesian sequential d-d optimal model-robust designs. *Computational Statistics and Data Analysis*, **47**, 655–673.

Simpson, J.R., Kowalski, S.M., and Landman, D. (2004). Experimentation with randomization restrictions: Targeting practical implementation. *Quality and Reliability Engineering International*, **20**, 481–495.

Smith, W.F. (2005). *Experimental Design for Formulation*. Philadelphia: SIAM.

Tack, L., and Vandebroek, M. (2001). (D_t, C)-optimal run orders. *Journal of Statistical Planning and Inference*, **98**, 293–310.

Tack, L., and Vandebroek, M. (2002). Trend-resistant and cost-efficient block designs with fixed or random block effects. *Journal of Quality Technology*, **34**, 422–436.

Tack, L., and Vandebroek, M. (2003). Semiparametric exact optimal run orders. *Journal of Quality Technology*, **35**, 168–183.

Tack, L., and Vandebroek, M. (2004). Budget constrained run orders in optimum design. *Journal of Statistical Planning and Inference*, **124**, 231–249.

Trinca, L.A., and Gilmour, S.G. (2001). Multi-stratum response surface designs. *Technometrics*, **43**, 25–33.

Vivacqua, C.A., and Bisgaard, S. (2004). Strip-block experiments for process improvement and robustness. *Quality Engineering*, **16**, 495–500.

Vivacqua, C.A., and Bisgaard, S. (2009). Post-fractionated strip-block designs. *Technometrics*, **51**, 47–55.

Wesley, W.R., Simpson, J.R., Parker, P.A., and Pignatiello, J.J. (2009). Exact calculation of integrated prediction variance for response surface designs on cuboidal and spherical regions. *Journal of Quality Technology*, **41**, 165–180.

Wu, C.F.J., and Hamada, M. (2009). *Experiments: Planning, Analysis, and Optimization*. New York: John Wiley & Sons, Inc.

Zahran, A.R., Anderson-Cook, C.M., and Myers, R.H. (2003). Fraction of design space to assess prediction capability of response surface designs. *Journal of Quality Technology*, **35**, 377–386.

Zhou, J. (2001). A robust criterion for experimental designs for serially correlated observations. *Technometrics*, **43**, 462–467.

Index

2-factor interaction effect, *see*
    interaction effect
2-level design, *see* two-level design
3-level design, *see* three-level design

A-optimality, 229
active factor/interaction, 26
adjusted standard error, *see*
    Kenward-Roger adjusted standard
    error
algorithm, *see* coordinate-exchange
    algorithm, 45
alias matrix, 9, 30–33
alias minimization, 44
aliasing, 20, 29, 31
average prediction variance, 88

balanced design, 1, 4, 6–8
best linear unbiased prediction,
    266–267
bias, 9, 18, 29, 31–32, 45
binary response, 192
blending property, 126
block effect, 91, 145, 152
    fixed, *see* fixed block effect
    random, *see* random block effect
block variance, 150
blocking, 132, 135, 161, 163, 185,
    240, 274
blocking factor, 136, 165, 211
blocking versus split-plotting, 240

BLUP, *see* best linear unbiased
    prediction
Box–Behnken design, 72, 185

categorical factor, 24, 69, 83, 201
categorical response, 192
categorized-components mixture
    experiment, 133
causation, 113–114, 116
center point, 223, 225–226
central composite design, 72, 97,
    136–137, 158, 185
coding, 15, 19, 24, 140
    effects-type, 24, 85
    factor levels, 15, 199
collinearity, 29, 83, 85, 118,
    124, 182
column factor effect, 261
completely randomized design,
    219–220, 222, 240, 254–256
composite optimality criteria, 93
compound optimality criteria, 93
concomitant variable, *see* covariate
    factor
confidence band, 16, 107
confirmatory experiment, 11
confounding, 31
constant, 21
constant variance, 4, 8
constrained design region, 92,
    101

Optimal Design of Experiments: A Case Study Approach, First Edition. Peter Goos and Bradley Jones.
© 2011 John Wiley & Sons, Ltd. Published 2011 by John Wiley & Sons, Ltd.

constraint, *see* inequality constraint, equality constraint, or non-linear constraint

continuous factor,   24, 69

contour plot,   108, 202

control-by-noise interaction effect, 249, 275

coordinate-exchange algorithm, 35–38, 44, 111, 130, 132, 215, 251, 273

correlated factor-effect estimates,   19, 34, 105, 263

correlated responses,   91, 217, 223, 235, 240, 243, 258, 269

cost,   1, 6–7, 216–217, 254, 257

covariance matrix of responses within a block,   153

covariate factor,   187, 190, 206–207, 216

Cox-effect direction,   131–132

criss-cross design,   258, 267

cubic effect,   95, 102, 108, 125

cuboidal design region,   92

curvature,   69, 83, 95, 124, 136, 224

D-efficiency,   77, 86–87, 226

D-optimal block design,   158, 160, 184

D-optimal design,   33–34, 73, 141

D-optimal split-plot design,   250

D-optimality criterion,   34–35, 69, 210

defining relation,   171

design matrix,   35

design point,   35

design region,   91, 127

desirability,   147–148, 161

drift,   187, 190

D_s- or D_β-optimal design,   210, 212–213

dummy variable,   83, 200–201

easy-to-change factors,   221–222, 230, 240, 242

effects-type coding,   24, 85, 183, 200

efficiency,   4, 6

efficiency factor,   143, 159, 184, 214

equal variance,   4

equality constraint,   123

error variance,   21

experimental region,   91, 127

factor scaling,   24, 122

factorial approach,   116

factorial design,   34, 44, 97, 171, 224, 256

feasible factor-level combinations, 101

feasible generalized least squares estimator,   156, 243

first-order effect,   21

first-order Scheffé model,   125, 128

fixed block analysis,   177, 179

fixed block effect,   153, 163, 177, 180, 185

fixed versus random block effects, 163, 178, 180–181, 185

Fraction of Design Space plot,   69, 77–78, 87, 92–93, 231

fractional factorial design,   34, 44, 117–118, 165–166, 171

full cubic model,   95, 102, 108

full quadratic model,   82

G-optimal design,   90, 92

generalized least squares estimation, 135, 145, 149, 154, 185, 219, 223, 235, 238, 243–244, 253, 258, 270

global optimum,   36

GLS, *see* generalized least squares estimation

hard-to-change factors,   71, 91, 216, 219, 221–222, 229, 240, 242, 248–250, 255

heredity,   43

heterogeneous experimental conditions, 91

hierarchy,   43

I-efficiency, 77, 90
I-optimal design, 69, 73, 90, 92–93, 226, 230
I-optimal split-plot design, 250
independent errors, 90
independent observations, 91
inequality constraint, 95, 100, 111
infeasible factor level combinations, 95
information matrix, 26
ingredient, 113, 123, 133
interaction effect, 9, 15, 17, 22–23, 31, 43, 45, 72, 106, 108, 207, 230, 262
intercept, 21, 123, 182, 228
IV-optimal design, *see* I-optimal design

Kenward–Roger adjusted standard errors, 157, 243
Kenward–Roger degrees of freedom, 179, 243, 273

lack-of-fit test, 105, 109–110
lattice design, 128
linear dependence, 113, 124
linear effect, *see* main effect
local optimum, 36
lurking variable, 116, 122, 133

main effect, 15, 21, 43
main-effects model, 21–22, 206
main-effects-plus-two-factor-interactions model, 22–23
mean squared error, 20, 25, 151–152
mixed model, 135, 153, 219
mixture constraint, 123
mixture design, 127–129
mixture experiment, 113, 117, 125, 132, 161
mixture-amount experiment, 127, 132
mixture-of-mixtures experiment, 133

mixture-process variable experiment, 114, 125, 132–133, 248
mixture-process variable model, 126–127
model matrix, 21–23
moments matrix, 89, 92
MSE, *see* mean squared error
multi-collinearity, *see* collinearity
multi-factor mixture experiment, 133
multistratum design, 254, 275

nonconstant variance, 7
nonlinear constraint, 112
nonorthogonal design, 34
nonregular design, 44
number of whole plots, 249

observational study, 113, 133
OFAT, *see* one-factor-at-a-time experimentation
OLS estimation, *see* ordinary least squares estimation
one-factor-at-a-time experimentation, 116–117, 221, 276
optimal design, 33–34
optimal run order, 212
optimal two-way split-plot design, 273
optimality criteria, 92
ordinary least squares estimation, 24, 30, 91, 135, 149, 163, 182–183, 209, 223, 234–235, 238
for split-plot data, 236
ordinary versus generalized least squares for split-plot data, 237
orthogonal blocking, 137, 141, 158, 165, 184–185
orthogonal design, 9, 12, 19, 26, 29, 33–34, 40, 45, 86, 262
orthogonal versus optimal blocking, 160–161, 172, 184
orthogonality and covariates, 211

p value, 27, 201
Pareto principle, 10, 41

Plackett–Burman design, 12, 17–18, 44

point-exchange algorithm, 45

power, 28, 152

prediction, 16, 73, 204

prediction interval, 107, 203–204

prediction variance, 1, 88, 92, 231

process variable, *see* mixture-process variable experiment

prototype, 248

prototype experiment, 253

pseudocomponent, 122, 129–130

pure error, 109–110

quadratic effect, 43, 69, 82, 84, 93, 95, 108, 224, 246

random block analysis, 178–179

random block effect, 135, 150, 153, 163, 178, 180, 185, 219

random design, 165

random error, 21

random intercept, 153

random run order, 91

random versus fixed block effects, *see* fixed versus random block effects

random versus systematic run order, 215

randomization, 215, 219, 222

region of interest, 91

regular design, 44

relative D-efficiency, *see* D-efficiency

relative I-efficiency, *see* I-efficiency

relative prediction variance, 77, 88

relative variance of factor-effect estimate, 25, 76, 93

REML, *see* restricted maximum likelihood estimation

replication, 103, 105, 109

residual error, 149, 152

residual error variance, 149

residual maximum likelihood estimation, *see* restricted maximum likelihood estimation

residual sum of squares, 109

response surface design, 69, 137

response surface model, 145

restricted maximum likelihood estimation, 145, 150, 156, 161, 185, 242, 273

robust process, 70, 72

robust product experiment, 248, 254, 275

root mean squared error, 20, 45

row × column interaction effects, 261

row factor effect, 261

run order, 212

sample variance, 150

scaling, *see* factor scaling

Scheffé model, 125, 128–129

screening experiment, 9, 21, 44, 69

screening for main effects and quadratic effects, 44

second-order effect, 22

second-order Scheffé model, 125, 129

serial correlation, 217

significance level, 27

significance test, 26–28, 157, 201

simplex, 127

singular design, 35

sparsity-of-effects principle, 10, 41

special-cubic Scheffé model, 125, 129

spherical design region, 92

split-lot design, 274

split-plot design, 91, 133, 216, 219–220, 222, 240–241, 247, 255, 275

split-plot model, 242

split-plotting versus blocking, 240

split-split-plot design, 254, 275

standard error, 19, 25, 157

strip-block design, 254, 258, 267

strip-plot design, 254–255, 258, 267, 276

sub-plot, 240–241

sub-plot effects, 242

sub-plot factor, 241–242

supersaturated design,   44
synergistic
    interaction effect,   43
systematic run order,   91, 215

t statistic,   27
ternary plot,   119–120
third-order effect,   108
third-order model,   102
three-level design,   83, 93
time as a covariate,   212
time trend,   211
    linear,   212
time-trend effect,   190, 212, 216
tolerance,   189, 203
trend-free run order, *see* trend-robust
    run order
trend-resistant run order, *see*
    trend-robust run order
trend-robust run order,   212–213, 216
trend-robustness,   214
two-level design,   21
two-level factorial design,   117
two-stage experiment,   248
two-way split-plot design,   91,
    254–255, 258, 267, 275–276

unbalanced design,   4–5
uncontrollable input,   165, 187, 206

unequal variances,   7
unexplained variation,   109

V, *see* variance–covariance matrix of
    responses
V-optimal design, *see* I-optimal design
variance component,   267
Variance Dispersion graph,   92
variance inflation,   34, 45, 105, 184,
    211, 214
variance inflation factor,   29, 143,
    160, 167, 184, 214
variance ratio,   5, 158, 251, 273
variance–covariance matrix of
    factor-effect estimates,   24–25,
    183, 209
variance–covariance matrix of
    responses,   155, 243, 258, 270,
    272
VIF, *see* variance inflation factor

whole plot,   240–241
whole-plot effects,   242, 249
whole-plot error variance,   249, 253
whole-plot factor,   241–242
whole-plot-by-sub-plot interaction
    effect,   242, 244, 249
within-block correlation coefficient,
    153

Printed and bound by CPI Group (UK) Ltd, Croydon, CR0 4YY

27/10/2024

14580285-0003